KU-567-041

GEOTHERMAL RESOURCES

Second Edition

GEOTHERMAL RESOURCES

SECOND EDITION

ROBERT BOWEN

Professor of Geology, Institute of Geology and Palaeontology,
Westfälische Wilhelms-University, Münster, Federal Republic of Germany

07 DEC 1989

UNIVERSITY OF STRATHCLYDE UNIVERSITY LIBRARY

ELSEVIER APPLIED SCIENCE
LONDON and NEW YORK

ELSEVIER SCIENCE PUBLISHERS LTD
Crown House, Linton Road, Barking, Essex IG11 8JU, England

Sole Distributor in the USA and Canada
ELSEVIER SCIENCE PUBLISHING CO., INC.
655 Avenue of the Americas, New York, NY 10010, USA

First edition 1979
Second edition 1989

WITH 59 TABLES AND 60 ILLUSTRATIONS

© 1989 ELSEVIER SCIENCE PUBLISHERS LTD

British Library Cataloguing in Publication Data
Bowen, Robert
    Geothermal resources—2nd ed.
    1. Energy sources: Geothermal energy
    I. Title
    621.44

    ISBN 1-85166-287-1

Library of Congress Cataloging-in-Publication Data
Bowen, Robert
    Geothermal resources/Robert Bowen—2nd ed.
        p.   cm.
    Bibliography: p.
    Includes indexes.
    ISBN 1-85166-287-1
    1. Geothermal resources.   2. Geothermal engineering.
    I. Title.
    GB1199.5.B68 1989
    333.8′8—dc 19

No responsibility is assumed by the Publisher for any injury and/or damage to persons or property as a matter of products liability, negligence or otherwise, or from any use or operation of any methods, products, instructions or ideas contained in the material herein.

Special regulations for readers in the USA

This publication has been registered with the Copyright Clearance Center Inc. (CCC), Salem, Massachusetts. Information can be obtained from the CCC about conditions under which photocopies of parts of this publication may be made in the USA. All other copyright questions, including photocopying outside the USA, should be referred to the publisher.

All rights reserved. No part of this publication may be reproduced, stored in a retrieval system, or transmitted in any form or by any means, electronic, mechanical, photocopying, recording, or otherwise, without the prior written permission of the publisher.

Printed in Great Britain by Galliard (Printers) Ltd, Great Yarmouth

D
621·4
BOW

Preface

Since the Arab oil embargo of 1974, it has been clear that the days of almost limitless quantities of low-cost energy have passed. In addition, ever-worsening pollution due to fossil fuel consumption, for instance oil and chemical spills, strip mining, sulphur emission and accumulation of solid wastes, has, among other things, led to an increase of as much as 10% in the carbon dioxide content of the atmosphere in this century. This has induced a warming trend through the 'greenhouse effect' which prevents infrared radiation from leaving it. Many people think the average planetary temperatures may rise by 4°C or so by 2050. This is probably true since Antarctic ice cores evidence indicates that, over the last 160 000 years, ice ages coincided with reduced levels of carbon dioxide and warmer interglacial episodes with increased levels of the gas in the atmosphere. Consequently, such an elevation of temperature over such a relatively short span of time would have catastrophic results in terms of rising sea level and associated flooding of vast tracts of low-lying lands.

Reducing the burning of fossil fuels makes sense on both economic and environmental grounds. One of the most attractive alternatives is geothermal resources, especially in developing countries, for instance in El Salvador where geothermal energy provides about a fifth of total installed electrical power already. In fact, by the middle 1980s, at least 121 geothermal power plants were operating worldwide, most being of the dry steam type.

The largest electricity producer from geothermal energy on Earth is the USA with a 2022-MW capacity as of 1985, this being expected to reach 4370 MW by 1992. Actually, by that year, world geothermal electrical installed power could reach 10 300 MW, representing approximately 0·5% of world total electrical installed power. This may seem rather insignificant, but the bare statistic rather misleads because, in the Third World,

v

contributions made by geothermal resources are important; for example, in the Philippines no less than 11·9% of total installed electric power in 1982 was produced geothermally. In addition, in countries like Iceland with highly adverse climates, geothermal energy is valuable both in space and process heating.

The proliferation of research and development programmes in the area of geothermics together with their practical implementation over the past quarter of a century necessitates the producing of many more specialists in geothermal energy than had been anticipated. Indeed, four training centres now exist in Iceland, Italy, Japan and New Zealand for this purpose and the author is indebted to one of them, the International School of Geothermics in Pisa, Italy, and in particular to its Director, Dr Mario Fanelli, as well as to Dr Enrico Barbier and Dr Mary Dickson, for supplying essential literature and photographs cited in the book. It is appropriate here also to thank John Wiley & Sons Ltd, UK, for permission to reproduce Figs 1.13, 2.5 and 7.2 from their publication, *Geothermal Systems: Principles and Case Histories*, edited in 1981 by L. Rybach and L. J. P. Muffler. I am also grateful to Mr Charles R. Imbrecht, who is Chairman of the California Energy Commission, USA, to the Pacific Gas and Electric Company, and Southern California Edison Company as they supplied geothermal photographs from that state. Mr Ed Macumber, Energy Division Lead of the Economic Development and Stabilization Board of the State of Wyoming at Cheyenne, and Ms Cindy Hendrickson of the Wyoming Travel Commission are sincerely thanked for sending me geothermal photographs from there. Icelandic photographs were kindly given by Ms Gil Middleton of Broomhill, Sheffield. The Athlone Press, UK, kindly permitted me to quote from *The Man with a Nose* by H. G. Wells.

The future looks even more promising when hot dry rock (HDR) experiments in the USA and Great Britain are taken into account. At Los Alamos National Laboratory, work has been going on for over a decade: 5 MW of geothermal energy has been produced and the artificial reservoir exploited for more than a year with a temperature drop of under 10°C. Economically, electricity so generated is highly competitive in price. The centre of activity was outside the Valles Caldera in north-central New Mexico and entailed two phases, one drilling to 3000 m into granite at 185°C to create hydraulic fractures at 2600 m by means of water under pressure, a second extending the work by constructing a larger, hotter and deeper system. This project at Fenton Hill informally cooperated with work of the Camborne School of Mines going on in the Carnmenellis granite at Rosemanowes, part of the batholith underlying the southwestern peninsula of the UK. Here explosive stimulation pre-treated the well prior to hydraulic fracturing and then the injection of water under pressure produced unsatisfactory results so that two other wells were drilled. Water was used together with a viscous gel because the connections with the new well were not optimal. A reservoir resulted and has been circulating continuously since August 1985.

The author formerly worked in the geothermal energy area in Iran and was particularly concerned with developing a resource at Mount Damavand, from which it was intended to insert 50 MW of electrical power into supplies for the capital, Tehran. Studies were being conducted also in Azerbaijan around two extinct volcanoes, Mounts Sahand and Sabalon. Accessibility was a problem here, not even to mention political danger partly connected with the close proximity of Turkey and the USSR. Support for the projects came from the energy ministry. Unfortunately, they had to be abandoned after the departure of the Shah and the arrival of the Ayatollah Khomeini in 1979. The first edition of this book was written back in England and periodic interest taken in potential development of geothermal energy in Eire, this in connection with the late Dr David Burdon, an old friend from earlier mutual collaboration in the United Nations.

The meteoric rise in geothermal data since 1979 justifies the appearance of this greatly enlarged and better illustrated second edition of *Geothermal Resources*. Commencing with an examination of the origins of earth heat, it proceeds to deal with geothermal systems and models, exploration and resources assessment, the exploitation and environmental impact (mostly benign) of geothermal fields, and the various uses to which geothermal energy can be put. There is a glossary of relevant terms and appendices on geothermal miscellanea, companies, organizations, places and world geothermal localities, and full Author and Subject Indexes are included.

The aim of the writer has been both to interest and to stimulate the reader in addition to offering a compact, but comprehensive, source of geothermal information for office and field use by earth scientists, engineers and isotope geologists, environmentalists, sociologists and land planners. I hope that this book fulfils this end.

ROBERT BOWEN
Münster

Contents

Introduction

'Now these things are so.'
HERODOTUS, 5th Century BC ('Father of History')

It has been known for a long time that the temperature of the Earth rises with depth below ground, heat flow from subterranean hot regions manifesting itself in this geothermal temperature gradient. It is enhanced by the continued generation of heat due to radioactive decay of uranium, thorium and potassium within the crust. But the natural heat flux driven by such temperature gradients is very low, the world average being around $50 \, mW/m^2$. This is three orders of magnitude lower than the mean solar input to northern Europe in winter; therefore such heat flows are much too small to be exploited directly. Nevertheless, over millions of years, they accumulate heat stores which can sometimes be mined.

However, the high concentrations of heat energy occurring in geothermal fields are another matter and these are much more attractive propositions from an economic point of view. For mankind, geothermal energy constitutes an integrative and practically self-renewing alternative energy source which as yet is far from fully exploited as compared with the conventional fossil fuel ones, despite its now-demonstrated capability of contributing significantly to the needs of many countries, particularly in the developing world. Geothermal exploration is necessary to find fields of commercial significance, and can identify them from related anomalies connected to the upwards movement of magma. Of course, factors such as temperature and enthalpy, the distribution of permeability and the depth of a suitable aquifer acting as a reservoir play a significant role in producing a spectrum of geothermal systems, including both convective and conductive varieties; the former include hydrothermal systems among which are almost all of those used until now to generate electric power commercially.

1

In 1970 the geothermoelectric power installed throughout the world was approximately 680 MW, over half of this being in Italy (384 MW). To the middle of the 1970s, there was a growth of approximately 7% annually; a similar development occurred with non-electrical usage of geothermal resources, the oil crisis of 1973 having acted as a great stimulus because of the necessity to reduce petroleum imports. By 1985 the global geothermo-electric power produced was approximately 4765 MW, a 600% increase in a mere 15 years. Almost two-thirds of this is in Asia. The inference that such power is crucial to the poorer countries is correct. Concomitant with all this, a greater understanding of geothermal resources has grown and more specialists are being trained in centres such as Pisa. There is an increasing number of case histories available and hot dry rock experiments are in progress, as also are investigations into geothermal aquifers, geo-pressurized reservoirs and even magma.

From that synthesis of data from the earth sciences called plate tectonics (see Chapter 1, Section 1.3.3) the distribution of geothermal resources on Earth becomes explicable, i.e. their locations relate to plate boundaries and parallel those of earthquakes and volcanoes as well as hot springs and geysers. The therapeutic value of some thermal sources was well known even in Roman times and natural heat was put to work in New Zealand ever since that country became inhabited, the Maoris having poetic legends to explain it. Geothermal anomalies and increases in geothermal gradients result from ascending heat embodied in mobile masses such as magma bodies. Such thermal sources as can be exploited are situated in rather shallow crustal levels of less than 10 km and constitute transient stores of heat.[1] It must be stated immediately that the word 'transient' is used here in a geological sense and refers to the geological time-scale.

In human terms, such energy sources are integrative and self-renewing. The associated magmas may be basaltic or silicic in type. The former arise from the material of the mantle by partial melting and either disperse their thermal energy through direct ascent to the surface and emplacement as dikes and sills or solidify near the base of the crust in such extensional regimes as the Basin and Range Province of the USA and hence increase the regional heat flow.[2,3] This may well induce hydrothermal convection along steep fault systems resulting from the extensional regime.[4] Silicic magmas are generated by the partial melting of mantle material and also by the differentiation of basalt, often associated with remelting of crustal material as well.[5] Because they are more viscous than basaltic magmas, they can be arrested several kilometres down in the crust.

According to R. L. Smith and H. R. Shaw in 1973, continental geothermal resources are more likely to occur in connection with silicic volcanism than with the basaltic variety.[6] Igneous intrusions possess variable temperatures which can be as high as 1200°C; therefore they can heat their emplacement environment considerably and produce a corresponding surface heat flow greater than several watts per square metre. Obviously, the effect is limited chronologically because of both conductive and convective cooling effects after intrusion occurs. Thus normally only Quaternary intrusions, i.e. those younger than a million years or so, in the shallow part of the crust, i.e. not deeper than 10 km, remain thermally active now.[7]

In overall terms, magmatic activity globally can be integrated into the plate tectonics hypothesis, which also permits the delineation of geographic regions with geothermal potential. The concept allows the inference that the upwelling of magma is most probable along the boundaries of plates, and geothermal resources associated with the intrusions of magma bodies will be found mostly along spreading mid-oceanic ridges, convergent plate margins (subduction zones) and intraplate melting anomalies (see Fig. 1). The rigid outer shell of the planet, i.e. the lithosphere, is split into seven large and numerous small plates which slide on a more viscous base, the asthenosphere, at velocities relative to each other of up to several centimetres yearly. The boundary between the asthenosphere and the overlying lithosphere is a zone of decoupling. Of course, there is interference at the plate boundaries. At the divergent ones (spreading ridges) new crust is created by upwelling molten material. At the convergent ones, one plate slides under another in a subduction zone. Along transform faults, plates pass each other horizontally and their margins are conserved. Where two plate boundaries or three plates meet is called a triple junction.

Apropos the spreading ridges, it is believed that up to one-third of planetary heat loss takes place along the system of submarine ridges roughly 40 000 km long and occupying a mere 1% or less of the terrestrial surface. Along these ridges lithospheric plates are sundered while basalt wells up from the mantle and fills in the gap to create new, oceanic lithosphere. Accompanying volcanism probably arises from the pressure release melting in the underlying mantle, as E. R. Oxburgh and D. L. Turcotte suggested in 1968.[8] Adjacent to the ridge axes, the neighbouring sea-floor slopes downwards, the hot lithosphere cooling with increasing distance and age from these features, a phenomenon accompanied by falling heat flow. As well as conductive cooling, there is hydrothermal convection in the oceanic crust and, if this is sufficiently permeable, cooling of the uppermost 6 or 7 km may take place. It was estimated that the

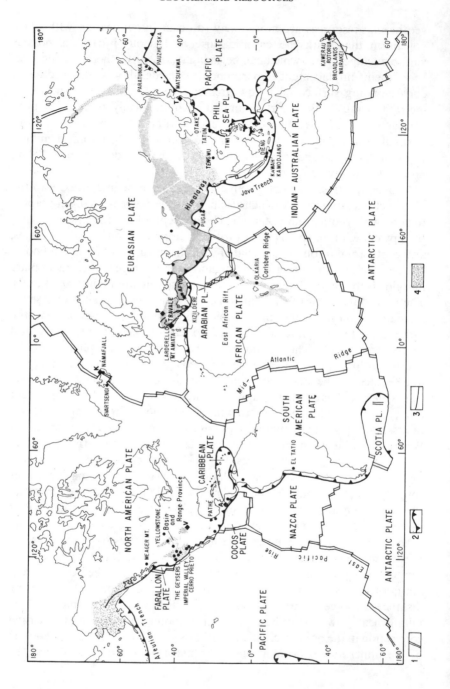

hydrothermal mass flow through the sea-floor from contemporary to 2 Ma in age is around 10^{11} g/km of ridge annually.[9]

Although there are some differences from a submarine oceanic ridge, Iceland constitutes the best place to see part of a ridge above sea level. The heat there is transferred to the surface by volcanism, conduction and hydrothermal activity, the output from the neovolcanic zone crossing the country totalling 56 MWt/km of ridge length. Of this, 21 MW/km come from volcanism, the same quantity from conduction, and 14 MW/km are contributed hydrothermally.[10]

Spreading at a much slower rate also occurs by the breaking up of continental lithosphere and manifests itself as rifts, below which pressure relief permits upwelling of molten material. The associated geothermal activity is lower-key and, in 1976, C. R. B. Lister estimated that the probability of finding a large-scale hydrothermal system is only 0·025 per km rift length and per cm/year spreading rate.[11] Continental rifting requires that a zone of weakness exists in the upper part of the crust and becomes associated with extensional tectonics, thereafter undergoing plastic deformation consequent upon rise in the thermal contribution from depth. This process induces faults to form and a central faulted block may result with an accompanying basal crustal mantle feature impregnated by basaltic material arising from the upper portion of the mantle because of the release of lithostatic pressure.[12] Of course, processes of this type are episodic, but geothermal activity persists for a long time. Probably the most famous continental rift structure on Earth is the East African Rift, which contains geothermal areas in Kenya and Uganda (discussed in Chapter 4), but there are many others of smaller size, e.g. the Basin and Range Province of the western USA, the Baikal Rift and the Upper Rhine Graben in central

---

FIG. 1. Lithospheric plate boundaries provide the framework for the global distribution of major geothermal systems. Plate boundary types: Spreading ridge (1), subduction/trench (2), transform fault (3). Shaded areas (4): Plate interior undergoing active extensional, compressional or strike-slip faulting. Base map and plate boundaries after Hamilton, W. (1976) and Panza, G. F. and Mueller, St. (1979); geothermal systems (dots) after Muffler, L. J. P. (1976a,b). Systems discussed in detail in Summary of Section I: Present status of resources development. In: *2nd UN Symp. on the Development and Use of Geothermal Resources, San Francisco*, pp. xxiii–xliv. A (Ahuachapán/El Salvador), K (Krafla/Iceland), P (Pannonian Basin/Hungary), T (Takinoue/Japan), V (Valles Caldera, Jemez Mts/USA). (Reproduced from Fig. 1.13 in Chapter 1, Geothermal systems, conductive heat flow, geothermal anomalies, L. Rybach. In: *Geothermal Systems: Principles and Case Histories*, ed. L. Rybach and L. J. P. Muffler, 1981. By permission of the publishers, John Wiley & Sons Ltd.)

Europe. The extension, subsidence and shoulder uplift of the last was discussed in 1986 by T. Villemin and others.[13]

Deep basins on continental crust are divisible into two major kinds, namely those with normal faults and with basement tilted either in the same direction (S-type) or in opposite directions (A-type), according to E. V. Artyushkov in 1987.[14] A large amount of stretch is involved in the former under a large angle of block tilting. This may be typical also in narrow S-type basins which do not exceed the thickness of the pre-stretched crust. It is only in basins of these two varieties that extension suffices to explain the subsidence by stretching. They may be called 'rifts'. Extended basins in the Basin and Range Province exemplify them. Deep basins of the A-type with a small angle of block tilting and deep basins of S-type width exceeding the thickness of the pre-stretched crust cannot be produced by stretching. They can be termed 'grabens'. Famous instances include the Fen-wei in central China and the Viking in the North Sea. In sum, Artyushkov proposed that deep basins created by stretching constitute rifts and narrow basins in which extension produced only a minor portion of the subsidence comprise grabens.[14] Deep internal processes and continental rifting were discussed in detail in a special issue of *Tectonophysics* in 1987.[15] In addition, in 1987 another special issue of the same journal discussed the main and regional characteristics of continental rifts.[16]

Subduction zones constitute regions in which oceanic lithosphere created at spreading ridges and moving laterally away from them are consumed by sliding under another plate, which is usually of 'continental' structure. The track of the descending plate is characterized by earthquakes occurring in brittle materials and comprises the Benioff zone. The material being subducted is in a state of both thermal and gravitational instability, being cooler and denser than the adjacent asthenosphere. Some frictional heating of the upper part will occur, but the earthquakes take place in the coldest (interior) part. Anyhow, substantial heating could occur only if the descent of the downwardly moving plate is resisted by large shear stresses (several hundred megapascals (MPa)) and these are very improbable given the rheological properties of the asthenosphere. In fact, P. Bird in 1978 reported stress values under 20 MPa from the topmost 100 km of subduction zones.[17] There are geothermal zones related to subduction zones in Kawah Kamodjang in Indonesia (thrusting of the India plate under the China one); in the Puga, Chumathang and Parbati Valleys, Himalayas, northwestern India (with same cause but in a complicated region of convergence of continental crust of each plate); and in El Tatio, Chile (connected with the subduction of oceanic lithosphere below the continental plate along the west coast of South America). An excellent recent survey of structures and processes in subduction zones appeared in 1985.[18]

Some observations on intraplate thermal anomalies are appropriate here. These are found in oceanic lithosphere in the Hawaiian Islands, the Azores and a number of volcanic chains and seamounts in the Pacific, e.g. Easter Island and the Cook–Austral Islands. In continental lithosphere, Africa possesses many Cenozoic volcanic centres clearly not related to plate marginal processes, e.g. the Tibesti, Hoggar and Bayuda ones. Domal uplifts occur in the first two and it is interesting that ancient rock art with pastoral motifs is found on the Tibesti Massif. Both it and the Hoggar record the last ice age in the shape of cryonivational zones of slow mass movement. The Nile describes a great bend around the region of Bayuda volcanics, an area of central Sudan. In fact, the Central African Rift System ends here, thereafter bearing southwest through the Jebel Marra complex to Mount Cameroun.[19]

The magma sources probably lie deep under the lithosphere and, while the latter can move over them, they go on acting as heat creators from fixed positions relative to the mantle, inducing mantle plumes. Because these do not move much relative to each other, they constitute potential reference frames for plate motions. Chemical plumes can be formed from the inhomogeneous distribution of radionuclides or from convective upwelling. Such perturbations may extend as far as 700 km laterally, produce heating up to 400°C and promote a heat flow anomaly of the order of as much as 70 mW/m². Associated effects in the overlying lithosphere include expansion due to heat, uplift, thinning, fracturing and eventual intraplate volcanism. That taking place in Africa is alkaline in type with differing degrees of silica-saturation and -undersaturation.[20]

It is possible to explain mid-plate volcanism in terms of the thicknesses of plates and their velocities because, according to L. Rybach in 1981, penetrative magmatism goes on with a time-scale one to several orders of magnitude greater than heat conduction.[21] In the cratonic portions of continental lithosphere having thicknesses around 200 km and heat flow of less than 45 mW/m², the rather slow movement of plates (under 2 cm annually) tends to suppress the upwards ascent of sub-lithospheric thermal anomalies and thus to suppress mid-plate volcanism also. Thinner lithosphere and slower plate movement are necessary in order to develop volcanism and thermal anomalies at the surface, and the time-scale will be 10 million years or so.[21]

However, there is another possible explanation for intraplate volcanism, namely that magma rises through propagating lithosphere fractures and thermal plumes may indicate the onset of rifting.[22] For instance, in Africa, uplifted domes such as the Tibesti and Jebel Marra may initiate a progression through rifted ones such as Kavirondo, East African Rift, to spreading rift types such as the Afar. The latter comprises a basaltic–trachytic strato-volcano and it is situated at the intersection of the Central African Rift System with a northwesterly-directed line of structures and volcanics passing through the Tibesti area and the Jebel al Haruj to the Jebel es Soda volcanics and the Nefusa Complex near Tripoli, Libya. The Jebel al Haruj is one of the most voluminous basalt complexes in Africa and covers 45 000 km², two post-Oligocene episodes having taken place. Radioisotope dating indicates that volcanicity here went from 6 to 0·4 Ma ago. The Jebel es Soda lies about 150 km northwest of it and covers 6000 km². Radioisotope ages from its oldest basalts range from 12·3 to 10·5 Ma (Middle to Late Miocene) in age. Turning to North America, western USA, mid-plate volcanism arises from the interacting of three lithosphere plates, namely the Pacific, the North American and the Farallon/Juan de Fuca. Probably thermal events and volcanic episodes relate to alterations in tectonic extension of regional stress fields.[23]

Finally, allusion may be made to thermal (gneiss) domes, these being found in metamorphic terrains and comprising a granitoid core within rocks of decreasing metamorphic grade and with horizontal extension up to many tens of kilometres. Obviously, a heat source must have created them and it is possible that localized release of carbon dioxide-rich fluids from the mantle took place. Support for the idea is given by the 'juvenile' carbon dioxide occurring in many spring discharges, e.g. at Urach in the Federal Republic of Germany.

Supporting data for the plate tectonic hypothesis derive from investigations into seismic events (earthquakes), volcanism, hot springs and geothermal waters. These phenomena occur around the world but are usually concentrated in certain regions. Geologic observation established the probability that the Earth has undergone earthquakes for at least several hundreds of millions of years and indeed historical records from China allude to these phenomena as early as 1800 BC, when they were attributed to supernatural causes. As late as AD 1750 a contributor to the *Philosophic Transactions of the Royal Society* excused himself to those 'apt to be offended at any attempts to give a natural account of earthquakes'. The Greeks considered that they are produced by such events as subterranean explosions and Aristotle classified them into six types, according to the kind of movement occurring, for instance elevation of the ground or its lateral shaking.

Earthquakes indicate that the planet is as stiff as steel which can behave as an elastic body when affected by short-term forces such as seismic waves, but, over the longer term, the Earth must behave as a fluid; for instance, the oblateness of the spheroid corresponds exactly to the period of rotation.

On occasion, earthquakes can arise from fault movements, e.g. the dextral San Andreas in California, adjacent to which the western sector of that state is moving northwards at an average rate of approximately 5 cm annually, each minor phase of this being marked by an earthquake. The region forms part of the circum-Pacific, arc–trench system known as the 'Ring of Fire' and also characterized by volcanoes. All of these features define mountain ranges, mid-oceanic ridges and major faults with considerable horizontal displacements. The location of earthquake foci implies the existence of a set of such faults underneath the Pacific marginal trench–island arc system dipping towards continents and extending down to depths as great as 700 km or more.

On land, probably the largest and certainly the most famous fault is the San Andreas, for which a project exists to drill down to the roots. Unfortunately, M. Barinaga in 1988 recorded that this has become a casualty of the failure of the National Science Foundation to obtain an expected budget increase.[24] The deepest scientific drill-hole in the USA, at Cajon Pass, was intended to reach 4·8 km, but the work will cease after completion of the second of three planned stages at a depth of 3·6 km. The aim was to test whether the fault in question is under less stress than was currently believed. NSF funding for Deep Observation and Sampling of the Earth's Continental Crust (DOSECC), a non-profit-making consortium comprising 40 universities, was reduced from an anticipated $6 million to $3 million after a White House budget summit in the autumn of 1987. This sum was not enough either to complete drilling in 1988 or to make measurements in the new segment of the existing hole. NSF has diverted an extra $1·8 million from different projects in its Continental Lithosphere Program to permit completion of the second phase both of drilling and measurements, but the reallocation rules out anything for 1989 and thus inactivates the project unless more money is made available in 1990.

The necessary evidence that San Andreas is a low-stress fault is dependent upon making sensitive heat-flow measurements to assess friction in the fault. These become more reliable at greater depths, at which they are less affected by shallow faults in the region. Should the low-stress hypothesis be substantiated, it would demonstrate that there must be friction somewhere else resisting the forward movement of the plates, possibly at horizontal faults deep below the surface. It has been recognized that earthquakes due to fracturing of rocks under accumulated stress are confined to the lithosphere, and in this part of the crust and uppermost mantle 100 km or so thick parallel tectonic activity is mostly confined to narrow zones such as the Ring of Fire around the Pacific Ocean. Such tectonically active zones are attributable to the boundaries of the seven main and more numerous minor semi-rigid and inert plates with peripheries of constructive, destructive and conservative kinds mentioned above and discussed in detail in Chapter 1 (see Section 1.3.3 on plate tectonics).

Damage to society through earthquakes and volcanoes can be devastating. As early as 1760, John Michell in the UK published a memoir on earthquakes, associating them with wave motion within the Earth. In the late 18th and early 19th centuries, investigations were made into the geological effects and it was noticed that buildings on soft ground were more affected than those on hard rock. Instrumental seismology really began in the middle of the last century when a *sismografo elettro-magnetico* capable of detecting earthquakes was invented in Italy. By the last decade of the 19th century, seismographs had been developed in Japan and the three principal types of seismic wave identified, namely P, S and surface. By 1906 the core had been recognized by using seismic waves and the San Francisco earthquake occurred. Thereafter a global system of seismic observatories was set up, an inner core was differentiated from an outer one and seismic waves, both natural and explosion-induced, were utilized to determine the thicknesses of layers down to that named after A. Mohorovičić in 1909.

After the Second World War, a great extension to the spectrum of recorded seismic waves took place, at one end permitting frequencies of the order of 100 Hz to be measured in ground movements and at the other facilitating measurements to be made of surface waves with periods as long as 10 min, and also of free oscillations of the whole Earth with periods of approximately an hour. Thus the spectral gaps between seismic wave vibrations of the order of seconds and diurnal tidal oscillations of the planet were bridged. Also rugged accelerographs were designed which could record strong ground motions in large-scale earthquakes.

Explosion seismology was further developed, especially with reference to the oil industry. The advent of nuclear explosions made it possible to extend the seismic explosion method plus its experimental controls to problems of the deep interior of the Earth. The first atom bomb was detonated in New Mexico in 1945 July 16d 12h 29m 21s (UT) at latitude 33° 40′ 31″ N and longitude 106° 28′ 29″ W from a tower 30 m above ground surface with a yield of *ca.* 19·3 kt. The first underwater nuclear explosion occurred 30 m below the surface of the ocean near Bikini Atoll and was recorded at only eight seismic stations. In March 1954 a hydrogen bomb was exploded near Bikini. Readings of seismic waves from four such explosions in routine station bulletins in 12 countries facilitated the computation of origin-times and these proved to be accurate within 0·0, 0·4, 0·6 and 0·1 s, respectively, for the four explosions. In 1956 atomic explosions at Maralinga, central Australia, enabled reliable knowledge to be obtained in a region where it had not been possible to make inferences from natural earthquakes.

In 1958, when the technical basis for a nuclear test ban treaty was under discussion in Geneva, there were about 700 seismograph stations operating and most were not particularly accurate (timing errors of several seconds were common). In the following year, the USA recommended installing a worldwide standardized seismographic network and this was done, each comprising three short-period seismographs and three long-period seismographs; timing and accuracy were maintained by crystal clocks, a calibration pulse being placed on each record daily. By 1967 some 120 stations existed in 60 countries. Data from them produced significant advances in regard to earthquake mechanisms, global tectonics and the structure of the interior of the Earth. By the 1980s digital stations were introduced, as were seismic research observatories in boreholes 100 m deep and modified high-gain long-period surface observatories. In addition, gravimeters with digital recording and response to very long wavelengths were deployed globally.

The aim is to accumulate information from seismic waves covering a broad band of frequencies. Analysis of waveforms in terms of frequency spectra is now widespread, as are special computing algorithms based upon Gauss's 'Fast Fourier Transform' developed by Cooley and Tukey. Modern minicomputers and microprocessors with peripheral display equipment permit the quick display of waveforms and spectra, and illustrate the value of graphics methodology to seismology.

Extraterrestrial seismology involving Apollo 11, 12, 14, 15, 16 and 17 lunar seismograph stations and other seismic stations on Mars constitutes an exciting new research field.

Earthquake damage limitation is the most desirable outcome of the work mentioned above and the same can be said of volcanological investigations. In 1902, for instance, three cataclysmic eruptions in Central America and the Caribbean caused the deaths of at least 36 000 people. They took place at La Soufrière on St Vincent (7 May), Mont Pelée on Martinique (8 May) and Santa Maria in Guatemala (25 October). Actually, a consequence of these events was the organization of volcano observatories which, through systematic documentation of local changes before, in the course of and after eruptions, determine the present eruptive behaviour of a volcano. Such data, together with geological mapping and dating work to assess the past history of a volcano, offer a basis for predicting future eruptive activity. This reflects the fact that most eruptions are preceded by measurable geophysical or geochemical changes, of which the most diagnostic involve variations in magma-induced seismicity and deformation of the surface of the volcano.

But even with such information successful forecasts are uncommon. It has been suggested that a mathematical analogy may be drawn between the terminal phases of material failure and the precursory behaviour of a volcano as it builds up towards an eruption.[25] The approach appears to be appropriate for the few eruptions analysed, but its real predictive potential has not been tested yet.[26]

It is interesting that H. G. Wells revealed his fascination in geothermal energy and its development for the benefit of mankind in a short story called 'The Queer Story of Brownlow's Newspaper', which he wrote in 1932 and referred to in his *The Shape of Things to Come* a year later.[27] Long neglected, the story was included, together with other uncollected short stories of his, in *The Man with a Nose* edited and with an introduction by J. R. Hammond, a book published by the Athlone Press, 45 Bedford Row, London WC1R 4LY, in 1984 (ISBN 0-485-11247-7). Previously, the work had appeared in the *Strand Magazine*, Volume 83, 1932. Permission by the publisher to quote the following geothermally fascinating portion is gratefully acknowledged. Firstly, to clarify, the protagonist Brownlow had been away and, after his return, found that his newspaper had been delivered. While reading it, he was amazed to see that it was dated 1971 instead of 1932. Among other items was the geothermal one. Unfortunately, he lost the newspaper and so could only describe the contents to his friend! This is how he did so:

> The streamer headline across the page about that seven-mile Wilton boring is, to my mind, one of the most significant items in the story. About that we are fairly clear. It referred, says Brownlow, to a series of attempts to tap the supply of heat beneath the surface of the earth. I asked various questions. 'It was explained, y'know,' said Brownlow, and smiled and held out a hand with twiddling fingers. 'It was explained all right. Old system, they said, was to go down from a few hundred feet to a mile or so and bring up coal and burn it. Go down a bit deeper, and there's no need to bring up and burn anything. Just get heat itself straightaway. Comes up of its own accord—under its own steam. See? Simple.'

'They were making a big fuss about it,' he added. 'It wasn't only the streamer headline; there was a leading article in big type. What was it headed? Ah! The Age of Combustion has Ended!'

Now that is plainly a very big event for mankind, caught in mid-happening, 10 November 1971. And the way in which Brownlow describes it as being handled, shows clearly a world much more preoccupied by economic essentials than the world of today, and dealing with them on a larger scale and in a bolder spirit.

The last paragraph of the tale reveals Wells in a rather over-optimistic mood, but there can be no doubt that geothermal energy does constitute a significant resource and one being rapidly developed. Its value to the constricted economies of the debtor countries is considerable and growing, so an up-to-date review of the subject is useful. An earlier work by the present author was published in 1979 and the present book, technically a second edition of the earlier one, is actually entirely new. Seven chapters follow this introduction and provide a state-of-the-art survey of the origins of earth heat, geothermal systems and models, geothermal exploration, geothermal resource assessment, the exploitation of geothermal fields, the environmental impact of such activities and the uses of geothermal energy in society; a glossary of relevant terms and appendices on some geothermal miscellanea and companies of geothermal interest are followed by indexes on authors cited, topics discussed, places mentioned and geothermal localities worldwide. A number of plates and figures illustrate various aspects of the subject and, where appropriate, acknowledgements are given. Its philosophy incorporates the remark of R. H. Dana that 'Earth Science is a particularly alluring field for premature attempts at the explanation of imperfectly understood data'.

REFERENCES

1. MUFFLER, L. J. P., 1976. Tectonic and hydrologic control of the nature and distribution of geothermal resources. In: *Proc. 2nd UN Symp. on the Development and Use of Geothermal Resources, San Francisco, CA, USA*, pp. 499–507.
2. WYLLIE, P. J., 1971. *The Dynamic Earth*. John Wiley & Sons, New York, 416 pp.
3. LACHENBRUCH, A. H., 1978. Heat flow in the Basin and Range Province and thermal effects of tectonic extension. *Pure Appl. Geophys.*, **117**, 34–50.
4. LACHENBRUCH, A. H. and SASS, J. H., 1977. Heat flow in the United States and the thermal regime of the crust. In: *The Earth's Crust*, ed. J. G. Heacock, pp. 625–75. American Geophysical Union Monograph No. 20. Washington, DC, USA.

5. ROBINSON, P. T., ELDERS, W. A. and MUFFLER, L. J. P., 1976. Holocene volcanism in the Salton Sea geothermal field, Imperial Valley, California. *Geol. Soc. Am. Bull.*, **87**, 347–60.
6. SMITH, R. L. and SHAW, H. R., 1973. Volcanic rocks as geologic guides to geothermal exploration and evaluation. *EOS, Am. Geophys. Union Trans.*, **54**, 1213 (abstract).
7. HEALY, J., 1976. Geothermal fields in zones of recent volcanism. In: *Proc. 2nd UN Symp. on the Development and Use of Geothermal Resources, San Francisco, CA, USA*, pp. 415–22.
8. OXBURGH, E. R. and TURCOTTE, D. L., 1968. Mid-oceanic ridges and geotherm distribution during mantle convection. *J. Geophys. Res.*, **73**, 2643–61.
9. FEHN, U. and CATHLES, L. M., 1979. Hydrothermal convection at slow-spreading mid-ocean ridges. *Tectonophysics*, **55**, 239–60.
10. PÁLMASON, G., 1973. Kinematics and heat flow in a volcanic rift zone, with application to Iceland. *Geophys. J. Roy. Astr. Soc.*, **33**, 451–81.
11. LISTER, C. R. B., 1976. Qualitative theory on the deep end of geothermal systems. In: *Proc. 2nd UN Symp. on the Development and Use of Geothermal Resources, San Francisco, CA, USA*, pp. 459–63.
12. MUELLER, ST., 1978. Evolution of the Earth's crust. In: *Tectonics and Geophysics of Continental Rifts*, ed. R. Shagam, pp. 11–28. Geol. Soc. Am. Memoir 132, Boulder, CO, USA.
13. VILLEMIN, T., ALVAREZ, F. and ANGELIER, J., 1986. The Rhinegraben: extension, subsidence and shoulder uplift. *Tectonophysics*, **128**, 47–59.
14. ARTYUSHKOV, E. V., 1987. Rifts and grabens. *Tectonophysics*, Deep internal processes and continental rifting, ed. C. Froideaux and T. T. Kie, Sp. issue, **133**(3/4), 321–31.
15. FROIDEAUX, C. and KIE, T. T. (eds), 1987. *Tectonophysics*, Deep internal processes and continental rifting, Sp. issue, **133**(3/4), 1–334.
16. RAMBERG, I. B., MILANOVSKY, E. E. and QUALE, G. (eds), 1987. *Tectonophysics*, Continental rifts—principal and regional characteristics, Sp. issue, **143**(1–3), 1–252.
17. BIRD, P., 1978. Stress and temperature in subduction shear zones: Tonga and Mariane. *Geophys. J. Roy. Astr. Soc.*, **55**, 411–34.
18. KOBAYASHI, K. and SACKS, I. S. (eds), 1985. *Tectonophysics*, Structures and processes in subduction zones. Selected papers from the Interdisciplinary Symposium No. 1, the 18th General Assembly of the International Union of Geodesy and Geophysics, Hamburg, 22–24 August 1983, Sp. issue, **112**(1–4), 1–561.
19. BOWEN, R. and JUX, U., 1987. *Afro–Arabian Geology: A Kinematic View.* Chapman and Hall, London, 295 pp.
20. GASS, I. G., CHAPMAN, D. S., POLLACK, H. N. and THORPE, K. S., 1978. Geological and geophysical parameters of mid-plate volcanism. *Phil. Trans. Roy. Soc. London*, **A288**, 581–97.
21. RYBACH, L., 1981. Geothermal systems, conductive heat flow, geothermal anomalies. In: *Geothermal Systems: Principles and Case Histories*, ed. L. Rybach and L. J. P. Muffler, pp. 3–36. John Wiley & Sons, Chichester, 359 pp.
22. OXBURGH, E. R. and TURCOTTE, D. L., 1968. Mid-oceanic ridges and geotherm distribution during mantle convection. *J. Geophys. Res.*, **73**, 2643–61.

23. CHRISTIANSEN, R. L. and MCKEE, E. H., 1978. Late Cenozoic volcanic and tectonic evolution of the Great Basin and Columbia Intermontane regions. Geol. Soc. Am. Memoir 152, Boulder, Colorado, pp. 283–311.
24. BARINAGA, M., 1988. San Andreas drilling halted. *Nature*, **332**(6160), 105.
25. VOIGHT, B., 1988. A method for prediction of volcanic eruptions. *Nature*, **332**(6160), 125–30.
26. TILLING, R. I., 1988. Lessons from materials science. *Nature*, **332**(6160), 108–9.
27. WELLS, H. G., 1932. The Queer Story of Brownlow's Newspaper. In: *The Man with a Nose*, 1984. Edited and with an Introduction by J. R. Hammond. The Athlone Press, London.

CHAPTER 1

The Origin of Earth Heat

'Speak to the Earth and it shall teach thee.'

JOB, 12: 8

1.1 BEGINNINGS

The genesis of the Earth, its differentiation, the development of the core, mantle and crust, accretional heating and heat source distribution in the planet through time are discussed.

1.1.1 Genesis

To describe the derivation of terrestrial heat, it is necessary to consider the history of the Earth itself, which is intimately connected with that of the solar system as a whole. In his classic work on *The Planets*, the Nobel Prize Laureate, Professor H. C. Urey, in 1952 presented a simple model for this, one which was later developed by E. Anders in 1964 and A. E. Ringwood in 1966.[1-3] New information from continuing researches into meteorites has made it necessary to devise a more complicated model.

Data show that the solar nebula was chemically heterogeneous and probably did not develop under conditions of physico-chemical equilibrium. Indeed, the Earth itself is not in chemical equilibrium, retaining traces of its inhomogeneous growth despite convection and volcanism through time. The likeliest sequence of events entailed intermediate processes between the initial gas and dust cloud and the existing bodies in the solar system. Initially, there may have been a loss of gas and separation of dust leading to the formation of randomly mixed, segregated or zonated lumps of matter anything up to 10 m across from which small planetesimals of up to 100 km diameter could have arisen.[4]

1.1.2 Accumulation of Planetesimals

Subsequently, some planetesimals may have accreted smaller ones during a major growth episode around 100 million years in length. The effect is regarded as particularly important at the time when Jupiter-scattered planetesimals prevented a planet forming in the region of main-belt asteroids and hindered the growth of Mars. When the planets had been formed, the vast majority of the residual planetesimals are thought to have been removed rather rapidly (over half an eon or so) in the heavy bombardment episode of which evidence is seen in the vast lunar basins. Later, this mopping-up decelerated, a process accompanied by meteorite formation which resulted, at least partly, from relatively slow sunward migration of collision debris (from dead comets and the main-belt asteroids) past the planet Mars.

1.1.3 Differentiation of the Earth

The thermal history of any planetary body will depend upon a number of factors, among the most important of which are the initial temperature, the quantities and distributions of heat sources, the ability to store heat (heat capacity) and the mechanism of heat transport. In the case of the Earth, the last is crucial, namely how it can eliminate internally produced and stored heat through the terrestrial surface. As will be seen later (e.g. see Section 1.2.7), thermal convection in the mantle dominates the heat transport and its vigour is measured by using the Rayleigh number, which is inversely proportional to the viscosity.

It is believed that most, if not all, of the Earth melted as it grew. Heavy iron-rich liquid sank into the core and crystalline silicate gradually formed a slowly convecting mantle with silicate-rich liquids of approximately basaltic composition transporting heat-producing lithophilic elements such as potassium, thorium and uranium towards the surface. Motions in the core are thought to be quite separate from those in the mantle.

In the mantle, the creep properties of the rock are a function of temperature, a change in the latter of 100°C causing a variation of viscosity by a factor of 10. The phenomenon should lead to an efficient self-regulating mechanism linking heat loss and the internal temperature of the planet. Quantification of the concept has been achieved by using a 'parameterization' for the convective heat transport. This is an attempt to evade the complexities of the flow pattern and the temperature field of convection cells (see Section 1.2.4) by substituting a simple relationship between the Rayleigh number, Ra (inversely proportional to the viscosity), and the Nusselt number, Nu (a measure for the global efficiency of convective heat transport). Then Nu is proportional to Ra^β, in which the exponent β is usually assumed to be of the order of $\frac{1}{4}$ or $\frac{1}{3}$.

Clearly, β is a key parameter. In parameterized convection models, the values quoted above appear to be valid, having been confirmed for Rayleigh–Bénard convection, where the viscosity is constant. This has been done both empirically and by means of theoretical and numerical work. The parameterized models show that heat generation and heat loss cannot be in equilibrium due to the secular decline in radiogenic heat production and the large thermal inertia of the Earth.

In addition, they show that the planet must have been cooling at least since the Archean when the heat flow was perhaps not much greater than now (but see Section 1.3.3). However, the mantle temperatures are thought by some to have been on average around 200°C or perhaps even 300°C higher then than now. If this were indeed the case, then more advanced stages of partial melting may have led to the copious generation of komatiitic magmas. In addition, it is probable that the production of oceanic plates was then faster and their mean thickness much less, e.g. from 25 to 40 km. These plates derive their name from plate tectonics, the concept asserting that the lithosphere is divided into a number of plates which move independently of each other. The lithosphere itself is the rigid shell of the planet, embracing that region between the surface of the Earth and its asthenosphere, namely the crust and the uppermost part of the mantle. The word sima is sometimes applied to the lower portion of the crust, this being taken as predominantly composed of silica and magnesium. In contradistinction, the upper portion of the crust which underlies all the continents may be designated as the sial, comprising mainly silica and alumina. There is a detailed discussion of the plate tectonics hypothesis in Section 1.3.3 below.

One of the major topics of debate at present relates to the geometrical relationship, if any, of mantle convection to the density structure responsible for the known seismic discontinuities. This is focused on the degree to which documented density changes around 650 km present an impediment to convective communication between the core and mantle. A detailed examination of this boundary is given in Sections 1.2.4, 1.2.7, 1.2.9, 1.2.10, 1.2.12, 1.3.4 and 1.3.9. Of course, the relatively thin, rigid and weak continental lithosphere takes no part at all in convection. However, it is believed that a poorly defined seismic discontinuity exists in its deeper parts. This is thought to mark the transition from felsic to mafic rocks or from amphibolite to granulite facies and it has been termed the Conrad discontinuity.

1.1.4 Accretional Heating

During a convective overturn, every gram of liquid which cools *ca.* 1200 K to less than 373 K, i.e. to the assumed temperature of the Earth's surface, loses some 120 J of latent heat and about 1500 J of sensible heat. It is not possible to give a reliable estimate of the amount of heating from accretion and core formation, but the latter would be unlikely to exceed 3000 J/g for the bulk Earth.[4] The former depends on how accretion occurred and also on the nature of planetesimal impacts. Probably most such bodies disintegrated before hitting the planet and most of the kinetic energy would have dissipated as radiation from the debris. After accretion, there should have been about 2 km thickness of water near the surface of the Earth and extensive vesiculation of impact debris ought to have taken place. If accreting material arrived with a mean temperature of 1000 K and the mean temperature of the Earth was *ca.* 3000 K at the end of accretion, then approximately 2500 J/g was needed to heat the Earth. For growth over 60 Ma, the radius of the planet on average would increase only 10 cm annually. Bodies 100 km across should excavate holes some 10 km or more in depth and up to 1000 km across. If the planet grew from such bodies weighing $2 \times 10\,000$ billion billion† g, a body would hit only every 20 years on average. It is impossible to give any quantifiable schema, but it may be assumed that most of the Earth's surface was cool for most of the period of accretion with only localized ephemeral surges from large impacts, and also that volcanism dissipated the impact heat as near-surface material participated in several impacts.[5] Of course, whether a global magma ocean developed at any time is highly speculative.

1.1.5 Heat Source Distribution During Accretion

Impact heating was concentrated in the outermost 100 km, where it rapidly dissipated through volcanism. The heat-producing elements potassium, uranium and thorium are concentrated here too and most probably primarily in the outermost 20 km (see Table 2.2).[6] The interior has only one main heat source, namely the gravitational energy of sinking iron-rich liquid and, if core formation went on continuously throughout growth, then this energy is taken up by the specific heat required to convey the deep interior from the <1400-K level when the Earth was very small to the roughly 4000-K level desired at the end of growth for a completely molten core. Also the adiabatic curve is steeper than the solidus and, therefore, it might be inferred that most of the Earth could have remained solid during

† Billion $= 10^9$.

FIG. 1.1.   Lava flows at Etna. Reproduced by courtesy of the International School
of Geothermics, IRG-CNR, Pisa, Italy.

accretion even if there were multi-level convection. Today the surface heat
flow of 4×10^{13} W is larger than the production of heat from radioactive
decay because of the cooling of the Earth (possibly one-third of the heat
flux) and delay in transport of heat to the surface.[7-9]

It is not necessary to assume that the planet has a content of uranium
exceeding 20×10^{-9}.[9] Indeed a quite permissible and lower value of
$(12-18) \times 10^{-9}$, i.e. within the range for devolatilized chondrites, is feasible.
Extrapolating back and allowing a six-fold depletion of potassium in the

bulk Earth, then the heat flow in the pre-Archean should have been about 1.3×10^{14} W, according to J. V. Smith.[4] From the work of R. St J. Lambert in 1980, it is apparent that this would have corresponded to the entire planetary surface then possessing a heat flow similar to that of Iceland now.[10] Hence it may be concluded that volcanism triggered by the escape of heat from the interior of the Earth could have destroyed early crust as this foundered under masses of new lava, such a process being aggravated locally by impact events from the same population of bodies which formed the lunar basins.[5,11,12] (Fig. 1.1).

There is no need to envisage complete volatilization of the surface of the Earth at the end of accretion, despite the probability that early episodes involving impact-induced fission must be considered possible. In fact, ice rather than hot water may have been present at this stage because the early Sun may have had its low brightness offset by a greenhouse effect due to an enhanced CO_2 concentration in the atmosphere. High-speed impacts may have triggered localized disruptions and accretion would have produced degassing of the planet through volcanism, a requirement for satisfying the constraint from argon isotopes, a primitive atmosphere rich in nitrogen and carbon monoxide being permissible.[13,14]

The above schema is the basis of the heterogeneous model of this planet. Its significance is circumscribed, to some degree, by the realization that most of the relevant information required to give an accurate picture of the far distant events under consideration undoubtedly has been lost during frequent collisions and melting in the early stages of its history. To complete it, the subject of plate tectonics must be examined as well and this is done in Section 1.3.3 (see below). Bearing this in mind, however, it is appropriate initially to make some additional remarks regarding the development of the Earth.

1.2 DEVELOPMENTS

These include a variety of topics relating to details of the core, mantle and crust of the Earth, subduction and melting processes, megaliths, plum pudding mixture, the role of eclogite, material transfer through the double-boundary thermal zone separating the upper and lower parts of the mantle, and a temperature profile.

1.2.1 Origins of Core, Mantle and Crust
The planet probably continuously differentiated into a core, mantle and

crust while hot planetesimals were being captured. This process entailed loss of an original uniform density because the existing core is much denser than the mantle and the crust. The accepted figures are $11 \cdot 5 \, g/cm^3$ in the inner solid core, $10 \cdot 2 \, g/cm^3$ in the outer liquid core, $ca.$ $5 \cdot 7 \, g/cm^3$ in the mantle as a whole and about $2 \cdot 7 \, g/cm^3$ in the crust (see also Table 1.1). The rather abrupt density change between the latter and the upper mantle comprises the famous Mohorovičić discontinuity (or MOHO).

In addition, the separation of core, mantle and crust represented a decrease in potential energy of the order of 10^{38} ergs, energy which would have been released in the form of heat whether this separation was slow or quick.

The earliest material was probably arid and reduced, later material being more oxidized and containing both water and carbon dioxide. Obviously, not only meteoritic evidence but also facts obtained from the existing Earth must be utilized in order to fill in some more details. Thus density and compressibility profiles confirm the existence now of a solid inner core and a liquid outer one, a mantle and a crust.[15]

A large number of properties of both the crust and the upper mantle can be inferred by correlating plate tectonics (discussed in detail later, see Section 1.3.3) and the phase equilibria for peridotite. H. S. Yoder Jr in 1976 suggested that a type of melting and cooling cycle must have taken place through Earth history in order to remove heat.[14] This could have resulted from thickening of the continents during the Proterozoic, which would have concentrated convection in the sub-oceanic areas so as to achieve a rather efficient cooling mode. Nevertheless, crustal patches survived and related to continental growth. Today oxygen and silicon are the most abundant chemical elements, together with significant quantities of aluminium, iron, potassium and magnesium, the commonest compounds being silicates and metallic oxides. So as to evaluate feasible conditions in the juvenile Earth, it is advisable to examine the properties of the existing Earth, whenever appropriate comparing them with those believed to have prevailed during the past history of the planet.

1.2.2 Oceanic Crust

The ephemeral oceanic crust is covered by sediments on top of alkali-poor basalts and gabbros which extend out from mid-oceanic ridges. These rocks arise from the leakage of magma out of chambers which are supplied by the melting of rising peridotite. The depleted residues (of harzburgite type) are thought to act as a kind of conveyor belt for the sediments and basalts. Incidentally, the chemical composition of harzburgite in this depleted form is 46·5% MgO, 43·6% silica, 8·8% FeO, 0·6% Al_2O_3 and 0·5% CaO.

Oceanic islands accrete from alkali-rich basalts rising from restricted areas of the mantle. Consequently, it is clear that, in the Earth today, the source regions of mid-ocean ridge basalts (MORBs) and ocean island basalts (OIBs) are chemically quite distinct.

MORB may be defined as the oceanic crust of that part of the ridge system with bathymetric depths exceeding 2 km. As regards their petrogenesis, MORBs seem to have formed by the partial melting of a pyrolite source region which underwent episodic extraction of small quantities of alkali magmas, enriched in incompatible elements, over a time interval exceeding a billion years (see also Section 1.3.3). The chemical composition of pyrolite is 44·6% silica, 38·1% MgO, 9·1% FeO, 4·3% Al_2O_3, 3·5% CaO and 0·4% Na_2O. There appears to be a near-uniformity of the MORB source reservoir attributable to mixing as a result of convection in this reservoir; but see also Section 1.2.6 below on 'plum pudding' mixture.

On the other hand, OIBs are not uniform, either in terms of incompatible element concentrations or of isotope ratios. The group includes anomalous sections of the mid-ocean ridge system such as Iceland and the Azores. Its isotopic heterogeneity is explicable by simple mixing between a MORB source reservoir (the upper mantle) and a relatively pristine and undepleted reservoir (the lower mantle). However, in certain cases of greater complexity, e.g. Tristan and St Helena as well as the Azores, this does not suffice and the inhomogeneity may have resulted from subducted oceanic crust and entrained sediments which have not been mixed back into the upper mantle reservoir. Such observed isotopic heterogeneities may have taken billions of years to develop. In this event, storage of subducted material in the mantle probably cannot have occurred in view of the strong mixing associated with mantle convection. An alternative proposal is that isotopic heterogeneities in OIB developed in the continental crust and mantle comprising the continental lithosphere.

1.2.3 Continental Crust

The continental crust comprises a highly heterogeneous mixture of both igneous and sedimentary rocks, some of which have undergone metamorphism since their accumulation through the past 3·8 Ga.[15] The mean continental age appears to be approximately 2·1 Ga based upon the Rb–Sr system and 3·3 Ga based upon the samarium–neodymium system. However, other estimates incorporating geological and geochemical constraints cover a greater time span extending from as little as 1·38 Ga to as much as 3·5 Ga.

Geothermal and volcanic areas are composed largely of igneous rocks, ←
either in their original condition or highly metamorphosed. In fact, granite is the commonest continental rock. Often it was intruded as a concentrated crystal mush or perhaps even in the solid state to comprise a batholith. Batholiths are intrusions of considerable thickness and lateral extension, with resultant heat flow at the terrestrial surface varying with time. A batholith will have an area exceeding $100 \, km^2$ and a diameter which is assumed to increase with depth, although this is hard to judge because no visible floor has ever been found for any particular one.

Other continental basements demonstrate, through the abundance of their Precambrian gneisses, that ductile laminar flow is common in the lower part of the crust. However, it has been argued that, because heat-producing elements and water have been lost to overlying rocks, its base is stable, as R. C. Newton et al. indicated in 1980.[16] Often basaltic rocks emerge through fractures in continental crust, but it is uncertain whether the resultant depletion of the upper mantle produces a stable 'raft' under each of the continents.[17]

1.2.4 Melting During Subduction

Of course, oceanic rocks can be compressed against continental margins or slid beneath them in the subduction process, during which both water and carbon dioxide are liberated as oceanic basalts and sediments become heated. The process reflects convergent movement, the subducting (oceanic) plate thrusting underneath the over-riding continental one. Most of the total volume of subducted oceanic lithosphere comprises mature plates anywhere from 60 to 100 km in thickness and between 50 and 100 Ma old, such plates having much greater thermal inertia than younger, thinner plates. When such mature plates reach the 650 km discontinuity, they will have interiors hundreds of degrees cooler than the surrounding mantle and even display seismic activity to this depth. Also they will possess mean viscosities several orders of magnitude higher than surrounding mantle.

It seems that, unless the subducting lithospheric slab is especially cool, it is very difficult to transport water deeper than about 100 km, although large blocks of carbonate isolated from water may well endure long enough to carry carbon dioxide into the mantle. Actually, at the moment, the degree to which melting occurs in a descending slab and the overlying mantle ledge is far from clear. No doubt a cold descending lithospheric slab beneath island arcs would constitute a major source of gravitational energy, density contrast with the enveloping asthenosphere causing sinking. Sea floor would be dragged along as well, mantle material upwelling so as to fill in gaps created at a ridge axis.

R. L. Armstrong in 1981 assumed that the higher temperature of the primal crust and upper mantle must have initiated extensive melting of such descending slabs as well as partial consumption of the continental margin.[18] That this assumption is valid is implied by the decreasing abundance of ultrabasic rocks from the Archean to the present, a phenomenon explicable only on the basis of a decreasing temperature for the upper mantle. Melts in the Earth later produced gigantic mountain chains such as those on the western margins of both North and South America. Their elevations are probably limited by the creep strength of the lithosphere, this being, to some extent, temperature-dependent. The situation in the mantle is similar in that, as noted above in Section 1.1.3, the creep properties of the rock are a strong function of the temperature, a temperature change of 100°C causing variation in viscosity by a factor of 10.

It is assumed that similar calc-alkaline volcanism took place in the infant Earth and later remelting processes produced potassium-rich granites surmounting stable continents. In the past quarter century, a great deal of experimental progress has been made in elucidating the petrogenesis of both basaltic and calc-alkaline volcanic rocks. Most petrogenetic models mainly relate to explaining the major and trace element compositions of these associations in terms of varying degrees of partial melting of mantle source rocks, succeeded by such processes as fractional crystallization and crustal contamination.

In association with this, the geochemists have provided much information regarding the uranium–thorium–lead, rubidium–strontium and samarium–neodymium isotopic systematics of relatively young volcanic rocks. The relevant data indicate highly complex chemical differentiation processes in the mantle taking place over time-scales of the order of 1–2 Ga, i.e. far earlier than the actual production of the volcanics themselves. The problem is how to reconcile the relatively recent partial melting and differentiation processes with the extremely old differentiation processes recorded by isotopic systematics within the framework of a generally acceptable model for the dynamic and geochemical evolution of both the mantle and the crust. Attempts have been made to do so.[19]

In the first 500 Ma or so of the life of this planet, volcanism was probably much more intense, so that continental-type crust probably would have been composed of only a few small scum patches not exceeding some hundreds of kilometres across. These probably concentrated above descending boundaries of polygonal convection cells. Centrally these would have possessed upwelling nuclei, between which and the peripheries there would have been cylindrical zones of adiabatic ascent from deep levels.

If such cells really existed, they would not have been able to cover the terrestrial spheroid because of Euler's relation; where some were present, there would have been no mantle wedges over descending oceanic crust and, of course, the magmatic processes would have been rather different from those associated with asymmetric convection.

1.2.5 Mineral Transformations and Slab Sinking (Megaliths)

However magmatism actually works at continental margins, there can be no doubt that the transformation of Ca–Al-bearing minerals such as plagioclase feldspar in the near-surface rocks into much denser minerals such as garnet occurring in foundering rocks could induce sinking. Now a descending slab of oceanic basalts and sediments should transform primarily into a garnet–pyroxene rock (eclogite) which is denser than all types of peridotite in the upper mantle, but not in the lower mantle where magnesium silicates transform into the extremely dense perovskite structure.

A simple model would involve an eclogite slab descending to the base of the upper mantle in the present-day Earth. Buckling of such slabs causes the development of 'megaliths' of relatively cool, but deformed, former oceanic lithosphere. This is discussed in detail later: see Section 1.3.4.[19] In the primal Earth, the garnet-carrying base of crustal scum could have foundered into the mantle as it sank under the superincumbent load of volcanic masses and debris from huge impact events.

1.2.6 'Plum Pudding' Mixture

Looking at the present continental crust, the interesting phenomenon of 'plum pudding' mixture has to be considered. Alkali basalts from depths not in excess of about 80 km and occasional kimberlite fragments from under 200 km in depth tear off rock fragments in the course of ascent to the surfaces of continents. They then form this mixture of metamorphosed lherzolites (olivine–two-pyroxenes–garnet/spinel–mica) and less abundant eclogites. The 'plum pudding' is veined by volatile-rich liquids and overlain by harzburgites.[20,21]

If whole-mantle convection took place, then this mixture may cause heterogeneities in both MORB and OIB. The chemical composition of basalt (MORB) is 49·7% silica, 16·4% Al_2O_3, 13·1% CaO, 10·1% MgO, 8% FeO, 2% Na_2O and 0·7% TiO_2. If, as some claim (see Section 1.2.2), MORB is more or less homogeneous, then the heterogeneities associated with OIB could be explained by invoking whole-mantle convection with a continuous stratification.

1.2.7 Convection in the Mantle

The nature of convection in the outermost 400 km of the Earth's mantle may be determined by the rheology of the mineral olivine, (Mg, Fe)2SiO₄, the most abundant and weakest phase. In fact, R. S. Borch and H. W. Green II in 1987 demonstrated that creep data for this mineral, when normalized by the melting temperature, can be extrapolated directly to the pressure, temperature and chemical environment of the upper mantle by employing the appropriate solidus and geotherms.[22] Of course, these were assumed, but the use of empirical rheologies provides much stronger constraints on viscosities in the upper mantle than those inferred by geophysicists from crustal plate motions and isostasy. For instance, many geophysical models assume Newtonian dislocation creep in the mantle, whereas experimental results suggest that upper mantle creep involves non-Newtonian dislocation creep mechanisms.

As S. J. Mackwell stated in 1987, no creep data have been directly measured for the spinel phase of olivine and so it is not clear whether the arguments of Borch and Green can be extrapolated to the mantle below the olivine–spinel transition.[23] In addition, as there are not enough data on the nature of the water-weakening effect in olivine, it is not yet established that high viscosities in the upper mantle necessarily imply a low content of volatiles. Nevertheless, the idea that the strength of olivine-rich materials depends upon the melting temperature affords a mechanism for determining the effects of changes in the physical and chemical environment on the behaviour of olivine-rich materials in the Earth simply by assessing the changes in the solidus of the material in a laboratory. Borch and Green presented data collected over a wide range of pressures and inferred effective viscosities from them which are claimed to be consistent with geophysical and tectonic constraints for a dry upper mantle.[22] The systematics are regarded as suggestive of an effective viscosity for the mantle which rose over the age of the Earth as outgassing progressively raised the solidus.

Borch and Green effected a set of axial compression experiments at constant strain rate over a pressure range of *ca.* 2 GPa, a wider range than any previously used.[22] The studies were conducted using a modified Griggs apparatus in which specimens enclosed in platinum capsules were surrounded by a molten alkali halide, greatly reducing the friction problems connected with solid-pressure-medium investigations. The material utilized was a synthetic harzburgite (depleted peridotite) comprising 97% olivine and 3% orthopyroxene with a grain size of *ca.* 50 μm. The significance of the rheology of olivine in regard to convection in the upper mantle has been discussed above. Borch and Green, in their model, gave an effective viscosity of 2×10^{21} Pa s, corresponding to a strain rate of 0.6×10^{-15} s^{-1}.[22] It is very difficult to compare effective viscosities with those assessed by geophysical measurements by assuming Newtonian behaviour ($n = 1$), but such a value for the two parameters in question (viscosity and strain rate) are of the right order of magnitude for plate tectonic motions.

M. W. McElhinny[13] and many others have discussed convection throughout the mantle, but assumed that a thermal boundary layer separates convection in the upper mantle from that in the lower mantle, and that this lies at a depth of around 650 km (see also Section 1.3.9). The upper mantle is taken as consisting mainly of peridotites depleted through various degrees of melting to yield the crust and it is rejuvenated, at least partly, by infiltrating liquids which leak out of descending material. A crucial assumption is that most sinking eclogite is mechanically mixed with adjacent depleted peridotite during the forced downward transportation of both which produces a depleted peridotite, but this is not generally accepted.

The gas loss of the outgassed mantle is extremely high, as much as over 99.6%, the isotopic signature of the lower mantle being entirely controlled by the un-degassed part of the mantle, even if the mantle proportions of the upper and lower mantle have changed through geological time (46–54% 4.5 Ga ago, 35–65% today). It is interesting to note that gas concentrations in these differ by a factor of 200 or more, which demonstrates that the mass exchange between the two reservoirs are minute (otherwise mixing with lower mantle material would have diluted all the observed isotopic excesses of the upper mantle).

1.2.8 The Role of Eclogite in the Upper Mantle

In fact, the literature on upper mantle mineralogy is bedevilled by the semantic factor since the petrological usage of the terms eclogite and peridotite is *sensu stricto* restricted to depths less than 150–200 km, at which the pressures do not rise above 45–60 kbar. At greater depth it seems that pyroxene constituents more and more tend to enter the garnet crystal structure while the $(Fe, Fe)2SiO_4$ component transforms from olivine structure to first the β-phase and then the spinel structure. It is not too much to say that until the role of 'eclogite' in the upper mantle is properly understood that part of the planetary interior cannot adequately be described.[4] In addition, the depths of the source regions from the various magma types are extremely unclear and indeed controversial.[24-27]

1.2.9 The Lower Mantle

As regards the lower mantle, the simplest supposition is that it also consists of peridotitic material, specifically the 'pyrolite' composition invented to model the upper mantle before extraction of basaltic crust.[28] Above, however, it was pointed out that often it is assumed that the lower mantle is compositionally different from the upper one because of inhomogeneous accretion and later mineralogical processes. In particular, it is believed to have a bulk composition rich in metasilicate ($MSiO_3$, where M is metal) rather than orthosilicate (M_2SiO_4) for pyrolite and peridotite. Since perovskite is very rich in magnesium and low in iron, it could fill the properties requisite for averaged lower mantle.[5] In fact, L.-G. Liu in 1979 proposed gravitational descent of perovskite during the growth of the Earth.[29]

In the Earth today there are at least two large-scale seismic discontinuities in the mantle, at depths of approximately 450 and 650 km, respectively, the deeper one separating an overlying transition zone from the underlying lower mantle. The major seismic discontinuity at 650 km entails phase changes at this depth. It may be that the discontinuity in question arises from an isochemical phase transformation of Mg_2SiO_4 spinel to $MgSiO_3$ perovskite plus $(Mg, Fe)O$ magnesiowüstite, occurring within a mantle of roughly uniform composition. This would allow whole-mantle convection to take place freely between the upper and lower mantles. Another explanation is that the discontinuity is connected to a change in chemical composition from an overlying upper mantle dominated by Mg_2SiO_4 olivine and spinel minerals to a relatively silica-rich lower mantle comprising essentially $MgSiO_3$ perovskite.

This hypothesis would imply essentially independent convective systems in the upper and lower mantles separated by a thermal boundary layer at 650 km. As regards the $(Mg, Fe)1-XO$ variety of magnesiowüstite (defect rock-salt structure), the problem is that the physical properties at high pressure of a bulk composition rich in $MgSiO_3$ resemble those of $(Mg, Fe)2SiO_4$ with a small Fe/Mg ratio and it seems that additional structural complications are to be expected, as M. Madon et al. indicated in

1980.[30] It is practically certain that the density for any particular bulk composition is not a simple function of pressure, temperature and mean atomic number. Also there is no obvious evidence obviating the possibility of a chemical change at the 650-km discontinuity.

1.2.10 Material Transfer Through the Double-Boundary Thermal Zone

Transfer of material through the double thermal boundary zone separating the upper and lower mantles may occur by perturbation, i.e. plumes of light material may escape from the lower to the upper mantle, but not a continuous stream of convecting material.[31] If the existing 650-km boundary is the result of a chemical discontinuity between olivine- and pyroxene-rich compositions, it is important to consider the effects of the advancing fronts of high-pressure minerals on convection during the growth of the Earth.[32]

Another significant feature refers to the distribution of iron among the mantle minerals. Iron is strongly partitioned into magnesiowüstite from perovskite, and overturning in the lower mantle might cause transfer of iron metal to the core and an associated upward transport of compensating ferric oxide component. However, this process would be inefficient in the solid state if limited by solid-state diffusion. Hence processes involving partial melting and perhaps mechanical deformation should be considered. Transfer of ferric oxide from the lower to the upper mantle would probably be inefficient. One inference is that possibly the iron distribution in the mantle is an inherited one deriving from original inhomogeneous accretion processes without much reference to subsequent events.

The heterogeneity of the upper mantle today is evident from isotopic analyses carried out on samples from the Ronda Ultramafic Complex in southern Spain, which represents a piece of the upper mantle tectonically emplaced into the crust. Within this relatively small body there exists the entire range of neodymium isotopic compositions and much of the range of strontium compositions found in rocks obtained from the sub-oceanic mantle.[33]

1.2.11 Inner and Outer Cores

The inner, presumably solid, core occupies about 16% of the total planetary volume and is thought to be rich in iron and nickel by analogy with meteorites; growth from the liquid outer core releases latent heat to drive the magnetic dynamo.[34] Large quantities of one or more light elements are necessary in order to reduce the density of the outer core from that of pure iron. The outer core, through which no transverse seismic waves can pass, is thought to be liquid, lying between depths of approximately 2900 and 5100 km. The seismic discontinuity between it and the lower mantle is termed the Gutenberg (or Wiechert–Gutenberg or Gutenberg–Oldham) discontinuity, where there is an abrupt decrease in the velocity of longitudinal waves and a change from solid to liquid phase. Although sulphur may be present by analogy with meteorites and 3 wt % in the bulk Earth would give sufficient for the core, it can be argued that the planet would not capture the 3–6% sulphur in unequilibrated chondrites if it captured a mere one-sixth of the potassium and rubidium, as has been postulated.[4,34] However, this matter is still *sub judice*. This is because of the evidence available for variation in oxidation state, mechanical separation of condensates and metastable processes. By analogy with iron meteorites, phosphorus, carbon and nitrogen may comprise minor constituents also. By analogy with enstatite meteorites, silicon may be present as well. Oxygen has been proposed as a major constituent and would not be reconcilable with the idea of a reduced interior of the Earth presented here.[34]

1.2.12 Temperature Profile

In the Earth now the temperature profile can be measured directly only on peridotite fragments obtained from the upper 200 km. Therefore proposals of deeper values are extrapolations based on an adiabatic gradient for convection coupled with estimates of thermal conduction through boundary layers at 650 and 2886 km, respectively. The combined profile so obtained and presented by J. V. Smith in 1982 (modified from Ref. 34) neatly lies below an extrapolated curve for the beginning of the melting of arid peridotite and permits sufficient time for the cooling of the planet by several hundreds of degrees Kelvin since continents became stable.[4,34] If a heating–cooling cycle at the Earth's surface be considered, for a mixture of peridotite, iron metal and iron sulphide at this surface, an Fe, FeS eutectic liquid should form at approximately 1250 K and blobs ought to sink in a hot early Earth so as to initiate a core. If wet, the peridotite would probably have melted at a slightly higher temperature so as to give a wet basalt. This might have escaped from a depleted arid residuum. Arid peridotite would have melted again as the temperature rose and thus released a range of basic to ultrabasic magmas rich to very rich in olivine and pyroxene components. If it be assumed that every part of the Earth melted a number of times in a chaotic surface zone as bodies churned this up, and also that the water and carbon dioxide became concentrated there, then deep-seated processes must have been constrained by the phase relations for arid peridotite and near-surface processes by those for hydrated and carbonated rocks.

It is apparent that early magmatism and cooling in the Earth can be represented by a cycle. This involves fertile peridotite (namely peridotite containing a basaltic component) rising adiabatically to a point where melting began, then the escape of basic to ultrabasic magmas to the surface leaving a residue of olivine–pyroxene (harzburgite) floating upon denser fertile peridotite (garnet–lherzolite) not yet melted. Cooling, hydration and carbonation of magmas sinking under a volcanic pile or becoming buried by impact blanketing takes place. Partial melting at 20–40 km depth yields wet calc-alkaline magmas escaping to the surface and 'eclogite' residues which sink because of the high density of garnet. Also there is a mechanical mixing of 'eclogite' and harzburgite residues to generate peridotites which are partly depleted by the quantity of material in the transient crust. Such partly depleted peridotites have been cooled by heat loss at the Earth's surface and tend to sink into the mantle to become temporarily capable of evading further melting as they fall below a critical adiabatic curve.[4]

1.3 INFERENCES

The concept of plate tectonics is examined together with crustal generation, stabilization of continental crust and the evolution of the planet together with the geochronological aspects of geothermal gradients and earth heat. Some interesting new developments are summarized in the final section on closing the gaps in knowledge.

1.3.1 The Early Crust
Clearly, in the early stages of the creation of the solar system, the heterogeneous accumulation of hot bodies to create the Earth probably resulted in the gradual formation of a crystalline barrier between a core which continually enlarged through denser, descending iron-rich bodies heated by gravitational energy and an initially ephemeral and chaotic crust cooled by volcanism and intermittent impact events. This is consistent with the ongoing emission of the atmophilic nuclide helium-3 from the planet.[35] The fact that helium escapes into space is advantageous in estimating the gas flux from the Earth's mantle. Due to the relatively short residence time of helium in the atmosphere (around one million years) compared with the geological time-scale, it can be assumed that the Earth's mantle outgasses as much helium as is lost from the atmosphere to space and that a steady

state exists. Given this assumption and no significant mass fractionation for helium isotopes when they escape into space, the helium-3 flux from the continents is about 2–3% of that from the mid-oceanic ridges. Certainly part of that outgassed from the continents derives from the mantle.[35,35a,36]

1.3.2 The Later Crust

The complex later crust can be envisaged as growing through two mechanisms, namely directly by melting of the upper mantle and indirectly by remelting of that material foundering under volcanic piles and the debris from impacts.[4] The composition of this early crust would most likely have been a mixture of ultrabasic, basic, calc-alkaline and possibly some potassium-rich granitic rocks. Poissibly until about 3·8 Ga ago all crust would have been recycled into the upper mantle, continents forming thereafter partly by the accretion of oceanic crust and partly through contributions of basaltic magma directly from the mantle.

It has been suggested that the style of mantle convection may well have altered from polygonal (cf. Section 1.2.4) through symmetrical linear finally to asymmetrical linear, although this is purely speculative.[4] However, it is likely that the early oceans and crustal nuclei were small.[37] If this was the case, then the transition from the Archean through the Proterozoic to the present situation with large plates must have been very complex, as, for instance, D. H. Tarling indicated in 1980.[38]

1.3.3 Plate Tectonics

There can be little doubt that this hypothesis is currently the most satisfactory, or at least the most widely accepted, explanation of the major architectural features of the planetary crust and their distribution. It derives from the attempt to extend transform fault theory to a sphere which involved division of the planetary surface into 20 blocks of various sizes separated by three types of boundary. These are oceanic rises where new crust is created, trenches where crust is destroyed and the above-mentioned transform faults at which neither process occurs. For mathematical reasons, the blocks in question were assumed to be rigid. Since the crust is too thin to possess the required strength, the blocks, now renamed plates, were believed to extend some 100 km or so down to the low-velocity layer of the mantle, i.e. the asthenosphere, this approximating a hydrostatic geoid. The relatively rigid upper 100 km region constitutes the lithosphere or tectosphere. The lithospheric plates in question have noticeably irregular motions and also are produced and consumed at uneven rates. Thus the configurations and positions of their boundaries are constantly altering.

Consequently, at the present time, the Atlantic Ocean is widening and the Pacific Ocean is shrinking. In the Pacific, the subduction zones on its two sides are migrating towards one another. Hence the subducted slabs are not only descending parallel to their dips, but they are also moving transversely. At least some of the slabs move backwards relative to the plates to which they are attached, this constituting retrograde motion.

Essentially, the driving forces for plate tectonics appear to relate to the negative buoyancy of such subducted ocean lithosphere. This has both cooled and contracted for an average of 80 million years, during which time it was on the surface. When subduction takes place, oceanic lithosphere is believed to be 2 or 3% denser than surrounding mantle and torques imparted by the sinking slabs may implant a mantle-wide stress field largely responsible for the inducement of plate tectonics. Several alternative suggestions are discussed later in this section, but here it is germane to mention that the slabs themselves comprise laminations, of which the most significant are the 6- or 7-km layer of MORB crust overlying a 30-km layer of harzburgite left over after basalt extraction. The parental material of these two products may represent the bulk composition of the upper mantle, i.e. pyrolite.

The oceanic zones of crustal formation by upwelling of mantle material, i.e. the oceanic rises, are a result of the phenomenon called sea-floor spreading and constitute constructive plate margins in the form of mid-oceanic ridges. Although sea-floor spreading has been enthusiastically adopted by some workers, it must be pointed out that the idea remains entirely speculative, particularly as regards its actual mode of operation. It is extremely difficult to account for some associated features, for instance the characteristic zigzag parts of a ridge–rift system. Convection does not appear to occur along lines and is especially unlikely where these are sundered by many offsets such as those affecting the major oceanic ridges. In addition, the assumption is necessary that injection and outflowing of lavas from the ridge axis is a symmetrical process which would be extremely improbable in nature. Thus boundaries for claimed magnetic polarity reversal are unlikely to coincide with the natural lava outflows mentioned. Paleontological evidence, e.g. the presence of Late Miocene foraminifera in the crestal rift of the Mid-Atlantic Ridge, conflict with the possibility of sea-floor convectional spreading and indeed argue for a cymatogenic nature for the main sub-oceanic ridges during Cenozoic time. (Cymatogeny is a mode of large-scale vertical deformation of the continental crust in which the fundamental structure induced at intermediate depths is a steeply tilted tectonic gneiss, with or without minor igneous activity. It manifests on the planetary surface by arching, and some arches are very large, for instance the Andes. Exaggerated cymatogeny becomes orogeny.)

The oceanic zones of crustal consumption, i.e. the trenches, are termed destructive plate margins and the crustal material is forced down under island arc/oceanic trench systems or mountain belts along the subduction zones (alternatively referred to as Benioff zones), sometimes to depths of 700 km or more. Conservative plate margins are those along which neither construction nor destruction is occurring, the plates laterally sliding past each other along transform faults which frequently offset the oceanic ridge axis.

Earlier, the characteristic planetary features alluded to at the beginning of this section were explained on the basis of a postulated continental drift which involved relative movement between continental blocks. This phenomenon is now envisaged as occurring between the plates, of which some are much larger in area than any continent, for instance the Eurasian plate. By contrast, others are quite small, for instance the many contributing to the so-called 'mayhem tectonics' of the southeastern Pacific Ocean. Features such as transcurrent faulting represent lateral movements and the mid-oceanic ridges are the results of plates separating from one another with crust arising from the 'join', trenches being found where subduction takes place. The vast number of confirmatory phenomena, e.g. volcanic tracts like the Pacific 'Ring of Fire' (which has been active through the Mesozoic and Cenozoic eras of the Phanerozoic Eon), areas of geysers and hot springs reflecting geothermal activity, earthquakes, orogenic belts and mountain ranges together with great fault zones, e.g. the San Andreas system in California, can be related to such motions of plates; see Fig. 1.2.

Geysers constitute perhaps the most spectacular of hydrothermal events, the word 'geyser' itself deriving from the old Icelandic verb *gjose*, to erupt. They involve reservoirs of hot water which, intermittently as well as explosively, eject all or a good part of their contents. Where they occur, there may be associated boiling pools, dry steam jets and mud pots. Interestingly, although there are many thousands of hot springs in the world, there are only some 400 geysers. The differences between these two categories lies in the periodic thermodynamic and hydrodynamic instability of the latter. Eruption takes place when a portion of the stored hot water becomes unstable. Thereafter an abrupt generation of steam occurs within the actual geyser, but close to its opening on the terrestrial surface. This induces an explosion because water, as steam, occupies over 1500 times as much volume as in the liquid state, i.e. it possesses the same ratio as gases generated by a solid explosive. After ejection of excess heat

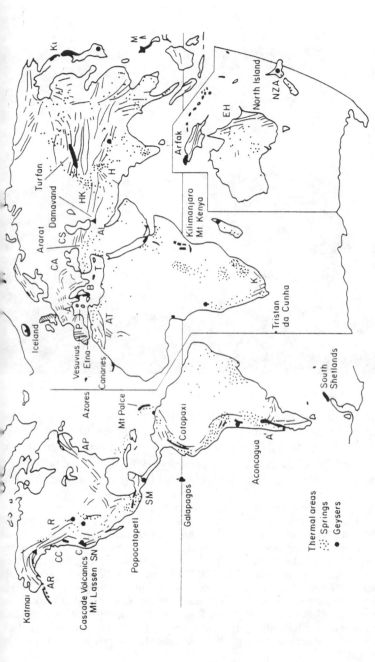

Fig. 1.2. Thermal springs, geysers, volcanic regions and folded mountain zones of the Earth which concord with the plate tectonics hypothesis as well as with observed major earthquake belts, and in particular that circling the Pacific Ocean and known as the Pacific Ring of Fire. Folded mountain regions: A, Alps; A′, Andes; AL, Alborz; AP, Appalachians; AR, Alaskan Ranges; AT, Atlas; B, Balkan Ranges; C, Cascades; CA, Carpathians; CC, Canadian Coastal; CS, Caucasus; EH, Eastern Highlands; H, Himalayas; HK, Hindu Kush; K, Karroo; NZA, New Zealand Alps; P, Pyrenees; R, Rockies; SM, Sierra Madre; SN, Sierra Nevada; T, Taurus. Volcanic zones named on map.

37

FIG. 1.3.

and water, the geyser returns to its previous stable state and commences a
new cycle of increasing instability. Most geysers in the world are located in
the Yellowstone National Park, Wyoming,† in the northwestern USA
actually about 200 of them, and they comprise 10% or so of all hot-spring
phenomena. In Yellowstone the hot springs are found in approximately 100

†Tragically 4·2 million acres were destroyed by forest fires affecting Yellowstone
National Park, Wyoming, and other western states of the USA, in 1988. To cope in
an area half the size of Switzerland, 30 000 fire fighters were involved at a cost of
US$350 million. Unfortunately, at least seven died. From this experience it appears
that attempts to control such fires may be counter-productive and only the sheer
magnitude of the disaster induced the US National Park Service to reverse its
normal stand of letting fires burn unless they threaten lives or property. This idea is
based upon ecological views that fighting fires causes dry fallen wood to accumulate
in dangerous quantities on forest floors. Regular, natural, 'let-burn' fires obviate
such phenomena as well as adding fresh nutrients to soil. Anyhow, as seen in 1988,
once full-scale forest fires start it is probably impossible to suppress them fully until
they have run their course.

FIG. 1.4.

FIG. 1.5.

FIG. 1.6.

FIG. 1.3.   Old Faithful Geyser Basin. Old Faithful is the most famous of the 10 000 hydrothermal features, including 200 geysers, bubbling mud volcanoes and brilliant pools, harboured by Yellowstone National Park, Wyoming. Reproduced by courtesy of Wyoming Travel Commission.

FIG. 1.4.   Grotto Geyser, Yellowstone National Park. This is one of the more active and most unusual geysers in Yellowstone National Park, named for the curved overhanging shape of its cone. The strange formation is the result of 'geyserite', a mineral deposit left by evaporating geyser water forming around the trunks of trees that grew in the region before the geyser became active. Grotto is located in the Upper Geyser Basin near Old Faithful. Reproduced by courtesy of Wyoming Travel Commission.

FIG. 1.5.   Riverside Geyser, which throws steaming water obliquely from its lower crater into the Firehole River. It is characterized by an unusual cone and eruptions lasting half an hour. At intervals of $5\frac{1}{2}$–8 h Riverside displays its upper crater spray. Reproduced by courtesy of Wyoming Travel Commission.

FIG. 1.6.   Morning Glory Pool, Yellowstone National Park, Wyoming. It derives its characteristic colour from the algae that grow in the warm waters. Reproduced by courtesy of Wyoming Travel Commission.

clusters. The geysers include such famous ones as Old Faithful, Giant, Giantess, Grand, Daisy, Riverside, Artemesia, Grotto, Lion and Beehive, which emit to heights of 30–60 m. Examples are shown in Figs 1.3–1.6. Old Faithful is probably the most famous geyser in the world and possesses a 4 m high, 50 m × 70 m sinter mound in a prominent position at one end of the Upper Geyser Basin, Wyoming. The actual vent of the geyser opens to a 1·5–3 m hole representing essentially a widened part of a fracture extending across the top of the mound. Premonitory splashing signals an eruption which continues with a series of ever higher emissions reaching a peak in a few minutes. The total time during which the water play goes on may last from 2 to 5 min. Eruptions are separated by periods of anything between half an hour and 100 min. In all, about 70 are active and all are surrounded by their characteristic heavy sinter deposits (see Fig. 1.7).

In Iceland there are a mere 40 active geysers, including the geothermal field called Geysir (see Fig. 4.2). Others occur in the North Island of New Zealand, especially at Whakarewarewa near Rotorua and some 45 km northeast of Wairakei. They are also found in Kamchatka in the northeastern USSR and in Japan. In addition, there are isolated geysers in

FIG. 1.7. Thermal springs in the USA. The Yellowstone National Park in Wyoming constitutes probably the largest and hottest region in the United States, while the geysers produce over 2000 MW of electric power.

Chile, Mexico, through Africa, in the Azores, in Indonesia and various islands in the Pacific Ocean, in mainland China, and in the Basin and Range Province of western USA, as well as in Unmak Island, Alaska. All main geyser areas are in volcanic regions and contain large amounts of rhyolite from which the geysers spout, although a few emerge from andesites and even basalts.

The general acceptance of plate tectonics nowadays (*pace* the fixists) is determined by the feasibility of separating kinematics from causes following ever-expanding knowledge of the horizontal movements of the plates through geological time as well as of their creation and subduction. It is important to note that the evidence comes almost wholly from the oceanic portions of plates. Indeed, were the plate tectonics hypothesis to be tested solely on land, it would be practically impossible to substantiate it. However, this does not imply that the continental records eliminate large-scale horizontal motions, simply that the actively-deforming belts are thousands of kilometres across and appear to indicate a semi-continuous type of deformation. Thus it would be unreasonable to expect observations on the continents to substantiate the inference that most of the tectonics on their surfaces are governed by the relative movements of a few rigid plates, themselves thousands of kilometres across. Epeirogenic plate movements have been examined as well, for instance in the Late Pleistocene along the continental shelf of North America by C. B. Officer and C. L. Drake in 1982.[39]

Clearly, there remains much ignorance on the matter, although it is possible to outline the general concept of heat-driven convective motions in the mantle. In addition, as indicated above, it is quite possible that the slightly greater density of subducted oceanic lithosphere *vis-à-vis* surrounding mantle permits descending slabs to impose torques in such a manner as to establish a mantle-wide stress field mainly responsible for the phenomenon of plate tectonics. However, the dynamics involved are far from well understood, especially in terms of the mechanisms controlling length-scales or the partitioning of heat transfer between the large-scale flow (plate motion) and the other scales of heat transfer which may be in operation. Alternative mechanisms include the following: viscous coupling may occur with the lower, mobile asthenosphere or pulling may result from localized subduction, this drawing oceanic crust towards it. In the case of small plates, they may react to large plates moving alongside them. It appears probable that, in the Earth now, the mechanisms of plate tectonics, whatever these are, appear to be responsible for approximately two-thirds of the present-day heat loss from the planetary surface.[40,41]

Great changes must have taken place during the history of the Earth since the Archean. Then the heat generated was probably higher than now (possibly only slightly—cf. Section 1.1.3), but perhaps by as much as a factor of three. There seems to be no general consensus on this particular matter. However, mantle temperatures were probably higher then, perhaps by as much as 300°C, and possibly the production of oceanic plates was more rapid, these having mean thicknesses of perhaps 30 or 40 km, i.e. less than now. As noted earlier in Section 1.1.3, magmatic regimes at Archean spreading centres were unlikely to have resembled those now existing and the higher mantle temperatures probably promoted greater degrees of partial melting and thus the copious generation of komatiitic magmas, the refractory residuum left after their extraction comprising mainly magnesian dunites. Such komatiitic magmas probably fractionated nearer the crust–mantle interface by the extensive crystallization of olivine, leading to the development of an oceanic crust comprising mostly tholeiitic basalts. Consequently, Archean oceanic lithosphere may be envisaged as consisting of layers of basalt, dunite cumulates and dunitic refractory residua.[42]

1.3.4 Generation of Crust

Any scenario involving a continuous and diminishing generation of crust associated with a decrease of recycling into the upper mantle after *ca.* 3·8 Ga could be tested, theoretically at least, by examination of the chemical compositions of rocks as a function of time. The results obtained could throw more light on the sequence of changes in the planet during that time interval which were alluded to in Section 1.3.3. In fact, however, this is hardly feasible because of the near-impossibility of allowing for the multitudinous effects of intervening chemical alteration arising from metamorphism. Nevertheless, some observations may be made.

It is permissible to explain decrease in abundance of ultramafic volcanic rocks (komatiites) with time by cooling of the upper mantle. In addition, sedimentary rock correlations seem to show that the early crust was more mafic than the later one which became enriched in granodiorite, potassium-rich granites becoming prominent by the end of the Archean. Despite some resemblances between high-grade metamorphosed regions dating from that era and rocks in modern continental margins, the constituents of the Archean lower continental crust do not accord, in any detailed manner, with the modern rocks.

This may reflect a process whereby subducting oceanic crust endured more melting when the mantle was hotter and perhaps also relates to changes in the recycling mechanism and the reservoirs. Many difficulties have arisen in regard to attempts to relate high-grade regions to the Archean greenstone belts because there are insufficient geochemical measurements and associated geological work in the field to date. However, they were certainly products of the oceanic domain and their present-day high frequency suggests an excessive original abundance. They have been recycled at oceanic rates before becoming incorporated into stable continental crust and at continental rates thereafter, i.e. during the subsequent 2·2 Ga or more. Therefore their very survival is evidence of their former abundance. It follows from this that the ocean domain was the dominant tectonic regime on the young Earth.

It is reasonable to assume upper-mantle heterogeneity through geological time and that crustal recycling aided in its rejuvenation. Naturally there has been plenty of time for extensive redistribution of elements in it during the four billion years or more which are involved. A preliminary model for upper-mantle seismic anisotropy was presented in 1986 by L. H. Estey and B. J. Douglas.[43] This was based upon the idea that the main features of the phenomenon could have resulted from mineral alignment of the principal constituent minerals achieved through dislocation motion (glide and climb).

The relevant reorientation mechanism utilized the dominant high-temperature glide systems of olivine and the pyroxenes, taken to be /100/(010) and /001/(100), respectively, to constrain the mineral alignment and therefore the large-scale seismic anisotropy of two petrological models. These are pyrolite (olivine-rich) and piclogite (clinopyroxene- and garnet-rich). In fact, the Lehmann discontinuity is claimed to be the result of a compositional change from pyrolite to piclogite, i.e. perhaps a change in anisotropy. However, the lateral migration of subducted slabs in the mantle tends to prevent closed cells forming as well as to stimulate mixing which may act in the direction of its chemical homogenization; cf. Section 1.3.8.[44]

If, with D. M. Shaw in 1980, it is assumed that up to half of the mantle has not contributed to the growth of the crust, then the lower mantle would have to be regarded as practically inert apart from small leakages of volatile elements from time to time.[45] Among these is hydrogen: it is especially hard to evaluate its loss from the Earth and the consequent degree of oxidation of the crust and upper mantle. In order for late accretion of oxidized material to have occurred, there should not have been a highly reduced atmosphere and, in the event that one arose, it would have swiftly lost ammonia and methane by photochemical reactions. In fact, all geological evidence from early rocks points to a weakly oxidized atmosphere before the development of abundant free oxygen from biological processes. Even prior to the emergence of oxygen-producing species there is evidence for biological activity, as, for instance, E. M. Cameron and R. M. Garrels indicated in 1980.[46]

Finally, it should be mentioned that a number of authors have suggested that the rate of growth of continental crust with time may have been episodic rather than uniform. Thus S. Moorbath in 1978 tentatively identified major episodes of crustal growth within the following time intervals, namely 3·8–3·4 Ga, 2·9–2·6 Ga, 1·9–1·6 Ga, 1·2–0·9 Ga and 0·6–0 Ga.[19] A very extensive episode of crustal growth is believed to have occurred around 2·6 Ga ago. Such episodes could be explained in terms of the interactions between convective overturns of the lower mantle and megaliths of subducted oceanic lithosphere, suspended somewhere near the 650-km discontinuity (cf. Section 1.2.12). Obviously, the density different-ials between such megaliths at this level and the surrounding mantle would be very delicately related to temperature differences between the two. If the megaliths were cool enough and were not separated into basaltic and harzburgite components, they might have sunk well below the level in question. This could have been facilitated by a major convective overturn of slightly superadiabatic lower mantle, which would increase the temperature of the mantle beneath this discontinuity (perhaps by as much as 400–500°C).

It is possible that the generation of continental crust may depend upon the cumulative rate of subduction of oceanic lithosphere, a process influenced by the above-mentioned interaction of megaliths with the 650-km discontinuity. High rates of subduction would be initiated if such megaliths sink deeply into the lower mantle prior to unmixing. This might take place after a major convective overturn of slightly superadiabatic lower mantle. Consequently, periods of rapid growth of continental crust may have coincided with major convective episodes in the lower mantle, as A. E. Ringwood indicated in 1982.[47]

As seen earlier, the Earth is thought by some to be cooling and, in fact, the mantle may have cooled by some 200°C or more during the past 4·5 Ga. Then megaliths could have sunk much deeper into the lower mantle than is possible now. Subduction would have proceeded more rapidly altogether, resulting in a more rapid average growth of continental crust between 3·5 and 2·0 Ga ago than took place during the last billion years.

1.3.5 Stabilization of Continental Crust

A model for this approximately 3–2·5-Ga-old event involving flushing out of water by carbon dioxide from the mantle was suggested by R. C. Newton and his associates in 1980.[48] For this an enormous supply of carbon dioxide would have had to be available. In this connection, despite the fact that dolomite–peridotite is stable to greater depth than mica–peridotite, there are inherent difficulties in transferring carbon dioxide from the atmosphere into the dolomite–peridotite of the mantle. J. V. Smith has proposed that elemental carbon in a weakly reduced part of the upper mantle may have been oxidized through mixing with appropriate oxidizing material transported downwards from the crust or maybe even upwards from the lower mantle.[4] It must be borne in mind that hydration and carbonation of some volcanic rocks probably removed a lot of water and carbon dioxide, at least temporarily, until burial to a depth of *ca.* 10 km caused devolatilization. On the other hand, further supplies of these compounds may have been contributed by comets.[49]

Intermediate volatility elements such as the halogens and the alkalis are valuable in providing constraints on crystal–liquid differentiation and recycling, because the known mineral hosts break down into the earliest melts. The halogen cycle is controlled primarily by amphibole, mica, apatite and melt, only about a quarter of the halogens remaining in the upper mantle. This may represent the quantity from subducted crust which is transiting the upper mantle prior to returning to the crust.

The alkali cycles are also interesting and the Na/K ratio probably increases considerably with depth since sodium-bearing clinopyroxene transformed into heavy sinking garnet as the planet grew. As noted earlier, high-density garnet may well induce parallel sinking of 'eclogite' residues. As regards eclogite, it may be germane to mention that in 1985 two special issues of *Chemical Geology* were devoted to the chemistry and petrology of eclogites,[50] and also their isotope geochemistry and geochronology,[51] and should be consulted for additional information. It is not impossible that some potassium entered sulphide in a reduced early Earth and was transported into the core, but basaltic melt would carry potassium upwards. Anyhow potassium is not needed in the core in order to explain the heat budget.[4,52]

Some interesting work on the recycling of continental crust through geological time was reported by E. Dimroth in 1985.[53] He estimated rates of magma emplacement, crustal growth and erosion for Archean greenstone belts and compared them with values obtained from modern accreting plate margins. It was suggested that the rate of magma emplacement into continental crust is larger than the rate at which the continental crust grows. Clearly, such an imbalance, if proven, would evidence either rapid recycling of continental crust, i.e. a return flow of crustal material into the mantle, or the episodic nature of magmatism. As regards this latter point, it is clear that, on a small scale, magmatism certainly is episodic, but this does not appear to be the case globally. Thus, for instance, the Pacific 'Ring of Fire', as already mentioned in Section 1.3.3 on plate tectonics, was active throughout both the Mesozoic and Cenozoic. Consequently, the recycling hypothesis is considered the more likely explanation of the data.

1.3.6 The Evolution of the Earth
This is characterized by a combination of cyclic, i.e. recurrent, and superimposed unidirectional phenomena. The short-term cyclic events leave residual records only and the long-term ones are usually composed of a set of such residual records. Thus the existing continental crust is the consequence of an agglomeration of remnants derived from repeated orogenic episodes. The dynamics of such processes are inherent in the concept of plate tectonics alluded to earlier, in Section 1.3.3. From this hypothesis it is believed that tectonic realms associated with the oceanic crust are geologically ephemeral structures with a rather low probability of preservation. On the continents, however, such tectonic realms are characterized by expanded records. From this it is feasible to assume that there is a long-term frequency for both the generation and the destruction of continental tectonic realms, as J. Veizer and S. L. Jansen indicated in 1985.[54]

1.3.7 The Geochronological Aspect of Geothermal Gradients
The planet is losing internal heat and this could be an irregular process. For instance, crustal metamorphism implies previous abnormal geothermal conditions which must have triggered perturbations of steady-state geotherms. Ancient events of this type are difficult to identify, but exposed glaucophane schist belts may represent fossil subduction zones and embody low-pressure and high-temperature phenomena. Kyanite–sillimanite facies reflect medium-pressure to high-temperature metamorphism and perhaps indicate past continental collisions. Andalusite–sillimanite facies pressure and temperature conditions are believed to be generated after intracontinental thinning and do not imply any antique volcanic areas. As many metamorphic traces have been found in Archean terrains, probably the process entailed similar conditions then to those of Proterozoic and Phanerozoic events of the same type.[54] The mineral assemblages in such metamorphic rocks sometimes indicate pressure and

temperature conditions of their formation from which the ancient palaeogeotherm and the palaeo-heat flow can be derived.

Today, away from anomalously perturbed geothermal regions such as subduction zones, the lithospheric thermal structure is possibly near equilibrium. Therefore its conductive heat loss is calculable relative to a steady-state geothermal gradient, employing the one-dimensional heat conduction equation for one uniform layer:

$$T_z - T_u = qz/K = A(z^2)/2K \tag{1.1}$$

This can be generalized for successive layers for different temperatures at the upper surface of a layer, T_u, of thickness z, thermal conductivity K and internal heat generation A, q being the heat flux through the upper surface and the temperature at depth z being T_z.

Global geothermal gradients on billion-year time-scales and also on hundred million-year ones were discussed by A. B. Thompson.[55] On the former, he indicated the advisability of distinguishing long-term global geothermal gradients characterizing the overall cooling of the interior of the Earth from those relating to short-term regional thermal episodes resulting from crustal and lithospheric deformation phenomena. Processes resembling modern plate tectonics may well have taken place in the Archean, hence continental thickening then could have resulted from the collision of plates. If this is true, then specialized Archean tectonics, e.g. basalt underplating and sub-crustal accretion, were probably only localized in importance. It has been suggested that, by the latter part of the Archean (3·5 to 2·7–2·5 Ga ago), most of the present-day continental mass had differentiated. Then proto-continental material may have covered approximately one-third of the planetary surface, being mostly accreted by ocean–continent collision and arc-amalgamation.

A. B. Thompson in 1984 mentioned that Archean (4 to 2·7 Ga ago), Proterozoic (2·7 to 0·6 Ga ago) and Phanerozoic (0·6 Ga to now) time intervals seem to be marked by three kinds of geotectonic conditions which are preserved in the continental crust. The first involves high-grade gneisses and greenstone/granite terrains extending from the Archean into the Proterozoic. 'Greenstone' is a term applied to non-schistose varieties of greenschist, the latter being a schistose metamorphosed rock, of which the green colour is derived from the abundance of chlorite, epidotite or actinolite contained in it. By comparison, the term 'greenstone belts' refers to normally strongly folded sequences of mafic and ultramafic volcanic rocks of Archean age forming synclines in surrounding granitic or gneissose rock. The second relates to stable continental cratons and apparently the end of the above-mentioned gneiss and greenstone/granite terrains in the Proterozoic. In this time interval, three global thermal pulses took place at 2·7–2·5, 1·9–1·6 and 1·2–0·9 Ga ago, the first being evidenced by isotopic studies showing major magmatic activity (see S. Moorbath in 1978).[19] The third comprises thickening of continents concentrating convection in sub-oceanic areas to produce an efficient cooling mode and extra tectonothermal evolution during the Proterozoic extending into the Phanerozoic.[54] No doubt sea-floor spreading and break-up of macro-continental masses would have occurred and formed the precursors to the existing, 200 million years old, configuration of continents and oceans.

It has been suggested that sub-continental and sub-oceanic steady-state conduction geotherms can be identified for the five main time intervals in the history of the Earth and those under the continents are considered to lose steepness from the Hadean (4·6 to 4 Ga ago) through to the Phanerozoic. In the latter, crustal thickening and the exhuming of metamorphic rocks probably occurred. Clearly, the appearance of metamorphic rocks on the terrestrial surface entails tectonic mechanisms capable of transporting vast continental pieces (exceeding $100 \, km^3$) or producing repeated continental thickening so that erosion and isostasy can do the job. It was inferred that, since cation, anion and isotopic closures are strongly temperature-dependent in metamorphic and some of the intrusive magmatic rocks, mineral assemblages formed in these record conditions which prevailed during uplift rather than at the maximum depth succeeding a tectonic event. Consequently, deduced geotherms are strongly time-dependent on 10^6- to 10^7-year time-scales, even though the total time of perturbation tops the geotherm and the exhumation process may exceed 10^8 years in length.

1.3.8 Earth Heat

Some general conclusions may be drawn from the foregoing discussions. The first is that the formation of the planet and its separation into core, mantle and crust constituted complex events which are only partly comprehensible because of gaps in the record of Earth history and inadequacies in technical knowledge. Secondly, the upper mantle may be in a steady-state condition because the apparent need for a short ($< 600 \, Ma$) residence time for lead and other incompatible elements in it makes it necessary that the rate of entrainment of lead into this convecting feature at one boundary layer be balanced by its removal at another, thus implying the steady-state model mentioned.[56] Of course, the idea of short residence times and a steady-state behaviour of the upper mantle for the most highly incompatible lithophile elements would impose bounds on the fluid dynamic behaviour of the mantle and the nature of mantle melting necessary for extraction of incompatible elements into the continents. In addition, the interpretation of trace element abundances and abundance ratios in the upper mantle must differ from that commonly accepted for the past 20 years or so. Certainly, the Earth is an enormous energy source. On average, $1.2\ \mu cal/s$ are believed to flow through every square centimetre of its surface and, in geothermal areas, the figure is higher by an order of magnitude at least; for example, at Yellowstone (USA) the average is 20–30 $\mu cal/s\ cm^2$. If these data are correct, then, since the surface area of the Earth is $5.1 \times 10^{18}\ cm^2$ and there are 3.15×10^7 seconds in a year, the total heat flow is 2.4×10^{20} calories annually (which is equal to 10^{28} ergs per annum). If this heat flow remained constant during the past 1 Ga, then the total heat flow during that time interval would have been 4×10^{10} calories per square centimetre of the surface of the Earth. As regards lithophile elements (potassium, uranium and thorium), highly incompatible heat-producing and related elements are contained within available repositories, i.e. the oceanic and continental portions of the lithosphere.

However, since the mean lifetime of the oceanic lithosphere is relatively short, ca. 100 million years or so, that is to say short even in terms of the residence times of potassium, uranium and thorium in the upper mantle, this particular reservoir is insignificant as regards the long-term storage of such elements. On the other hand, the continental lithosphere has a mean elemental storage age for its crustal part of 2·0–2·5 Ga for samarium and neodymium, as S. L. Goldstein and his associates pointed out in 1984.[56] The conclusion is that this must constitute the principal repository. It is germane to add that the uranium and thorium concentrations in the upper continental crust are approximately 1000 times greater than those in the upper mantle. Some inclusions found in kimberlite diamonds and xenoliths seem to be, at least partly, 2·0–3·5 Ga old and it is feasible that continental lithospheric mantle may have a mean age not much different from the actual continental crust itself.[56]

The entrainment rates necessary in order to satisfy the upper mantle residence time estimates entail a mass influx into the upper mantle of approximately 1.3×10^{10} kg per annum and 1.3×10^6 kg per annum for potassium and uranium, respectively. Of course, these are presumably balanced by equal mass effluxes of these elements into continental lithosphere, which means that the mass fluxes of these elements into the continental crust must be directly related to their entrainment rates. In 1987 R. K. O'Nions referred to a preferred model in which such entrainment occurs at 670 km and the rate of addition of continental lithosphere for these and related elements is totally controlled by such entrainment and the fluid dynamic behaviour of the boundary layer between the upper and lower mantle.[57]

The question of the material transfer of chemical compounds, e.g. ferric oxide, across this boundary layer has been discussed in Section 1.2.10 above. It was demonstrated by S. J. G. Galer and his associated in 1986 that a mean elemental age of nearly 2·0 Ga for all of the incompatible elements in the continental lithosphere can be reached with an entrainment rate of 2–5% per Ga of the upper mantle mass.[58] As O'Nions remarked, it is hard to resolve additions to the continental crust at the rates suggested by the model calculations, this reflecting the problem of extracting bulk properties of the continental reservoir from the heterogeneous samples which are actually exposed at the surface.[57] The optimum approach so far has been by means of the Sm–Nd system since ratios between the two isotopes involved show little variation and are not much disturbed by sediment formation and recycling. Analyses of the sedimentary record imply a mean age of 2·3 ± 0·3 Ga for that part of the continental crust which contributed to the sedimentary mass. In fact, the relationship between crustal residence ages and stratigraphic ages is compatible with the rates of crustal addition necessitated by this type of model.

The transport of heat and volatiles is an important topic. Within the convecting upper mantle there are very low thermal and chemical diffusivities and, as a result, they constitute rather ineffective means for effecting such transport on geological time-scales. Advection is the main method of both chemical and heat transport. In 1983 R. K. O'Nions and E. R. Oxburgh made a comparison between estimates of radiogenic heat and heat fluxes from the mantle.[59] This suggested that the radiogenic helium flux of 4×10^9 atoms per m^2/s necessitated only 5 ppb of uranium and equivalent thorium in the upper mantle for it to be maintained.

However, such a quantity of uranium and thorium, including a contribution from potassium (assuming that K/U = 10 000) would contribute a mere 5% or so of the oceanic heat flux. The extreme imbalance between radiogenic helium and heat fluxes from the mantle noted above means that either there are completely different origins for the helium-4 and the heat or that there is some major difference in their loss mechanisms.[58] Other ongoing investigations into heat flow could provide illuminating information regarding both the mantle and its overlying crust. For instance, G. Vasseur and R. N. Singh in 1986 noted the high variability of continental heat flow data which reflects the varying distribution, at various space- and time-scales, of heat-producing elements, erosion and hydrothermal circulation within the crust as well as thermal convection within the mantle.[60] They used a statistical approach in order to account for random horizontal variations in radiogenic heat sources together with models simulating vertical variation. When the horizontal scale length of heat generation distribution is not much greater than the vertical one, lateral heat conduction drastically modifies the statistical properties of heat flow distribution and the apparent thickness derived from a heat flow generation plot is sharply reduced. The procedure was applied to actual data recorded over New Hampshire, USA.

An interesting discussion of thermal evolution models for the Earth was given by U. R. Christensen in 1985.[61] He asserted that recent numerical work on heat transport by variable viscosity convection showed that dependence of heat flow on the mantle temperature may be less than anticipated so that plate velocities and heat flow in the Archean may have been less than 50% higher than now. Of course, if this idea is incorrect, then much larger differences can be expected. If it is correct, the present-day Urey ratio comes out as 0·5 (50%) or slightly less and geological, geochemical and palaeomagnetic evidence, although it is admitted that this is subject to some uncertainties, is taken as favouring evolution models based on a weak coupling of heat loss to interior temperature.

The Urey ratio is defined as the relationship between heat production and heat loss. Various values have been given to it (from 0·5 to 0·9) and, in some of the thermal evolution models, low values may have resulted from using inappropriate model parameters.[61] If it be assumed that all of the heat is radiogenic in origin and that the Earth has a Urey ratio of ca. 0·5 (50%), then there must be a region in the planet where a major disparity in the transport of helium and heat exists. The Urey ratio can be estimated from the abundance of radioactive elements in the Earth according to geochemical/cosmochemical models. Of the various values assigned to it, 0·5 is taken as optimum. Then the disparity in question could be at either the upper or the lower boundary layers of the upper mantle convecting layer. If the lower boundary was responsible, the planet would have to lose heat much more effectively than helium and the consequence would be the storage of radiogenic helium in the mantle at the expense of heat.

Some observational data support the idea that helium reaching the basaltic part of the oceanic crust is efficiently degassed mainly through the hydrothermal processing of ridges. R. K. O'Nions in 1987 indicated that there are two lines of reasoning pertaining to this in relation to the incompatible behaviour of helium in the course of mantle melting.[57] One refers to the fact that helium in basaltic glass is normally three or even four orders of magnitude more abundant than it is in olivines, according to M. D. Kurz and W. J. Jenkins in 1981.[62] The other relates to the observation that hot spot centres such as Hawaii and the Loihi Seamount are degassing argon-36, presumably deriving from the lower mantle in contrast to the near-absence of this nuclide in spreading ridge basalts.[63] Actually, the Loihi Seamount basalts possess almost two orders of magnitude more more argon-36 relative to argon-40 than ridge basalts which have to be efficiently degassed so as to avoid 'pollution' of the upper mantle with plume-derived argon-36, as C. J. Allègre and his associates noted in 1983.[64] Consequently, the differential transportation of both heat and helium seems likelier to occur either at or below the lower thermal boundary layer. R. K. O'Nions and E. R. Oxburgh in 1983 suggested that heat moves between the lower and upper parts of the mantle primarily by conduction across the thermal boundary layer, which is believed to impede the transport of helium to a much higher degree.[59] Now, in the event that this is true, i.e. that the diffusivity of heat through the boundary exceeds that of helium, the disparity between the surface helium and heat fluxes reflects differential behaviour arising at the lower boundary to the upper convecting mantle.[56] The diffusive transport of helium through the upper mantle mineral phases is probably much slower than advective transport and it is thought that helium diffusion in the minerals of the continental crust is too small to remove helium from the crust on the relatively short time-scales involved, i.e. 10^8 years. Helium is known to move advectively in the upper part of the crust in association with groundwaters and other crustal fluids, but, as might be expected, the transport mechanism in the deeper, hotter and more ductile lower crust is rather obscure. However, D. R. Porcelli and his associates in 1986, among others, provided evidence that, in certain circumstances, helium-3 may be carried, as a tracer component, by a carbon dioxide-dominated fluid.[65] A comparison of the $C/^3He$ ratio in mid-oceanic ridge basalts (MORBs), the East Pacific Rise and Galapagos vents showed a rather uniform ratio near 10^9.[57,66]

I. Barnes and others in 1984 recorded CO_2 emanations from numerous active tectonic regions and sometimes the $\delta^{13}C$ of the carbon dioxide is between $-4‰$ and $-6‰$, this being the assumed range for mantle carbon.[67] From a detailed evaluation of the Massif Central (France), the work of A. Matthews and his associates in 1986 showed that this carbon is, to some extent, from the mantle, and also that $C/^3He$ ratios range from about 10^9 to 10^{11}.[68] These data necessitate mantle fluids traversing metamorphic basement there and they should reach the surface with a ratio within an order of magnitude of the estimate for the initial mantle value.

Mantle melting below the area in question, and emplacement of melts into the crust from which both carbon dioxide and helium-3 degas, seem to constitute the main means of helium transportation. Citing this, R. K. O'Nions regarded it as too soon to infer a wider applicability to CO_2 emanations in other environments which are actively undergoing extension involving melting and CO_2 emplacement into the deeper parts of the crust.[57] However, if this becomes possible, it could well be that estimates of helium-3 and carbon fluxes could be utilized in order to set limits upon the total quantities of mantle-derived carbon and basaltic melt which enter portions of the crust which are suffering metamorphism. Clearly, therefore, the distribution of helium-3 in the active portions of the continental crust has provided indications as to how sedimentary basins may have formed and has also provided clues regarding the nature and fluxes of the various fluids which enter the crust at sites of active tectonism. This stems from the long-term isolation of primordial helium in an only partially degassed reservoir mostly convectively isolated from the upper mantle. Of course, this isolation was not total and permitted sufficient helium-3 to penetrate to the upper mantle for subsequent small-scale leakage through continents to be usable as a tracer of mantle volatiles.

It appears probable that crustal regions exist which were chemically isolated throughout the history of the planet. As it has not proved feasible to reconcile chemical and isotopic observations with a mantle freely convecting throughout, support is thereby given to models of layered mantle convection. The upper mantle may possess many of its trace element and isotopic characteristics as a result of interactions taking place at both the upper and lower boundary layers. At the latter, primordial helium-3 and argon-36 as well as other rare gases become entrained from the lower mantle, more especially at sites of instabilities producing plume or jet generation. Since incompatible elements, e.g. lead, thorium, uranium, caesium and rubidium, appear to have relatively short residence times (< 600 Ma), their abundances must be maintained by the introduction of material at one or another of the thermal boundary layers to the upper mantle. Once again, entrainment from the lower mantle would seem to be the most viable source and its actual rate would only have to be some 0·02–0·05 of the upper mantle mass/Ga.

Unfortunately, the long-term chemical and isotopic development of the continental lithospheric boundary is far from satisfactorily understood at present. This is particularly true of the yet-to-be-defined localities where incompatible elements become extracted from the mantle into the continents. As R. K. O'Nions indicated in 1987, it may be that views on the relationship between island arcs and continent generation are not

TABLE 1.1
Heat production from different layers of the Earth

	Total for spherical shell	Radioactivity per unit volume	Mean density (g/cm^3)	Heat production per unit mass $(erg/g \; year)$
Continental				
Crust	20	200	2·7	104
Upper mantle	8	8	3·35	3·34
Lower mantle	3	1	5·15	0·27
Oceanic				
Upper mantle	28	28	3·35	11·60
Lower mantle	3	1	5·15	0·27

applicable to both major and incompatible trace elements.[57] Probably the lower thermal boundary layer produces a considerable degree of fractionation between the radiogenic heat and helium generated in the lower mantle. Even if it is the case that, as mentioned above, the diffusivity of the boundary layer for heat is greater than that for helium, it still seems to be responsible for the imposing of a time constant of ca. 2·0 Ga on heat loss from the mantle. The inferred imperfect chemical isolation of the lower mantle and the leakage of helium through this region of the Earth yields, at the surface, a first-class tracer of fluid movement through the continental lithosphere. Data on heat production from the various structural units of the planet taken from F. D. Stacey (1969) are listed in Table 1.1.[69]

1.3.9 Closing the Gaps
Clearly, a great deal remains to be discovered about the planet before a more complete understanding of such inherent processes can be attained. However, the gaps are constantly lessening in number as new knowledge develops.

For instance, at the end of 1987 it was reported that, using a new technique of seismic exploration, California Institute of Technology geophysicists identified, some 644 km under the Asian continent, slabs of the crust perhaps constituting remnants of the Pacific plate that slipped underneath it.[44] If substantiated, this supports the idea that crustal material sinks only part of the way into the intermediate portion of the mantle, thereafter possibly being reheated, later to re-emerge at the mid-oceanic ridges. This might be in agreement with the remarks of Z. Garfunkel and his associates in 1986 that slab migration is usually retrograde (cf. Section 1.3.3 above), i.e. opposite to the direction of motion of the plates to which they are attached.[70] A result of lateral slab migration is that material is displaced away from one side of the slab and an equal volume flows in from the other side. A mass flux is thereby generated in the mantle and this is comparable in size with the flux involved in the overturn of the oceanic lithosphere. Clearly, the flow induced by slab migration might constitute an important part of the large-scale mantle circulation associated with plate motions.

Two-dimensional numerical models with retrograde slab migration explicitly built in as a boundary condition demonstrate that migrating slabs, in contrast to non-migrating ones, are at an angle to streamlines in the surrounding mantle. Thus the parallelism of Benioff–Wadati zones and streamlines cannot be employed in order to discriminate between mantle flow models. Such laterally migrating slabs do not separate streamlines which turn in different directions at depth and also do not turn backwards beneath the plates to which they are attached. Instead such slabs become part of the circulations under the over-riding plates and, even at depth, they continue moving away from their oceanic ridge sources. Migrating slabs do not separate mantle convection cells. Numerical simulations show that slab migration slows the flow under the attached plate by diverting part of it to the circulation beneath the over-riding plate. The enhanced flow under this latter plate drags it more forcefully towards the subduction zone by increasing both the magnitude of the basal drag and the length of the over-riding plate subject to a trench-directed drag, this comprising a 'trench suction' force. Retrograde slab migration increases this phenomenon, and could be significant in initiating and maintaining back-arc spreading. Through long geological time intervals migrating slabs may allow large volumes of the mantle to seep out and alter flow patterns as well as dispersing material to great distances. All this inhibits formation of regular closed cells and mixes the mantle, so tending towards chemical homogeneity (cf. Section 1.3.4).

Also in 1987 attention was given to the relationship between large-scale asphericities of the observed non-hydrostatic geoid and the seismically inferred lateral density variations in the mantle of the Earth. Thus A. M. Forte and W. R. Peltier derived poloidal-flow internal-loading Green functions for incompressible fluid models of the mantle comprising either a single constant-viscosity spherical shell or two adjacent spherical shells having different viscosities.[71] Such functions, as well as spheroidal and toroidal Love numbers, have been discussed recently by J. B. Merriam in 1986.[72] From them were derived kernels connecting the surface divergence, the geoid and the surface topography fields to the lateral density heterogeneity implied from seismic tomographic imaging techniques. These kernels supported the argument that both the surface divergence and geoid fields, consisting of harmonic degrees 2–5, may be in reasonable fit with only an 8-fold viscosity increase at a depth of 1200 km. However, coupling from poloidal to toroidal flow necessary to understand surface velocity spectra may perhaps permit the observations to be comprehended in terms of a viscosity increase at depth which is smaller than that required by non-hydrostatic geoid data in the context of pure poloidal models. By employing observed kinetic energy in the surface plate motions as a constraint, the authors were able to infer a value for the steady-state upper mantle viscosity of $(2 \cdot 0 \pm 0 \cdot 5) \times 10^{21}$ Pa s. This differs from a lower value of 1×10^{21} Pa s obtained from analyses of signatures of the glacial isostatic adjustment process.

Also, regarding the upper mantle, an interesting report on a peridotite mass in Hokkaido, Japan, was made by M. Obata and N. Nagahara in 1987.[73] Here the layering is characterized by wavy and oscillatory patterns, in terms of both bulk-rock and mineral chemistry. From the linear trend in bulk-rock chemistry noted it was concluded that such layering resulted from segregation of partial melt in the mantle. The authors presented a physical model of melt segregation in the gravity field. Since permeable flow is strongly coupled with solid creep and compaction, magma waves may be generated through a fluid dynamic oscillation in a partially molten rising mantle. In this way a vertical chemical stratification could be produced, starting from an initially homogeneous mantle. This hypothesis would necessitate that the most depleted part, dunite, is a residue of partial melting and the least depleted part, plagioclase lherzolite, represents a melt accumulation region. The partial melt concerned may have been of picritic composition.

Mantle investigations are proceeding apace. For instance, the nature of the 650-km seismic discontinuity and its implications for mantle dynamics and differentiation were discussed by A. E. Ringwood and T. Irifune in 1988.[42] Chemical buoyancy relationships in the course of the subduction of young and thin oceanic plates trap them in a gravitationally stable layer at a depth between 600 and 700 km which, to some extent, isolates the convective systems of the upper and lower mantles. On the other hand, in the subduction of more mature, thick plates, their upper, differentiated layers may well break through this barrier and become entrained in the convective circulation of the lower mantle. The consequent hybrid convection model would offer a potential explanation for the properties of the seismic discontinuity at about 650 km depth and also for the chemical evolution of the mantle. On the fate of subducted lithosphere, P. Olson in 1988 made some interesting comments on possible mantle geotherms and sites of active chemical differentiation, noting that differences of bulk composition between the upper and lower mantle could explain the seismicity terminus at 650 km (although there is absolutely no proof that any such compositional difference in fact exists).[74]

Allied topics cover such interesting themes as continental break-up through geological time. Thus Z. Ben-Avram and R. P. Von Herzen in 1987 alluded to the connection between this phenomenon and heat flow in the Gulf of Elat (Aqaba), Israel.[75] The idea that heat flow in continental rift zones might evidence temperature anomalies at depth, these either originating or resulting from continental break-up, was earlier proposed by M. P. H. Bott in 1982.[76] Following this up, Ben-Avram and Von Herzen made appropriate measurements of heat flow at five sites in the major basins of the Gulf of Elat off the northern Red Sea.[75] The gulf in question is actually located at the southernmost end of the Dead Sea Rift, which is a transform plate boundary. A probe was utilized which permitted multiple penetration of the bottom during a single deployment and the thermal conductivities were determined by means of needle probe measurements upon sediment cores. A mean heat flux of approximately 80 mW/m^2 was recorded and the values appear to increase southwards. This latitudinal gradient may relate to the general trend of gradual thinning of the continental crust of the gulf towards the Red Sea, where oceanic crust exists. These heat flow data and other geophysical information are quite in accord with a propagation of rifting from the Red Sea northwards along the Dead Sea Rift, where the fractures are mainly shear in type. Therefore the gulf comprises a transition between two tectonic regimes, i.e. extensional Red Sea to the south and transform Dead Sea to the north.

Light may be shed upon the subduction process following the discovery in 1988 of an active mud volcano by scientists in the French submersible Nautile approximately 400 km east of Trinidad and Tobago at a depth of almost 5 km in the Caribbean Sea; investigation of this, as well as leading to a better overall understanding of the phenomenon of subduction, could improve short-term prediction of volcanic eruptions and earthquakes.[77] The exploration was led by X. Le Pichon of the Laboratory of Geology at the Ecole Normale Supérieure and J. P. Foucher of IFREMER, a French organization for civilian oceanography. The feature comprises an egg-shaped mud mound about 1 km in diameter and 20–30 m in thickness, and has its surface ribbed with concentric lines. Temperature measurements imply that the volcano is not extinct since, whilst the water temperature at such depths is only 2°C on average, the temperature 2 m under the mud in its centre reaches 21°C. Such mud volcanoes may be caused by the subduction process, sediments saturated with water undergoing high-pressure squeezing. This would be consistent with observations made in this case where the volcano is located 30 km or so from the Barbados ridge. Interestingly, the surface of the mud mound teems with animal life; for example, clam colonies abound and there are also worms, sponges and sea anemones concentrated in places characterized by extrusions of minerals in liquids emerging from the sediments. The main food source appears to be methane-eating bacteria. As pressure rises more fluids and sediments will be extruded, therefore monitoring the behaviour of faunal colonies living and feeding on mud volcanoes might be a feasible approach to predicting earthquakes. Thus, while it is quite clear that the foregoing discussions on the heat of the Earth will have to be continuously modified, the main outlines given above embody a reasonable steate-of-the-art summary of what really occurred so as to create and sustain it during the past 4·5 Ga.

REFERENCES

1. UREY, H. C., 1952. *The Planets*. Yale University Press, New Haven, 245 pp.
2. ANDERS, E., 1964. Origin, age and composition of meteorites. *Space Sci. Rev.*, 3, 583–714.
3. RINGWOOD, A. E., 1966. Chemical evolution of the terrestrial planets. *Geochim. Cosmochim. Acta*, 30, 41–104.
4. SMITH, J. V., 1982. Heterogeneous growth of meteorites and planets, especially the Earth and Moon. *J. Geol.*, 90, 1–48.
5. RINGWOOD, A. E., 1979. Origin of the Earth and Moon. Springer-Verlag, Heidelberg, 350 pp.
6. TAYLOR, S. R., 1979. Lunar and terrestrial potassium and uranium abundances: implications for the fission hypothesis. *Geochim. Cosmochim. Acta*, Suppl. 11, 2017–30.
7. DAVIES, G. F., 1980. Review of oceanic and global heat flow estimates. *Rev. Geophys. Space Phys.*, 18, 718–22.
8. SLEEP, N. H., 1979. Thermal history and degassing of the Earth: some simple calculations. *J. Geol.*, 87, 671–86.
9. SCHUBERT, G., CASSEN, P. and YOUNG, R. E., 1979. Subsolidus convective cooling histories of terrestrial planets. *Phys. Earth Planet. Inter.*, 38, 192–211.
10. LAMBERT, R. ST J., 1980. The thermal history of the earth in the Archean. *Precamb. Res.*, 11, 199–213.
11. GRIEVE, R. A. F., 1980. Impact bombardment and its role in proto-continental growth on the early Earth. *Precamb. Res.*, 10, 217–47.

12. FREY, H., 1980. Crustal evolution of the early Earth: the role of major impacts. *Precamb. Res.*, **10**, 195–216.
13. MCELHINNY, M. W. (ed.), 1979. *The Earth: its Origin, Structure and Evolution.* Academic Press, New York, 598 pp.
14. YODER, H. S. JR, 1976. *Generation of Basaltic Magma.* National Academy of Sciences, Washington, DC, 265 pp.
15. WINDLEY, B. F., 1977. *The Evolving Continents.* John Wiley, Chichester, 619 pp.
16. NEWTON, R. C., SMITH, J. V. and WINDLEY, B. F., 1980. Carbonic metamorphism, granulites and crustal growth. *Nature*, **288**, 45–50.
17. JORDAN, T. H., 1981. Continents as a chemical boundary layer. *Phil. Trans. Roy. Soc., London*, **301**, 359–73.
18. ARMSTRONG, R. L., 1981. Radiogenic isotopes: the case for crustal recycling on a near-steady-state no-continental-growth Earth. *Phil. Trans. Roy. Soc., London*, **301**, 443–72.
19. MOORBATH, S., 1978. Age and isotopic evidence for the evolution of continental crust. *Phil. Trans. Roy. Soc., London*, **A288**, 401–13.
20. DAWSON, J. B., 1979. Kimberlites and their xenoliths. Springer-Verlag, Berlin, 252 pp.
21. GURNEY, J. and HARTE, B., 1980. Chemical variations in upper mantle nodules from southern African kimberlites. *Phil. Trans. Roy. Soc., London*, **297**, 273–93.
22. BORCH, R. S. and GREEN II, H. W., 1987. Dependence of creep in olivine on homologous temperature and its implications for flow in the mantle. *Nature*, **330**, 345–8.
23. MACKWELL, S. J., 1987. Creep in the mantle. *Nature*, **330**, 315.
24. MCKENZIE, D. P. and WEISS, N., 1975. Speculations on the thermal and tectonic history of the Earth. *J. Roy. Astr. Soc. Geophys.*, **42**, 131–74.
25. ALLÈGRE, C. J., BRÉVART, O., DUPRÉ, B. and MINSTER, J.-F., 1980. Isotopic and chemical effects produced in a continuously differentiating convecting Earth mantle. *Phil. Trans. Roy. Soc., London*, **297**, 447–77.
26. ANDERSON, D. L., 1981. The early evolution of the mantle; a geochemical and geophysical model. *Eos*, **62**, 418.
27. O'NIONS, R. K. and HAMILTON, P. J., 1981. Isotope and trace element models of crustal evolution. *Phil. Trans. Roy. Soc., London*, **A301**, 473–87.
28. RINGWOOD, A. E., 1975. *Composition and Petrology of the Earth's Mantle.* McGraw-Hill, New York, 618 pp.
29. LIU, L.-G., 1979. On the 650-km seismic discontinuity. *Earth Planet. Sci. Lett.*, **42**, 202–8.
30. MADON, M., BELL, P. M., MAO, H. K. and POIRIER, J. P., 1980. Transmission electron diffraction and microscopy of synthetic high pressure $MgSiO_3$ phase with perovskite structure. *Geophys. Res. Lett.*, **7**, 629–32.
31. YUEN, D. A. and PELTIER, W. R., 1980. Mantle plumes and the thermal stability of the 'D' layer. *Geophys. Res. Lett.*, **7**, 625–8.
32. FEARN, D. R. and LOPER, D. E., 1981. Compositional convection and stratification of Earth's core. *Nature*, **289**, 393–4.
33. REISBERG, L. and ZINDLER, A., 1986. Extreme isotopic variations in the upper mantle: evidence from Ronda. *Earth Sci. Planet. Lett.*, **81**, 29–45.
34. SMITH, J. V., 1981. The first 800 million years of Earth's history. *Phil. Trans. Roy. Soc., London*, **A301**, 401–22.

35a. OZIMA, M., 1975. Ar isotopes and earth-atmosphere evolution models. *Geochim. Cosmochim. Acta*, **39**, 1127–34.

35. TOLSTIKHIN, I. N., MAMYRIN, B. A., KHABARIN, L. B. and ERLIKH, E. N., 1974. Isotope composition of helium in ultrabasic xenoliths from volcanic rocks of Kamchatka. *Earth Planet. Sci. Lett.*, **22**, 75–84.

36. SHIMUZU, M., 1979. An evolutional model of the terrestrial atmosphere from a comparative planetological view. *Precamb. Res.*, **9**, 42–7.

37. WINDLEY, B. F. (ed.), 1975. *The Early History of the Earth*. John Wiley, Chichester, 619 pp.

38. TARLING, D. H., 1980. Lithosphere, evolution and changing tectonic regimes. *J. Geol. Soc. London*, **137**, 459–67.

39. OFFICER, C. B. and DRAKE, C. L., 1982. Epeirogenic plate movements. *J. Geol.*, **90**, 139–53.

40. SCLATER, J. G., JAUPART, C. and GALSON, D., 1980. Heat flow through oceanic and continental crust. *Rev. Geophys. Space Phys.*, **81**, 269–312.

41. ENGLAND, P. and BICKLE, M., 1984. Continental thermal and tectonic regimes during the Archean. *J. Geol.*, **92**, 353–67.

42. RINGWOOD, A. E. and IRIFUNE, T., 1988. Nature of the 650-km seismic discontinuity: implications for mantle dynamics and differentiation. *Nature*, **331**(6152), 131–6.

43. ESTEY, L. H. and DOUGLAS, B. J., 1986. Upper mantle anisotropy: a preliminary model. *J. Geophys. Res.*, **91**, 11393–406.

44. ANON., 1987. Scientists find slabs of Earth's crust. *International Herald Tribune*, Thursday–Friday, 24–25 December, No. 32 606, 52/87, p. 7.

45. SHAW, D. M., 1980. Development of the early continental crust. Part III: Depletion of incompatible elements in the mantle. *Precamb. Res.*, **10**, 281–99.

46. CAMERON, E. M. and GARRELS, R. M., 1980. Geochemical compositions of some Precambrian shales from the Canadian Shield. *Chem. Geol.*, **28**, 181–97.

47. RINGWOOD, A. E., 1982. Phase transformations and differentiation in subducted lithosphere: implications for mantle dynamics, basalt petrogenesis, and crustal evolution. *J. Geol.*, **90**, 611–43.

48. NEWTON, R. C., SMITH, J. V. and WINDLEY, B. F., 1980. Carbonic metamorphism, granulites and crustal growth. *Nature*, **288**, 45–50.

49. SILL, G. T. and WILKENING, L. L., 1978. Ice clathrate as a possible source of the atmospheres of the terrestrial planets. *Icarus*, **33**, 13–22.

50. SMITH, D. C., FRANZ, G. and GEBAUER, D. (eds), 1985. *Chem. Geol.*, Chemistry and petrology of eclogites, Sp. issue, **50**(1/3), 368 pp. (a collection of papers presented at the First International Eclogite Conference held in Clermont-Ferrand, France, 31 August to 2 September 1982).

51. SMITH, D. C. and VIDAL, PH. (guest eds), 1985. *Chem. Geol. (Isotope Geoscience)*, Isotope geochemistry and geochronology of eclogites, Sp. issue, **52**(2), 271 pp. (a collection of papers presented at the First International Eclogite Conference held in Clermont-Ferrand, France, 31 August to 2 September 1982).

52. STACEY, F. D., 1980. The cooling Earth: a reappraisal. *Phys. Earth Planet. Int.*, **22**, 89–96.

53. DIMROTH, E., 1985. A mass balance between Archean and Phanerozoic rates of magma emplacement, crustal growth and erosion: implications for recycling of the continental crust. *Chem. Geol.*, **53**, 17–24.

54. VEIZER, J. and JANSEN, S. L., 1985. Basement and sedimentary recycling—2: time dimension to global tectonics. *J. Geol.*, **93**, 625–43.
55. THOMPSON, A. B., 1984. Geothermal gradients through time. In: *Patterns of Change in Earth Evolution*, ed. H. D. Holland and A. F. Trendall. Springer-Verlag, Berlin, pp. 345–55.
56. GOLDSTEIN, S. L., O'NIONS, R. K. and HAMILTON, P. J., 1984. A Sm–Nd isotopic study of atmospheric dusts and particulates from major river systems. *Earth Planet. Sci. Lett.*, **70**, 221–36.
57. O'NIONS, R. K., 1987. Relationships between chemical and convective layering in the Earth. *J. Geol. Soc. London*, **144**, 259–74.
58. GALER, S. J. G., GOLDSTEIN, S. L. and O'NIONS, R. K., 1986. Limits on the extent of chemical and convective isolation in the Earth's interior. *Lunar and Planetary Abstracts*, **XVII**, 247–8.
59. O'NIONS, R. K. and OXBURGH, E. R., 1983. Heat and helium in the earth. *Nature*, **306**, 429–31.
60. VASSEUR, G. and SINGH, R. N., 1986. Effects of random horizontal variations in radiogenic heat source distribution on its relationship with heat flow. *J. Geophys. Res.*, **91**(B10), 10397–404.
61. CHRISTENSEN, U. R., 1985. Thermal evolution models for the Earth. *J. Geophys. Res.*, **90**(B4), 2995–3007.
62. KURZ, M. D. and JENKINS, W. J., 1981. The distribution of helium in oceanic basalt glasses. *Earth Planet. Sci. Lett.*, **53**, 41–54.
63. RISON, W. and CRAIG, H., 1983. Helium isotopes and mantle volatiles in Loihi Seamount and Hawaiian Island basalt xenoliths. *Earth Planet. Sci. Lett.*, **66**, 407–26.
64. ALLÈGRE, C. J., STAUDACHER, TH., SARDA, PH. and KURZ, M. D., 1983. Constraints on the evolution of the Earth's mantle from rare gas systematics. *Nature*, **303**, 762.
65. PORCELLI, D. R., O'NIONS, R. K. and O'REILLY, S. J., 1986. Helium and strontium isotopes in ultramafic xenoliths. *Chem. Geol.*, **54**, 237–50.
66. EDMOND, J. M., VAN DAMM, K. L., MCDUFF, R. E. and MEASURES, C. I., 1982. Chemistry of hot springs on the East Pacific Rise and their effluent dispersal. *Nature*, **297**, 187–91.
67. BARNES, I., IRWIN, W. P. and WHITE, D. E., 1984. Map showing world distribution of carbon dioxide springs and major zones of seismicity. Misc. Information Ser., US Geol. Surv., Map 1-1528.
68. MATTHEWS, A., HILL, R. I., FOUILLAC, C., O'NIONS, R. K. and OXBURGH, E. R., 1986. Mantle gases in the Massif Central, France. *Terra Cognita*, **6**, 257–8.
69. STACEY, F. D., 1969. *Physics of the Earth*, John Wiley & Sons, New York, 324 pp.
70. GARFUNKEL, Z., ANDERSON, C. A. and SCHUBERT, G., 1986. Mantle circulation and the lateral migration of subducted slabs. *J. Geophys. Res.*, **91**(B7), 7205–23.
71. FORTE, A. M. and PELTIER, W. R., 1987. Plate tectonics and aspherical Earth structure: the importance of poloidal–toroidal coupling. *J. Geophys. Res.*, **92**(B5), 3645–79.
72. MERRIAM, J. B., 1986. Transverse stress Green's functions. *J. Geophys. Res.*, **91**(B14), 13903–13.

73. OBATA, M. and NAGAHARA, N., 1987. Layering of Alpine-type peridotite and the segregation of partial melt in the upper mantle. *J. Geophys. Res.*, **92**(B5), 3467–74.
74. OLSON, P., 1988. Fate of subducted lithosphere. *Nature*, **331**(6152), 113–14.
75. BEN-AVRAM, Z. and VON HERZEN, R. P., 1987. Heat flow and continental breakup: the Gulf of Elat (Aqaba). *J. Geophys. Res.*, **92**(B2), 1407–16.
76. BOTT, M. P. H., 1982. The mechanism of continental splitting. *Tectonophysics*, **81**, 301–9.
77. ANON., 1988. Studying life around an undersea volcano. *The Times*, Science Report, Wednesday, 6 January, No. 62 970, p. 10.

CHAPTER 2

Geothermal Systems and Models

'And ye came near and stood under the mountain: and the mountain
burned with fire unto the midst of heaven'.

DEUTERONOMY, 4: 11

2.1 GEOTHERMAL SYSTEMS

In Chapter 1 the origin of the heat of the Earth has been discussed and it is
now possible to proceed to examine localized accumulations, that is to say
geothermal systems. The word 'geothermal' refers to the thermal energy of
the planetary interior and it is usually associated with the concept of
systems in which there is a large enough reservoir of heat to comprise
energy sources. It is important to indicate that only these are significant for
human needs. This is because, while the Earth comprises a vast heat source,
most of the heat is either too diffuse or too deeply buried to permit
economic exploitation.

Such systems may be both defined and classified on the basis of their
geological, hydrogeological and heat transfer characteristics. A separate
classification scheme orientated towards the economic aspects of
geothermal resource assessment for geothermal energy is given in Chapter
3 dealing with geothermal exploration, based upon nomenclature
commonly utilized in Italy. Leaving that for later discussion, it may be
stated here that, as a result of the fact that heat energy is stored mainly in
rock masses, an appropriate working fluid, i.e. water or steam, is required in
order that diffuse heat may be collected and thereafter transferred and
concentrated so as to create a geothermal reservoir. It is probably the
density difference between the downwardly descending and cold recharge
water and the hot geothermal water which induces the motion of the fluid in
question, the geothermal water ascending through buoyancy.

On *a priori* grounds, it is clear that high porosity (relating to the storage
coefficient) as well as high permeability (relating to hydraulic conductivity)

are essential parameters for economically important concentrations of geothermal heat to form. In 1981 L. Rybach pointed out that the degree of concentration of thermal energy in a particular geothermal reservoir can be demonstrated by simply comparing the average heat content (i.e. the above-the-surface temperature) of crustal rocks in the upper 10 km, namely 85 kJ/kg, with the enthalpy of saturated steam at 236°C and 3·2 MPa, namely 2790 kJ/kg.[1]

Enthalpy (heat content) may be defined as a thermodynamic function expressing the energy contained in a system with certain volume under certain pressure. It is used in geothermal work to denote the quantity of heat contained in the amount of water or steam produced, the relevant unit being the joule, J (0·239 cal). The proposed comparison gives an enrichment factor of more than 30 in the extractable fluid in a high-grade geothermal resource. Identification of such geothermal systems is possible because of the differences in thermal energy concentrations in relation to adjacent terrains, these comprising positive geothermal anomalies. They may include geological, hydrological and heat transfer phenomena as well as shallow, young, magmatic heat sources, the hydrothermal circulation of meteoric water in fault and fracture systems, and other factors of which the most important relate to ascending magma.

Since there is a vast variety of locally variable parameters involved, for instance changes in porosity and permeability, the depths of reservoirs, enthalpy/temperature and so on, obviously a great range of geothermal systems is to be expected. At one end could be artificial, man-made, heat extraction circulation boreholes in hot dry rocks and, at the other, the penetration of recharge meteoric water to considerable depths below ground at which heat is extracted from hot rock and afterwards transported back to the surface. In the latter case, the necessary working fluid is actually supplied by the system concerned. Between these extremes many other feasible situations may be found, for instance thick aquifers in sedimentary rocks covering large areas frequently in normal heat-flow regions, such bodies being suitable for tapping or pumping to extract the warm geothermal waters. After heat is removed, the cold waters are sometimes re-injected back into the relevant aquifer.

Some general remarks on heat flow would be appropriate here. Heat is transferable in a solid by three main processes, namely conduction, convection and radiation. At those temperatures pertaining to geothermal systems radiation is not applicable. Except for mass movements, conductive heat transfer can be described by the heat flow vector:

$$q = -K\nabla T$$

TABLE 2.1

The thermal conductivity, K, of different types of rock at room temperature[1]

Rock type	K $(W/m)^a$
Granite	2·5–3·8
Gabbro and basalt	1·7–2·5
Peridotite and pyroxenite	4·5–5·8
Limestone	1·7–3·3
Dolomite and salt	ca. 5·0
Sandstone	1·2–4·2
Shale (depending on water content)	0·8–2·1
Volcanic tuffs (depending on porosity)	1·2–2·1
Deep sea sediments (depending on water content)	0·6–0·8
Water	0·6

^{a} The unit of heat flow is now mW/m^2 (as compared with the old heat flow unit, where $1\,HFU = 41·8\,mW/m^2$).

where K is the thermal conductivity tensor and the negative sign signifies that heat descends the temperature gradient ∇T. In transient heat flow situations another thermal property, i.e. diffusivity κ, is involved, and $\kappa = K/\rho c$, where ρ is the density and c is the specific heat. Some typical values for K are given in Table 2.1.

The heat flow expresses at the surface the geothermal conditions prevailing at depth and it can be assessed either by measuring the temperature gradient, e.g. in a borehole, or by empirical determination of K on drill cores, cuttings or *in situ*. The methods used on land are described, together with necessary corrections, by (*inter alia*) J. H. Sass and R. J. Munroe in 1974.[2] It is relevant to add that consideration of several thousand terrestrial heat flow measurements led to an average conductive flow estimate of $59\,mW/m^2$, which corresponds to a capacity of *ca.* 10^{21} J/year released through the entire surface of the planet.[3,4] The location of heat resources is discussed in Chapter 3.

2.2 CLASSIFICATION

The following classification of geothermal systems is widely accepted.

2.2.1 Convective Geothermal Systems

These are characterized by natural circulation of the working fluid, the majority of the heat being transferred by circulating fluids and not by conduction. The convective process promotes increased temperatures in the upper part of the circulation system and a corresponding lowering of temperature occurs in the lower part. Convective geothermal systems can be subdivided into hydrothermal and circulation systems.

2.2.1.1 Hydrothermal Systems

These are characterized by both high porosity and high permeability. Almost all geothermal systems which have been developed to produce commercial electric power fall into this group.[5] Obviously, this is possible where the systems are shallow (under 3 km deep) and hence can be tapped by drilling. The actual heat source is in the form of shallow and young magmatic intrusions, which are to be found in specific geological environments such as spreading ridges, convergent plate margins, continental rifts and intraplate melting anomalies, all of which are discussed below.

There are two classes of hydrothermal system, namely vapour- and liquid-dominated.

(a) Vapour-dominated. Examples are Larderello in Italy and The Geysers in California, USA (see Figs 2.1 and 2.2). At the former, on 4 July 1904, electric energy was produced for the first time from geothermal steam, actually when Prince Piero Ginori Conti lit five bulbs with power supplied by a small dynamo driven by such steam.

However, as early as 1818, the French emigré Francesco Larderel had settled in Tuscany and started a borax-producing industry. Boric acid had been produced as early as 1702 by W. Homberg from borax, $Na_2B_4O_7$, by treating it with sulphuric acid, and was thereafter prescribed as 'Homberg sedative salts'. In 1827 Larderel conceived the idea of using natural steam to heat boron-enriched water of pools in the area, collecting the boric acid left after their evaporation. This obviated the need to employ wood from local forests, ever scarcer and dearer, as a fuel, and long flat sheets of lead were utilized in evaporation and concentration of boron contents of the original waters. These were called Adrian boilers, after their inventor, Adriano de Larderel, and geothermal steam flowing under them heated the overlying waters. The boric acid concentration increased from 2 to 4 g/litre on entry to 150 or 160 g/litre and, after cooling and refining, the boric acid was ground into a fine powder or minute particles. In fact, this was the first industrial utilization of geothermal energy.

Within such vapour-dominated systems two sub-types are sometimes distinguished, namely Larderello and Monte Amiata. The former embodies characteristics found also at The Geysers in California and Matsukawa, Japan. They have initial temperatures near 240°C in reservoirs at or below depths of 350 m, these being uniform and influenced by the maximum enthalpy of saturated steam (669·7 cal/g at 236°C). They have pressures below hydrostatic. Where intense surface activity occurs, waters are usually

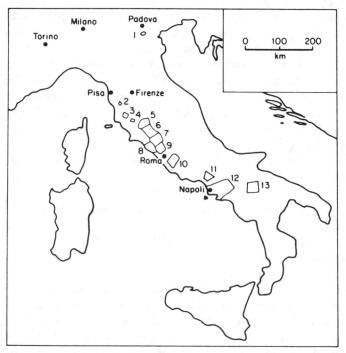

FIG. 2.1.   Italian geothermal regions. 1, Monte Berici; 2, Montecatini; 3, Larderello/Travale; 4, Roccastrada; 5, Monte Amiata; 6, Monte Volsini; 7, Monte Cimini; 8, Tolfa; 9, Monte Sabatini; 10, Colli Albani; 11, Roccamontana; 12, Campania Ovest; 13, Monte Vultura.

acidic, having pH values as low as 2, although neutralization with ammonia may take place. Sukphates may be high and chlorides low. Discharge rates for fluids are low, only tens to hundreds of litres per minute. Most of the heat appears to be in solid phases. The Monte Amiata sub-type resembles hot natural-gas fields to some extent and has, for comparable initial pressures (20–40 kg/cm^2), much lower temperatures around 150°C and much higher initial gas contents ($\geq 90\%$). Steam is a small initial constituent and, after production and decompression, the initial vapour of

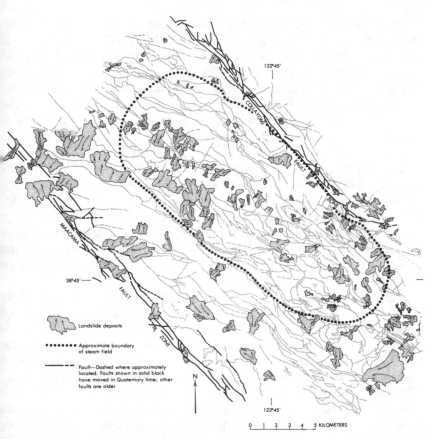

FIG. 2.2. Map illustrating landslides and faults in The Geysers geothermal field, California, USA. After McLaughlin, R. J., 1978. *Preliminary Geologic Map and Structural Section of the Central Mayacamas Mountains and The Geysers Steam Field, Sonoma, Lake and Mendocino Counties, California.* US Geol. Surv. Open-file Rept 78-389 (2 sheets), scale 1:24 000. Reproduced from Fig. 7.2 in Chapter 7, Environmental aspects of geothermal development, by M. D. Crittenden Jr, in *Geothermal Systems: Principles and Case Histories,* ed. L. Rybach and L. J. P. Muffler, 1981, by permission of the publishers, John Wiley & Sons Ltd.

high steam content is flushed out of the reservoir and replaced by relatively low-pressure steam of lower gas content arising from the boiling of water at moderate temperatures. A tendency for fluids to go from dry vapour to vapour and liquid water was observed at the Bagnore field of the Monte Amiata district. However, as seen in the second paragraph below, the practical significance of such detailed classification is small and in any case restricted to eventual uses of the resource.

(b) Liquid-dominated. Heber, East Mesa (see Section 2.5) and the Salton Sea (see Section 2.6.2), all in Imperial Valley, California, USA, and Wairakei and Ohaki–Broadlands in New Zealand, are examples. Liquid-dominated systems are much commoner than the vapour-dominated type.[6] Interestingly, until 1973, Wairakei was the only liquid-dominated geothermal system in the world to be exploited for electric power. It is controlled by the New Zealand Electricity Department and administered by the Ministry of Works. Its construction was based on a design of and supervised by the consulting engineering company of Merz and McLellan with Sir Alexander Gibb and Partners, both from the UK. Geologically, the site is characterized by a layer of pumice and soft rock 90 m thick, under which 30–60 m of impermeable dense fine sandstone and mudstone rock occur. Underlying these there is a porous mass of cemented and fissured pumice and rhyolite with, at 600 m depth, a massive ignimbrite layer. The aquifer permeability is *ca.* $10^{-13}/m^2$, which may be compared with an associated pore or matrix permeability which is orders of magnitude less at an estimated value of $\leq 5 \times 10^{-17}/m^2$.

Both vapour- and liquid-dominated systems can coexist in a reservoir and hence the distinction between them is really significant only in terms of utilizing the resource. In this connection the name given applies to the dominant mobile phase in a particular reservoir in the undisturbed state, the other phase being only partly mobile. For instance, vapour-dominated systems contain immobile or only slightly mobile water, whilst liquid-dominated ones may either contain only liquid water or a steam–water mixture. Vapour reservoirs may possess a liquid-dominated 'condensate layer' near the surface, as D. E. White and his associates pointed out in 1971.[7]

In all such systems nearly all the heat and mass are transported by the convection of liquid water and/or steam through highly porous and permeable rocks. Consequently, they are different in kind from dry rock before exploitation and geopressurized systems in which the dominant mechanism for the transfer of heat is conduction. The principal known hydrothermal systems are associated with seismically active areas having extensive faulting. It is interesting that, despite the fact that some hydrothermal systems are located in porous and permeable sedimentary/volcanic rocks, the majority are found in fractured rocks which may be almost impervious, the fractures functioning as conduits for the geothermal fluids. From this it is not surprising to discover that there is much circumstantial evidence indicating that faulted regions are often associated with geothermal systems in general as well as with hydrothermal ones in particular. Convective hydrothermal systems frequently produce surface phenomena from which they can be located and these include fumaroles

FIG. 2.3. A fumarole at Vulcano, Italy. Reproduced by courtesy of the International School of Geothermics, IIRG-CNR, Pisa, Italy.

(Fig. 2.3), boiling springs and so on, all of which are important factors in delineating the actual geothermal reservoir itself as well as in determining the inherent temperatures. Relevant temperature logging in boreholes sometimes shows geothermal gradients up to 200°C/km which demonstrate mainly conductive heat transfer in an impervious cap rock.[8] In some places, either low or even negative temperature gradients have been recorded: the latter probably result from the infiltration of cold meteoric water into the hydrothermal system. Where a geothermal gradient falls between 40 and 80°C/km, the region is termed a semi-thermal one, in contrast to a hyperthermal region in which the geothermal gradient exceeds 80°C/km.

Of course, the lateral extent of such phenomena will be an index of the size of the hydrothermal system involved. An alternative method for determination of this size is by means of gravity mapping where young silicic bodies at shallow depth are concerned, negative anomalies arising relative to adjacent terrain as a consequence of the steam fraction in highly porous reservoir rocks or because of a diminution in density induced through thermal expansion. In The Geysers, a vapour-dominated

FIG. 2.4.   The Geysers, California, USA: general view of operations. Reproduced by courtesy of the Pacific Gas and Electric Company, California, USA.

hydrothermal field, there exists a negative gravity anomaly. In some cases, this may be masked by factors such as silicification by self-sealing, and positive gravity anomalies may be substituted. Electrical methods can be utilized as a means of finding convective geothermal systems, despite the fact that the measured resistivity of a rock is affected by salinity and porosity to a greater extent than by temperature. Luckily, salinities are normally low in volcanic rocks where resistivity mapping is a feasible approach to the problem of locating geothermal reservoirs (see Fig. 2.4).[9-11]

2.2.1.2 Circulation Systems

These occur in low porosity–low fracture permeability environments in areas of high to normal regional heat flow. These can develop in areas not containing young igneous intrusions, arising from the deep circulation of meteoric water in the thermal regime of conductive and regional heat flow. Obviously, in order that such systems can exist, the related rocks must possess fractures and fault zones of adequate permeability to permit water to circulate. The temperature reached by the water will depend mainly on the size of the regional heat flow and also, to some extent, on the depth to which circulation is possible. In the absence of other considerations, it would be in shallow depths in areas of higher conductive heat flow that the higher temperatures will be attained.

There are numerous such circulation systems scattered around the world manifesting themselves as thermal springs, which occur in a vast variety of geological settings involving many different rock types, these being marked by low primary porosities and permeabilities. Thus discharges are normally near or indeed at the intersections of fault and fracture systems, but the associated recharge areas are of course much larger, sometimes by orders of magnitude. The residence times of thermal waters in such circulation systems can easily reach 1000 years or more.

2.2.2 Conductive Geothermal Systems

These include low-temperature, low-enthalpy aquifers, including geo-pressurized zones, and hot dry rock.

2.2.2.1 Low-Temperature, Low-Enthalpy Aquifers

Such aquifers are located in high-porosity and high-permeability sedimentary sequences, including geopressurized zones, in regions of either normal or slightly elevated heat flow. The conduction which is solely responsible for the geothermal regime is frequently in a steady state and the working fluid is either available (as in deep aquifers) in sedimentary basins or it must be added (as in the case of hot dry rock); see Section 2.2.2.3.

With deep sedimentary basins extended aquifers may be present, occurring at different depths in layers with high porosities and permeabilities, the associated temperatures usually not exceeding 150°C (a condition of low enthalpy). The water bodies may possess old, connate porewater, which can be highly saline, or they may be recharged from adjacent highlands. If artesian (confined) conditions exist in a particular aquifer, then the geothermal water will discharge freely to the surface when penetration by drilling takes place. The actual productivity of a specific aquifer can be defined by its transmissivity, i.e. the product of aquifer thickness and permeability.

Geothermal aquifers are considered to be a national resource of low-grade heat in the UK. Main events in the British programme involved the drilling of an initial exploratory borehole at Marchwood near Southampton in 1979 followed by a long-term pumping test in 1981. In the same year a secondary exploratory borehole was drilled at Larne and a production borehole was drilled at Southampton. A pumping test at the latter site demonstrated that the relevant reservoir is probably bounded by faults and would not support the flow rates necessary for a production schedule. Consideration was given to a smaller field trial. A third exploratory borehole was drilled at Cleethorpes in 1984, with preliminary tests. The evaluation of a hypothetical geothermal scheme at Grimsby was undertaken, based upon the results of the Cleethorpes borehole and using heat pumps. In 1985 studies were effected on the utilization of the Cleethorpes borehole for heating a glasshouse. A report by the Energy Technology Support Unit (ETSU) at Harwell was discussed by the Advisory Council on Research and Development (ACORD), and it was recommended that the Aquifer Programme should be brought to a tidy conclusion. The Secretary of State accepted this. In 1986 consultants completed the assessments of the economics of employing the intermediate-depth aquifer in the Cleethorpes/Grimsby area and the British Geological Survey finalized their work.[12]

2.2.2.2 Geopressurized Reservoirs

These comprise a rather unusual category of geothermal resource. They are, in fact, a special case of sedimentary aquifer in which the pore fluids sustain pressures exceeding hydrostatic pressure (the pressure of the water column), thus bearing a large part of the total overburden (lithostatic pressure). Instances have been given of pore pressures of the order of 100 MPa.[1] Such geopressurized reservoirs are isolated by overlying impervious and low-conductivity shales which also underlie them. Contrary to the situation in ordinary sedimentary environments in which both heat and porewater are expelled through diagenetic compaction, geopressurized reservoirs constitute heat traps of considerable efficiency. A good example is the northern Gulf of Mexico Basin in the USA, both on- and off-shore, where geopressurized zones occur in a depth range of 3–7 km and have substantial quantities of natural gas in addition to heat.[13]

2.2.2.3 Hot Dry Rock

Hot dry rock in a high-temperature, low-permeability environment is particularly interesting for its potential in areas not normally considered geothermally promising. Geothermal projects in such localities were commenced in the early 1970s, notably at Los Alamos in Nevada, USA, but also at the Camborne School of Mines in Cornwall, UK. Their aim was to develop technology which would produce an alternative energy source exploitable over a significant part of the terrestrial upper crust. In the relevant experiments, rock temperatures varying between 50 and 327°C were used. Further details are given in Chapter 5; the next stage for both projects should entail the development of new reservoir systems at depths and temperatures from which electric power can be generated commercially.

2.3 HEAT TRANSFER

The geophysical signatures of geothermal systems include hot springs and allied phenomena and the geochemical ones include the various geothermometers, e.g. silica, sodium–potassium, which will be described in Chapter 3 below. They have proved to be extremely valuable in determination of heat flow distribution patterns.[14,15] In 1980 a new gas thermometer was developed by F. D'Amore and C. Panichi.[16] They claimed that the chemical composition of gas mixtures emitted in geothermal regions can be utilized in order to evaluate deep thermal temperatures; their work is also discussed in Chapter 3.

The above manifestations result from heat in the upper part of the crust of the Earth and this is the region of interest in identifying and exploiting geothermal resources. The transfer of thermal energy is effected by means of conduction and, even in geothermally anomalous areas, heat flow through impermeable rock is only by conduction. Of course, the distribution of heat flux is affected by convective effects and those boundary conditions arising from topographic relief as well as by regional variations of heat properties such as thermal conductivity. Other influential factors are the variations, spatio-temporally, of actual heat sources relating to matters such as magma intrusions and the generation of heat by radioactive decay (Table 2.2).

As noted above (Section 2.2.1.1), practically all geothermal energy trapped until now for commercial purposes was derived from convective geothermal (hydrothermal) sources and, in such systems, near-surficial

TABLE 2.2

Heat from radioactive decay and its production in common rocks
[adapted from E. Bullard (1973)[17]]

Radioisotope	Heat generated (cal/g year)	Half-life (10^9 years)	Reaction
^{238}U	0·70	4·5	$U \rightarrow {}^{206}Pb$
^{232}Th	0·20	13·9	
^{40}K	0·21	1·31	$K \rightarrow {}^{40}Ca$ or ^{40}Ar
K ordinary	$2·7 \times 10^{-6}$		
^{233}U	4·3		
^{235}U	0·03	0·71	
U ordinary	0·73		

Rock type	Elemental concentration			Heat production ($\mu cal/g$ year)			
	U (ppm)	Th (ppm)	K (%)	U	Th	K	Total
Granite	4·7	20	3·4	3·4	4·0	0·9	8·3
Basalt	0·6	2·7	0·8	0·44	0·54	0·23	1·21
Peridotite	0·016	0·004	0·0012	0·0012	0·001	0·0003	0·013

water seeps to considerable depths through porous and permeable conduits and meets hot rock. Subsequently, the heated fluid, i.e. water and/or steam, ascends through buoyancy, and convection cells arise in which leakage to higher levels or above the surface is compensated by the inflowing of meteoric groundwaters. Sometimes such systems produce impressive surface manifestations such as geysers. The question of convective heat and mass transfer in these systems is discussed in detail later in this chapter.

2.3.1 Fractured Media

In hydrothermal systems, fractured media are involved and, of course, the transportation of both mass and energy in these is of considerable interest to those concerned with groundwater and petroleum, among others. Consequently, studies have been effected on the geometry of fracture systems and how they react to hydrodynamic gradients.[18] Obviously, the existence of fissures in rocks exerts a great influence upon fluid flow because of the much greater permeability; for instance, C. Louis suggested in 1970 that the matrix permeability of rock is significant only when continuous joints do not occur or for joint apertures not exceeding 10 μm across.[19]

Also fracture permeability is almost invariably anisotropic, the associated porosity being more sensitive to rock stresses and fluid pressures than the porosity of the actual matrix.[18]

In order mathematically to describe flow through such fissured rocks, it is important to bear in mind that the most significant variable to be considered is probably the spacing of discontinuities. In the case where the distance between these is comparable with the dimensions of the rock mass being investigated, then the detailed fracture geometry becomes critical.[20] Unfortunately, it is usually impossible to assess this geometry in geothermal systems and so the discussion is usually confined to situations of opposite character, i.e. those in which the discontinuities involved are extremely small in comparison with the size of the geothermal reservoir in question. The fissured rock can then be treated as a continuous medium with anisotropic permeability.

A continuum model for fluid flow through fractured rock masses was investigated by W. G. Gray and his associates in 1976.[21] They assumed that extensive fracturing exists and that it is so distributed as to permit a superficial discharge through the fractures as well as the pores to be considered. However, despite the fact that it is not unduly difficult to present balance laws for mass and energy transport, their model has little practical use because it is limited by the problem of describing the exchange of both heat and mass between fluid in the pores and fluid in the fractures. To obviate this drawback, it is feasible to represent the fissured rock mass under discussion by an equivalent porous medium which should have permeability which is anisotropic and thus variable in space. Among others, J. W. Pritchett and his associates in 1976 did so in regard to modelling the behaviour of the geothermal reservoir at Wairakei, New Zealand, where most of the aquifer permeability of about $10^{-13}/\mathrm{m}^2$ results from fractures, with an associated pore or matrix permeability of less than $5 \times 10^{-17}/\mathrm{m}^2$.[22]

2.3.2 Balance Laws for Fluid Flow

In practice, balance laws for fluid flow through porous media refer primarily to saline waters because these predominate in most hydrothermal systems, although to a variable degree, e.g. from 10% of that of seawater at Wairakei to ten times that of seawater at the Salton Sea.[5] They are associated with small quantities of non-condensible gases as well. Usually, the porewaters in question are regarded as pure water; suitable governing equations which express the balance of mass, momentum and energy in a deformable porous matrix were developed by numerous authors, among them D. H. Brownell Jr and colleagues, and S. K. Garg and J. W. Pritchett in 1977.[23,24] A totally interacting rock–fluid system was considered and the factors involved include the relation between pore collapse and fluid extraction rates as well as the re-injection of such collapsed pores by means of condensate injection.

It is also desirable to investigate the results of the variability induced by naturally existing mechanical forces, for instance the stresses produced by overburden, on the condition of the geothermal system. It was suggested that there are significant interactions between the fluid and the porous matrix, as could perhaps be anticipated on *à priori* grounds. Hence general interaction theories are required so as to assess the effect of the motion of the fluid on the motion of the matrix and vice versa.[5]

If systems in which the matrix is taken to be almost or completely stationary are considered, then the fluid motion in them will not be affected by it (except insofar as variations in the porosity occur). In them Darcy's Law is applicable and the several phases involved, namely the rock matrix, liquid water and water vapour, are probably in local thermal equilibrium. From this it is clear that only the energy balance of the relevant mixture (rock, liquid water and water vapour) is interesting in such situations. For simplicity, it must be assumed that (1) the rock porosity, ϕ, at a point x depends upon the local fluid pressure alone, and (2) that the liquid and vapour are in local pressure equilibrium in a manner such that the capillary pressure may be neglected.

The balance equations in question may be stated as follows.

Mass
Fluid (liquid/vapour mixture):

$$\frac{\partial}{\partial t}(\phi\rho) + \nabla(\rho_i q_i + \rho_g q_g) = 0 \tag{2.1}$$

Momentum
Liquid:

$$q_1 = -\frac{R_1 k}{\mu_1}(\nabla p - \rho_1 g) \tag{2.2}$$

Vapour:

$$q_g = -\frac{R_g k}{\mu_g}(\nabla p - \rho_g g) \tag{2.3}$$

Energy

$$\frac{\partial}{\partial t}[(1-\phi)\rho_r E_r + \phi\rho E] + \nabla[\rho_1 E_1 q_1 + \rho_g E_g q_g + pq_1 + pq_g]$$
$$= \nabla(K_m \nabla T) + [\rho_1 q_1 + \rho_g q_g]g \tag{2.4}$$

in which

E = specific internal energy for liquid–vapour mixture $(= [(1 - S)\rho_1 E_1 + S\rho_g E_g]/\rho)$;

E_i = specific internal energy for ith phase (i = r, rock matrix; i = l, liquid; i = g, vapour);

g = acceleration due to gravity;

k = permeability tensor;

K_m = thermal conductivity of the mixture (solid–liquid–vapour);

p = pressure;

q_i = volume flux vector for the ith phase (i = l, g);

R_i = relative permeability for the ith phase (i = l, g);

S = relative vapour volume;

t = time;

T = temperature;

x = space vector;

μ_i = viscosity of the ith phase (i = l, g);

ρ = density for liquid–vapour mixture $(= [1 - S]\rho_1 + S\sigma_g)$;

ρ_i = density for the ith phase (i = r, l, g);

ϕ = porosity.

Thereafter it is necessary to relate the four equations given above to appropriate constitutive factors and, in addition, to initial and boundary conditions. In the given balance equations, liquid and vapour viscosities, densities and internal energies as well as relative vapour volume, fluid density ρ, internal energy E, temperature T and pressure p are involved, but the specification of mixture thermal conductivity, K_m, entails acquisition of knowledge regarding both the liquid and the vapour conductivities, K_1 and K_g.

There are semi-analytical or tabular equations of state for water and, to avoid complications, it may be simply stated that, for the purposes of geothermal applications, it is permissible to write the following equation:

$$E_r = c_r T_r \qquad (2.5)$$

in which c_r is the constant-volume heat capacity of the solid.

The thermal conductivities of dry and fluid-saturated rocks have been examined in considerable detail, for instance by H. J. Ramey and others in 1974.[25] Those of most rocks decrease as the temperature rises. The thermal conductivities of fluid-saturated rocks are greater than those of dry rocks by a factor of anything up to five.

The permeability, k, is complexly inter-related to porosity, ϕ, as well as to both the fluid pressure and the temperature. In geothermal work, it is usual to take k as a function solely of ϕ, thus $k = k(\phi)$.

At the present time there is not enough information available for steam/water systems for the dependence of relative permeabilities on steam saturation and temperature to be assessed. Attempts to do so have been made, however, on the basis of expressions for oil/gas reservoirs.

Porosity, ϕ, normally is a function of the current state of stress in a rock mass, i.e. conditions induced by both rock and fluid stresses, the preceding stress history, the temperature and the type of rock involved. In geothermal work, it is sometimes assumed that the reservoir is subjected to uniaxial compression only and also that the overburden stays more or less constant. If this is justified, then the porosity is given by the following expression:

$$\partial\phi/\partial t = (1 - \phi)C_m \, \partial p/\partial t + 3(C_m\eta_s\mathcal{K}) \, \partial T/\partial t \qquad (2.6)$$

from S. K. Garg and his associates in 1977,[26] where C_m is the uniaxial compaction coefficient, $\eta(\eta_s)$ is the coefficient of linear thermal expansion for porous rock (rock grain) and $\mathcal{K}$ is the bulk modulus of porous rock.

The uniaxial compaction coefficient depends upon the pore pressure, as also does the bulk modulus, and both relate to the loading direction (increase or decrease in pore pressure) and the stress history. In the field, it has been found that even small changes in the pore pressure can affect the parameters in question very greatly indeed; for example, at Wairakei an average reservoir pressure drop of under 3 MPa is believed to have caused a 15-fold alteration in the uniaxial compaction coefficient.

S. K. Garg and D. R. Kassoy in 1981 presented a very important Boussinesq approximation, of which the simplifying arguments are presented below.[5] Although the equations cited above for two-phase flow are complex, a simpler approach is often permissible, for instance one pertaining to single-phase flow. Thermally driven flow in such a situation, i.e. in liquid geothermal systems, may be discussed. However, some assumptions are necessary. One is that the rock matrix is both isotropic and homogeneous physically, namely with regard to porosity, permeability and thermal conductivity. These parameters should be temperature-independent.

Secondly, water is taken as an incompressible liquid with both density and kinematic viscosity dependent upon temperature in the following manner:

$$\rho = \rho_0[1 - \alpha(T - T_0) - \beta(T - T_0)^2] \qquad v = v\rho\sigma(T) \qquad (2.8)$$

in which ρ_0, α, T_0 and β are constants; $\sigma(T)$ is a prescribed function of temperature; ρ is density; and v is the kinematic viscosity. Then pressure, work and viscous dissipation are regarded as negligible and fluid (solid) internal energy, $E(E_r)$, is given by

$$E = c_v T \qquad E_r = c_r T \qquad (2.9)$$

in which c_v (c_r) denotes the constant-volume heat capacity of the fluid (rock matrix).

Taking into account these assumptions, the balance laws [see eqns (2.1) to (2.4) above] can be restated thus:

Mass
Fluid (liquid):

$$\nabla q = 0 \qquad (2.10)$$

Momentum
Fluid (liquid):

$$(\nabla p/\rho_0 - g) + \alpha(T - T_0)[1 + \beta/\alpha(T - T_0)]g + v_0 \sigma/kq = 0 \qquad (2.11)$$

Energy
Rock–fluid mixture:

$$[(1 - \phi)\rho_r c_r + \phi \rho_0 c_v] \, \partial T/\partial t + \rho_0 q c_v \nabla T = \nabla(K_m \nabla T) \qquad (2.12)$$

The pressure, p, is not a thermodynamic variable, but is rather determined by the imposed boundary conditions.

Obviously, the best approach to investigating both heat and mass transfer in a particular geothermal system entails obtaining, analysing and interpreting field data, the results being applicable to creating a model which will be physically acceptable, i.e. it will accord with the relevant geology as well as with the chemical properties of both the rocks involved and water. Thereafter it may be possible to produce a mathematical model which is based on porous media flow and from this it would be feasible to derive quantitative estimates of those physical processes included in the conceptual model itself. Clearly, such a model ought to describe the source of water supplying the geothermal system and this will entail the mechanisms for water transport to depth, how heating is effected at depth, the ascent of buoyant, heated liquid, its dispersal into aquifers and the subsequent cooling of aquifer liquids through processes such as mixing with colder groundwaters, emission at the surface of the ground and conduction. It must be emphasized that, as yet, no one model based upon the flow of a liquid in a saturated porous medium can describe every facet of heat and mass transport which takes place in a particular geothermal system. Only specific aspects can be considered in essentially idealized modelling.

2.4 IDEAL MODELS

Appropriate non-dimensionalized equations can be employed to describe convective processes occurring in an enclosed liquid-saturated porous medium and they can indeed be derived from the foregoing equations (2.10) to (2.12).[5] In Ref. 5 the authors provided a detailed examination of this matter and mentioned that the equations in question also permit the Rayleigh number to be determined, i.e. the ratio of the convective heat transport to that conducted across the system. In the past, convection in saturated porous media has been examined, bearing in mind the flow of a fluid with constant properties in an isotropic and homogeneous medium. However, in idealized convection problems, flow in geothermal situations can be linked with considerations of physics.[26]

2.4.1 Linear Stability
In the simplest models, linear stability is applied; that is to say, the conditions necessary for convection to commence in a water-saturated isotropic and homogeneous medium between two horizontal and impermeable infinite plates kept at uniform, but different, temperatures are studied. When, as is actually typical of geothermal systems, $\Delta T' = O(10^2 \text{ K})$, the variation of viscosity with temperature becomes significant ($\Delta T'$ is the maximum impressed temperature difference); cf. the work by J. H. Keenan and his associates in 1969.[27]

Earlier, it was suspected by H. L. Morrison and his co-workers[28] that the critical value of the Rayleigh number in a variable-viscosity system is reduced from the constant-property value $4\pi^2$. Subsequently, linear-stability solutions were developed for a liquid having a viscosity which decreases with temperature. Crude estimates of eigenvalues demonstrated reduced Rayleigh numbers. Precise values for Rayleigh numbers were found later by D. R. Kassoy and A. Zebib in 1975.[29] Viscosity variation, $\mu(T)$, may be described by an empirical formula for liquid water in which, for linear theory, $T \simeq 1 - \tau z$,[29] where $\tau = \Delta T'/T_0'$ (the overheat ratio) and the factor z refers to two-dimensional planar calculations in the x–z plane and also relates to the local Nusselt number, Nu, which may be utilized in order to express heat transfer results.

It must be noted that, where $T_0' = 298 \text{ K}$ and $\Delta T'$ is less than 225 K, the viscosity decreases from top to bottom in the system by as much as eight-fold. Results also showed that the critical Rayleigh number decreases by a factor of about 4·5. Later a variation of thermal expansion coefficient was included in the calculation with temperature and an associated density increase produced a further reduction in the Rayleigh number.[30]

Refined and definitive stability calculations produced very accurate variable viscosity and expansion coefficient formulae.[31] The variations of the Rayleigh number with τ as found by L. W. Morland and his associates in 1977 differ slightly from those recorded by J. M. Straus and G. Schubert in the same year because their non-dimensional parameters μ and T are not quite the same.[31,32] Depth effects on the critical Rayleigh number were recorded for systems in which the depth exceeded 4000 m and $\Delta T' < 250$ K. However, the great majority of geothermal systems are shallower than this.

In order to account for the finite extent of such a real geothermal system, A. Zebib and D. R. Kassoy in 1976 investigated the stability problem in a rectangular parallelopiped with insulated vertical boundaries when $\Delta T'$ was less than 200 K.[33] The difficulty is that all such experiments have only a restricted value as regards actual geothermal systems in which the geological conditions are far too complex to accord with the assumed properties of any model, particularly with respect to constant temperature, homogeneous isotropic material properties and a steady-state conduction temperature profile.

With regard to this last requirement, F. T. Rogers Jr and H. L. Morrison in 1950 indicated that the conduction temperature distribution in the rest state is most probably unsteady in the geophysical context, e.g. injection of a sill under a cold, water-saturated horizontal layer of porous rock will produce a conduction temperature profile with large localized gradients.[34] Nevertheless, it has been shown that both heat and mass transports are sensitive to the viscous properties of the water.

2.4.2 Real Geothermal Systems

These will have complex lithologies which can be modelled by making the assumption of a distinction between their vertical and horizontal permeabilities. The effect of anisotropic permeability on linear stability in a horizontal slab configuration implies that viscosity decrease with depth resulting from rising temperature is counter-balanced by a permeability decrease resulting from compaction, as R. A. Wooding pointed out in 1976.[35] His calculations in New Zealand indicated that, in the Taupo Volcanic Zone (see Fig. 2.5), where the representative depth of the geothermal reservoir is taken as 3 km, an anisotropic case gave a spacing of *ca.* 1·1 ×

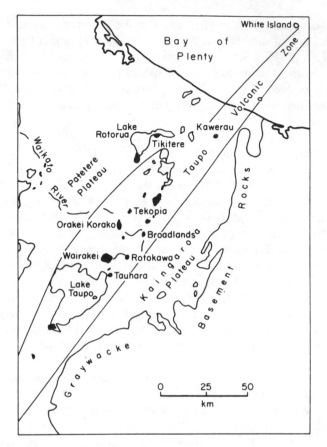

FIG. 2.5. The Taupo Volcanic Zone, New Zealand.

10^4 m, corresponding to spacing of well-defined zones of anomalously high heat flow there. The inference was that the pattern of liquid upwelling taking place is attributable to hydrodynamic processes only, these occurring in an anisotropic medium, internal geological structures such as faults playing no part at all. Of course, hydrological flow on a large scale may be induced by artesian pressure gradients in permeable rocks and, if a magma body is intruded under such a zone, then thermal gradients will arise in the system and can promote buoyant convection.

2.4.3 Non-linear Theory
Rather high, almost surficial, heat fluxes present near existing geothermal

systems result from heat transfer connected with liquid convection below ground, and the movement suggests that the Rayleigh number of such a system exceeds a critical value. In an idealized model representation, non-linear equations are applicable to solving the problems of magnitudes of heat and mass transfer, as well as the spatial distribution of velocity, temperature and pressure.

Of course, in a real geothermal field, it is feasible to obtain data comparable with calculated heat flux distribution figures and the variations of temperature and pressure with depth. In models involving a homogeneous, isotropic, porous matrix, the heat and mass transfer factors will be unlike those actually obtaining a real geothermal reservoir in which sedimentary stratigraphic characteristics can order main flow phenomena. The constant-property convection occurring in a system composed of an overlying permeable layer and an underlying impermeable one of equal thickness was considered in 1962 by I. G. Donaldson.[36] Movement of fluid in the upper layer stimulated a vertical distortion of the isotherms in the lower one. An extension of this exercise involves consideration of three layers, namely a permeable layer enclosed between upper and lower impermeable ones. M. L. Sorey in 1975 indicated that, at a given Rayleigh number, the Nusselt number was lower for this model than for the two-layered one alluded to above.[37] The heat transfer diminishes as that part of a system comprising low permeability and pure conduction zones increases.

Interesting work was reported by R. Rana and his associates in 1979.[38] They looked at convection in a rectangular reservoir made up of three horizontal layers and recharged at its upper surface. Whilst it is true that the main body of water enters and leaves such a system once only, both events taking place at the said upper surface, a barrier effect arises where the middle layer is impermeable and produces small zones of closed streamline patterns in the lowest layer. This may mean that lithological effects may be significant in some models of convective flow.

Compaction increases with depth, therefore this factor may affect the flow pattern: the matter was examined by R. J. Ribando and K. E. Torrance in 1976.[39] Permeability decreases exponentially with depth by a factor of 280. At Ra (Rayleigh number) = 200 and Nu (Nusselt number) = 1·76, the most active convection takes place in the upper part of the system. In the deep and impermeable part, conduction is the principal means of heat transfer, although small quantities of liquid may have penetrated. Thus the rest state isotherm pattern is practically undisturbed there.

The inference is that in the deep formations in a given geothermal system, peripherally involved with water convection, conduction will be the controlling factor in supplying heat to the overlying convection system.

Actual field observations show that real geothermal systems are often of a transitory nature. A calculation for convection in a rectangle with constant-temperature horizontal boundaries demonstrated that a time to steady state of $t \approx 10$, corresponding to 10^4 years, is valid in the case where $Ra = 800$.[40] In addition, the convection process becomes erratic and oscillatory if sufficiently large values of Ra are attained, e.g. where $Ra = 800$, cycles of oscillatory convection may be of the order of $t = 10$ or $10\,000$ years in the case referred to above. Such convection processes are not confined to two dimensions in geothermal systems, but rather operate in a three-dimensional mode with flows of this sort being probable in both infinite parallel-plate and box-like configurations; see, for instance, J. M. Straus and G. Schubert (1978).[41]

Weakly non-linear theory was employed by A. Zebib and D. R. Kassoy in 1978 in order to demonstrate that both two- and three-dimensional steady flow patterns were feasible in the box configuration in the case where each of the horizontal directions is an integer multiple of the height.[42] That both steady modes exist may justify the inference that initial conditions on the transient problem would determine which configuration actually appears.

In 1979 J. M. Straus and G. Schubert employed a three-dimensional Galerkin procedure in order to study the transient development of convection in a cube for $45 \leq Ra \leq 150$, where Ra is the Rayleigh number and the initial conditions were random.[43] They derived two- and three-dimensional steady solutions for all values of Ra, which underlined the significance of initial conditions in determining the steady-flow configuration. The Nusselt number results showed that, at $Ra = 97$, the heat flux of two- and three-dimensional modes is the same. The inference which may be obtained from such three-dimensional considerations for these idealized models of geothermal systems is that the flow configuration in one of them could well be a function of the geological and geophysical development of the relevant region.

2.4.4 Feasible Models

The following discussion relates to attempts made to understand heat and mass transfer processes taking place within typical liquid-dominated hydrothermal geothermal systems by using hypothetical, but physically feasible, models of which mathematical descriptions refer to the physical properties of real systems. Relevant parameters such as structural characteristics and boundary conditions should parallel situations occurring in the field.

2.4.4.1 Pipes

The first of these was intended to encompass the major features of a complete geothermal region. It was developed in 1942 by T. Einarsson, who based it upon a pipe system.[44] Whilst qualitative properties of the model in question were examined by various authors, among them J. W. Elder in 1966, an initial quantitative treatment was given by I. G. Donaldson in 1968.[45,46] In this latter model, cold water charges a vertical porous channel of given width, and heating takes place in a connecting horizontal channel wherein the bottom boundary temperature increases linearly. The centrally

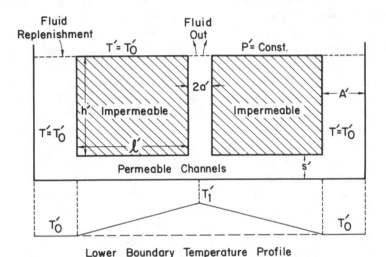

FIG. 2.6. The pipe model of I. G. Donaldson (1968). Reproduced by permission of John Wiley & Sons Ltd from *Geothermal Systems: Principles and Case Histories*, ed. L. Rybach and L. J. P. Muffler, p. 53, 1981.

placed, vertical, porous channel represents the upwelling zone of a whole geothermal region, convection resulting from the horizontal pressure gradient consequent upon the difference between the cold and hot hydrostatic heads of the two channels mentioned above. This vertical convection induces an upwards distortion of high-temperature isotherms and large increases in total surface heat transfer; see Fig. 2.6. Of course, some simplifying assumptions were made, for instance spatially invariant mean values for the velocity field and constant viscosity as well as uniform porous medium characteristics. If the central channel has a radius r and height h, a non-linear variation of the Nusselt number on the upper surface with the aspect ratio r/h occurred, an increase in r causing increased vertical convective transport through the upper surface of the said central channel and also upwards isotherm distortion in the impermeable blocks surrounding it. The latter change produces an enhancement in the surface conductive heat flux. It was noted that the upwards isotherm distortion caused by convection resulted in reduction from the pure conduction value of the temperature gradient in the lower 85% of the central channel, i.e. most of the temperature rise takes place in the uppermost 15%. In fact, many such geothermal field observations have been recorded. The liquids are emitted through hot springs or other exists at a very high temperature and, where the Rayleigh number is significantly larger than the range considered by Donaldson (Ra < 89), then the temperature of the surface environment has little effect on the hot spring efflux temperatures.[46] Where the central channel possesses a much lower permeability than that of the horizontal channel, the flow resembles that described already. Where the permeability in the whole system is large enough, then a flow instability arises and produces enhanced heat transfer by means of cellular convection superimposed upon the through-flow.

The pipe model is regarded as valuable in obtaining reasonable estimates of mass transfer through a system and overall surface heat flux, the first directly depending upon the properties of the rock system which connects the cold charge zone with the region of upflow. Of course, these are not known very well as a rule because of the lack of borehole information available from the region around the actual geothermal area. Also other factors such as, for example, the Rayleigh number, as well as the mass flow rate, influence the overall heat flux. This means that the latter parameter is not directly related to the mass flow rate. Consequently, the model cannot be employed to examine spatial characteristics of mass transfer and temperature in such an active geothermal area since insufficient details are to hand regarding fault zones, joint systems, aquifers and so on.

2.4.4.2 Convection Cells

A survey of work done entailing large-scale convection cells models was given by P. Cheng in 1978.[47] This was particularly concerned with modelling heat and mass transfer in a whole-island geothermal reservoir and an initial value problem was formulated for the convection process in a confined-island aquifer with a constant-temperature upper boundary (impermeable, of course) caused by the abrupt rise of a hot spot at the lower boundary (an intrusive). Others have examined the results which might stem from such an intrusion initiating a convection system when quickly emplaced in a deep, water-saturated and fractured rock system.[48] Flow was assumed to take place mainly in cracks arising from regional tectonic stresses and also from the intrusion itself, such cracks extending to depths of 15 km or so. Obviously, cooling of the pluton is more rapid where water circulates through the cracks in cases where they are derived by thermal stressing as temperatures fell.

In an instance given by D. Norton in 1978, an impermeable pluton was considered, this possessing a height of 4 km and a width of 6 km together with an initial temperature of 1573 K.[49] Emplacement occurred in a 10 km × 24 km reservoir in which $k = 10^{-14}/m^2$ in the uppermost 3 km, an order of magnitude less in the middle 3·5 km and again an order of magnitude less than that in the lowermost 3·5 km. The initial distribution of temperature corresponds to the normal geothermal gradient determined by a heat input of 50 mW/m² through the bottom boundary. On cooling to 973 K, the permeability of the said pluton reduces to the value of $10^{-17}/m^2$. The convection process is the cause of a ten-fold increase in the surface heat flux above the regional values over an 80 000-year time interval. At considerable depth convection is enhanced.

A rather similar study was effected by L. M. Cathles in 1977.[50] He calculated the heat and mass transfer process induced in a fractured and rectangular saturated reservoir of 10 km width and 5 km depth by the intrusion of a body 1·5 km wide and 5 km high, at 973 K. At first, the reservoir has a normal geothermal gradient based upon a prefixed basal heat flux of 63 mW/m² and a top temperature of 293 K. Vertical boundaries are insulated. Transient numerical calculations are effected for the porous media equations with uniform permeability and the realistic properties of pure water are included. Phase changes are allowed, but no regions of coexisting phases are feasible. True boiling or condensation takes place across a surface in the geometrical space. If the relevant fracture permeability is taken as $2·5 × 10^{-16}/m^2$ (so that Ra is approximately 250 and the uppermost surface is impermeable), the maximal conductive surface heat flux peaks at some 410 mW/m² during the succeeding 20 000 years. Cooling is finished in 100 000 years. If surface mass flux is allowed, then the net convective and conductive surface flux peaks at some 7120 mW/m² after 7750 years. This is probably practically wholly due to convection.

L. M. Cathles's hydrodynamic patterns show that fluid percolates down over a wide region and, when salinity effects are included, there is a phase transition at depth. At an early stage during cooling, i.e. after 5000 to 50 000 years or so, a supercritical region is formed above the pluton and fluid ascending through this cools to create a discrete condensation surface. Such a phenomenon would comprise a good geothermal resource lasting for anything up to 10 000 years. However, it would disappear before any appreciable surface expression of high heat flow became evident, and hence would be hard to locate unless an associated geophysical anomaly occurred. In the reservoir, the pressure field is almost hydrostatic, although anomalously low pressures were calculated at great depths (<10 MPa below the hydrostatic value) when the pressure is specified on the impermeable upper boundary. This is probably associated with the variability of average system density in a constant-volume configuration. It is more practical either to specify the pressure at depth or to make allowance for uplift of a confining surface water table acting as a barrier to prevent geothermal waters from emerging at the surface while allowing thermal expansion.

2.4.4.3 Hot Springs and Faults

Hot springs are highly efficient at transferring energy and their waters ascend from deep hot reservoirs and lose heat to adjacent rock formations as they rise. Should there be a hydraulic connection between the upflow zone and intersecting beds, then convective charging may well take place. However, where an impermeable barrier exists between the two regions, only conductive heat transfer is possible. Such a barrier can result from hydrothermal alteration on the edge of a vertical conduit.

M. L. Sorey modelled the situation in 1975.[37] He examined the cooling process in an ascending hot-water column deriving from the surface of a deep reservoir, the mass flow rate and the reservoir temperature being specified. Heat is removed from the conduit via an impermeable barrier into an adjacent and near-surface aquifer, water exiting at the surface through a hot spring at a temperature higher than that of the neighbouring area but lower than that of the original reservoir. The near-surface aquifer mentioned has a bottom temperature equal to that pertaining to the deep reservoir as well as a distant and insulated vertical boundary. At the upper surface, either a convective–radiative heat loss condition or a specified temperature value is specified.

In such a steady-state model, all heat conducted into the aquifer along the vertical (hot) boundary will eventually be lost through its upper surface. One instance involves a mass flow rate of 110 000 kg daily from a reservoir at 453 K lying in a planar vertical fault zone which is 10 m in width, 1 km long and 1 km deep. In the case where an adjacent aquifer is impermeable and with the surface temperature specified at 283 K, then the heat loss from the upflow zone is 544 MW and the water temperature at the surface is 349 K. This is a high-Rayleigh-number convection model in a fault zone and the temperature drops 55 K in the lowermost 88% of the 1-km-deep fault as compared with a decrease of approximately 50 K in the uppermost 12%. The surface heat flux next to the spring is some 3430 mW/m^2. The surface temperature specified is inclined to over-strain the energy transfer process in the upper part of the aquifer and promotes excessive surface heat flux.

In the case where the surface heat loss boundary condition is utilized, with a heat transfer coefficient of $4 \cdot 186 \times 10^{-2}$ W/m^2 K, then the surface spring temperature will rise to 364·8 K. At the aquifer surface, the temperature and heat flow varies from 326 K and 150 mW/m^2 at 5 m from the fault plane to 286·5 K and 150 mW/m^2 beyond 1 km. There is an even more marked distortion in regard to the vertical distribution of temperature, having a gradient of 0·06 K/m in the lowermost 92% of the fault as compared with 0·41 K/m in the uppermost 8%. Where the aquifer permeability is fixed at $k = 10^{-14}$/m^2, then the effective Rayleigh number is 200.

Heat addition from the fault zone affects naturally occurring processes of convection in the aquifer to some 500 m inwards from the vertical boundary. The spring temperature is 369 K at that point since the heat loss from the fault plane is reduced to *ca.* $4 \cdot 186 \times 10^5$ W from the earlier value of $5 \cdot 46 \times 10^5$ W. In the fault itself, the vertical distribution of temperature is even more skewed. The 55-K drop covers approximately 95% of the fault dimension, and a thermal boundary layer (*ca.* 5% of the fault dimension) includes the remaining drop, the corresponding gradients being 0·058 and 0·58 K/m. This shows the significant influence of convective mass and heat transfer in a geothermal system.

Other models have been proposed, for example by D. R. Kassoy and A. Zebib in 1978.[51] They looked into the cooling of an ascending column of liquid water in a vertical, narrow channel of saturated porous matter with impermeable walls along which the temperature increases linearly with depth. This modelled the fracture permeability system connected with a fault zone in an area of continual seismicity, hot liquid charging the fault zone at depth. This liquid stems from a highly fractured basement. If the fracture permeability is great compared with the intergranular value for the adjacent unfractured sediment (containing clays and/or being hydro-thermally altered), then the ascending liquid will traverse the fault zone without any mass loss and emerge as a surface hot spring.

Naturally, the specified temperature boundary condition implies a much lower thermal resistance of adjacent rock compared with that of the saturated material in the fault system and this is an idealization of the actual conditions. Solutions can be derived for large Rayleigh numbers. The surface temperature in the centre of the fault is larger than that of adjacent (impermeable) material. Most of the vertical heat transport occurs through convection of hot liquid and there is substantial conductive loss through the fault walls. This re-emphasizes the efficiency of heat transport by means of convection even at low flow rates of a mere 10^{-2} m/day. The argument is valid for high-Rayleigh-number systems wherein a particle of fluid loses a small quantity of heat by conduction as it traversed the system.

Another hot-spring model was proposed by D. L. Turcotte and his associates in 1977.[52] They looked at two-dimensional convection rolls found in a thin vertical slab of saturated porous media having impermeable and insulated vertical boundaries, this representing the fracture field connected with a fault zone. At the lowermost impermeable boundary, the heat flux is prescribed and surface water penetrating the convection cell through the uppermost surface does so with a prescribed ambient temperature, that emerging, i.e. the hot-spring emanation, having to satisfy a condition that the vertical gradient of the conductive heat flux disappears. Such a boundary condition allows an exit temperature higher than the rock surface value specified as ambient.

The situation was formulated in terms of distinct rock and water temperatures and solutions for the appropriate porous media equations developed numerically by employing input information from the Steamboat Springs in Nevada, USA. In this particular hydrothermal system, almost equal rock and water temperatures occur in the basal 90% as a result of the fact that the local characteristic conduction time is short in comparison with the analogous convection time. However, a temperature differential appears in the topmost 20% fuelled by the disparate conditions of cooling on the rock and the water at the uppermost surface. Unfortunately, actual measurements in the field suggest that the temperature difference in question is, in fact, less than anticipated.

2.5 EAST MESA ANOMALY, IMPERIAL VALLEY, CALIFORNIA, USA: A CONCEPTUAL MODEL

Here field data were utilized by K. P. Goyal and D. R. Kassoy in 1977 and the former alone in 1978 in order to derive a conceptual model for the East Mesa geothermal reservoir which they based upon a fracture zone (fault) of finite width extending downwards almost vertically through a cap containing clays through interbedded sediments in the reservoir and into the basement.[53,54] The fault is assumed to be charged at depth by liquid previously heated in the fracture system of the basement. Liquid ascends into the reservoir system of the fault, but cap clays prevent vertical transport and therefore water extruded from the fault by artesian pressure flows horizontally.

Mathematically, the fracture zone is treated as a vertical slab of porous material and the adjacent reservoir aquifer as a porous medium having only horizontal permeability. The cap rock is regarded as impermeable. Spatially uniform temperature boundary conditions are imposed on the cold surface of the cap as well as at the hot bottom boundary of the reservoir; on the lateral boundary, well away from the fault, the distribution of temperature is controlled solely by vertical conduction, the pressure distribution is hydrostatic and mass flux is allowed to conserve matter.

In the high-Rayleigh-number approximation, the liquid will rise adiabatically in the fault zone except for a thin area under the clay cap. The adjacent aquifer becomes charged along the whole length of the fault. On a horizontal scale commensurate with the depth of the fault, the cooling effect through the conductive clay cap of the cold surface is restricted to a narrow, growing thermal boundary layer in the aquifer adjacent to the interface. Under that the fluid moves horizontally at the high temperature of the source fluid in the fault.

If a point is selected far enough away from the fault zone, say at a distance of ten times the fault depth, then all of the aquifer is affected by the cold upper surface. Because adiabatic flow takes place in the fault, the majority of the drop in temperature occurs in the clay cap. There the conductive surface heat flux is maximized and gradually declines away from the fault zone. This results from the fact that an increasingly large fraction of the temperature change can take place in the growing thermal boundary layer in the topmost portion of the aquifer. Consequently, at depth, the temperature variation shows a large gradient in the cap succeeded by a rapidly diminishing gradient in the boundary layer. Ultimately, near the far boundary, the surface flux falls to the background value. At this boundary, simple conductive heat flux goes on and the temperature variation with depth is distributed over the whole cap and aquifer.

The maximum surface heat flux turns out to be greater than the background value by a factor of approximately four, which agrees quite well with the field observations at East Mesa. A similar theoretical and

FIG. 2.7.   East Mesa, California, USA. Reproduced by courtesy of the Southern
California Edison Company, California, USA.

observational agreement was obtained for the distribution of the surface
heat flux with distance from the fault zone. The inference drawn is that fault
zones can offer a mechanism for the charging of adjacent geothermal
aquifers.

In 1988 it was reported that Geothermal Resources International, San
Mateo, California, expanded its development activities to the East Mesa
area, where the company acquired the rights to two long-term contracts to
sell electricity to the Southern California Edison Company (see Fig. 2.7).

2.6 PRE-PRODUCTION MODELS OF HYDROTHERMAL SYSTEMS

In geothermal systems, the quantity of fluid to be produced and the total
usable energy must be calculable. The hydrothermal types comprise
flowing convective fluids heated at depth and ascending surfacewards
through reduction in density. They are both non-isothermal and dynamic
because of buoyancy effects. Since those transient processes related to the

commencement of convection in any geothermal system take place over long periods of time (exceeding 10 000 years), a steady-state approach is perfectly in order in describing those physical processes occurring in an unexploited system over, say, a few decades.

Nevertheless, to obtain adequate performance predictions for any particular hydrothermal field, pre-production temperatures and flow fields must be established. At first, pre-production flow will dominate in a system, except immediately around wells. As development goes on, the natural flow will become subject to perturbations caused through these wells (of injection and production types).

Equations given in Section 2.3.2 above are applicable to constructing a mathematical model capable of describing such pre-production flow and usually they are solved numerically. Recently, computer programs have been developed which can solve equations of heat and mass transport in geothermal reservoirs in one, two or three dimensions and some are very flexible in reference to fluid and rock properties, geometry and boundary conditions. One, MUSHRM, is able to handle multi-phase, multi-species fluid flow in one, two or three dimensions (e.g. water/steam, water with dissolved methane/free methane).[26] In it each computational zone in the finite difference grid can contain a different rock type and it is possible to include all practically possible boundary conditions; see also Section 2.6.2 (on the Salton Sea, California, USA).

During the first stages of exploration and initial development of a geothermal field, there will be inadequate data available regarding such parameters as the fluid and rock thermomechanical properties, rock porosity and permeability, surface heat flow and so on. Even the relevant stratigraphy may not be clear, although routine geophysical measurements are valuable in determining this stratigraphy. As information accumulates, computer-based simulation of the pre-production fluid and heat flow in the geothermal reservoir facilitates its incorporation into a wholly integrated geohydrological model which, although speculative, can be extremely useful.

Of the many essays at developing a quantitative model of the pre-production fluid and heat transport in a real geothermal system, those of M. L. Sorey in 1976 (see Section 2.6.1), T. D. Riney and his co-workers in 1977 (see Section 2.6.2) and J. W. Mercer with C. R. Faust in 1979 (see Section 2.6.4) are especially interesting.[55-57]

2.6.1 Long Valley Caldera, Sierra Nevada, California, USA

This elliptically-shaped depression on the eastern front of the Sierra

Nevada covers about 450 km^2 and contains a hot-water convection system as well as a number of hot springs. Field data show a total heat discharge of 2·89 × 10^8 W.[5] It has been estimated that anywhere from 190 to 300 kg/s of water at 438–555 K discharges from the reservoir upwards towards the hot springs, the highest temperature corresponding to the lowest mass flux.[55]

M. L. Sorey utilized a three-dimensional conceptual model which comprises five horizontal layers covering 4·5 × 10^8 m^2 and having a depth of 6 km, the five layers corresponding to the major rock units known to be present as a result of seismic and geological investigations.[55] The uppermost 1 km is composed of post-caldera sediments and volcanic rocks, and includes the groundwater system. Layer 1 is an impermeable cap (except along some portions of the rim of the caldera where recharge occurs and in the Hot Creek gorge area where hot water ascends along faults and discharges in the springs in the gorge) and the hydrothermal system is believed to lie in layers 2 and 3, which contain welded and fractured tuffs. Layers 4 and 5 are considered to be impermeable basement rocks, but they are thermally conductive.

Since magma exists below the 6-km level, as is indicated by both heat flow and seismic studies, this is simulated by a temperature distribution at the base of the model that is kept constant through time. Development of temperature in the system was computed under assumed mass flux boundary conditions of fluid recharge and discharge by numerically solving the fluid and heat transport equations employing a Boussinesq approximation. In the numerical model, each horizontal layer was divided into 82 grid blocks (nodes) and it was assumed that hot water discharges only over the surface of the node (including the springs in Hot Creek gorge) and through the southeastern rim, recharging occurring along the western and northeastern rim of the caldera. Fractures in the welded tuff associated with faulting are taken as incorporating the channels requisite for flow in the hydrothermal system. Since only a single deep well (*ca.* 2 km) was drilled in the caldera, there are practically no data regarding the thermo-mechanical properties of the reservoir rocks, namely on porosity, permeability and specific heat. The groundwater flow is from the western and northeastern rim to zones of discharge at lower elevations in the Hot Creek gorge and, underground, through the southeastern rim of the caldera. An extra impulsion to flow results from the density differences between the hot and cold parts of the flow system, a buoyancy effect. The effective permeability of the reservoir was given by Sorey as (3–5) × 10^{-14}/m^2 by means of specifying pressures based on water-table altitudes in the recharge and discharge areas and also through adjusting the permeability distribution in order to give the desired mass flux of water (*ca.* 250 kg/s).

The model in question is speculative, but provides a better comprehension of the heat and mass transport in the system. It has been employed to derive the depths of fluid circulation for which the underlying magma chamber could supply the necessary heat flow of *ca.* 630 mW/m² over the area of the caldera for various time intervals. For instance, simulation for 35 000 years demonstrated that the existing heat discharge could have been sustained by fluid circulation to between 1·5 and 2·5 km depth. Simulated rock temperatures at 1·5 km depth under the Hot Creek gorge area approach 473 K and, interestingly, cooler temperatures east of this (consequent upon recharging along the northeastern rim) are consistent with those actually reported from the above-mentioned sole deep test borehole drilled in the caldera. In April 1987, at a symposium on this region held at the Lawrence Berkeley Laboratory, research was compiled to describe the active hydrothermal and recent magmatic systems there. However, indications of a possible magma body were regarded as ambiguous.

2.6.2 Salton Sea, California, USA

This geothermal field is a saline and high-temperature geothermal system at which the San Diego Gas and Electric Company constructed a nominal 10-MWe Geothermal Loop Experimental Facility (see Fig. 2.8). Nine geothermal wells have been drilled into the main rock unit, a sandstone with shale lenses overlain by an impermeable shale cap rock. In the former, there is believed to be an upper and a lower reservoir separated by a rather thick shale layer. All layers dip northwest and are parallel to the famous Brawley fault zone. From insufficient data it was estimated that the horizontal permeability of the upper reservoir beds exceeds $5 \times 10^{-13}/\text{m}^2$. Here sands predominate and have a porosity exceeding 30%.

A two-dimensional model for part of the area was made by T. D. Riney and his associates in 1977.[56] The dipping and thickening of the upper reservoir was handled by including the component of gravity along the dip direction and also by varying the properties of the rock so as to allow for variations in thickness. Both the Brawley and the Red Hill faults were regarded as obviating any fluid flow across the lateral boundaries, which is perhaps not justified in view of the fact that, while the fluids emerging from wells on opposite sides of the Brawley fault seem to have different origins, this is not recorded in the case of the Red Hill fault; see Figs 2.9 and 4.5.

It is interesting that temperature at the mid-plane of the upper reservoir

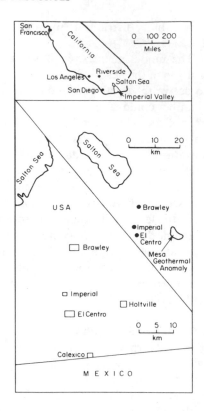

FIG. 2.8. The geographical setting of the Salton Sea and Imperial Valley in southern California, USA.

is much greater at the northwestern end and, from temperature–depth profiles at both ends and a brine equation of state, the corresponding hydrostatic pressures are found to be, at the mid-plane, 8507 kPa at the northwestern end and 3802 kPa at the southeastern end. Taking into account the temperature variation and angle of dip along the length of the reservoir in question, it appears that, in the absence of lengthwise flow, the value of the hydrostatic pressure at the northwestern end mid-plane is required to be 8824 kPa. The lengthwise pressure drive of about 317 kPa seems to produce an influx of approximately 323 K groundwater from the southeastern end and this would cool the upper reservoir if the hot brine infusion from the lower reservoir were completely sealed off by the intermediate shale layer.

All of these boundary conditions and reservoir characteristics were built into a reservoir simulator called MUSHRM intended to simulate a pre-production model for a part of the geothermal field containing the

FIG. 2.9. The Salton Sea, California, USA. Geothermoelectric power plant on the shore of the Salton Sea in southern California taps hot brine which is highly corrosive. Most of it is re-injected, but some is removed by trucks. Reproduced by courtesy of the Southern California Edison Company, California, USA.

Geothermal Loop Experimental Facility. Calculations were effected until an appropriate concordance with the mid-plane pre-production temperatures in the upper reservoir was derived. Initially, a one-dimensional version was applied to the dipping and thickening upper reservoir and incorporated the provision that, for each zone, there is an infusion of hot brine at the rate necessary to obtain the corresponding projected mid-plane pre-production temperature. Total rates of influx and also the lengthwise variation of the influx, as computed from one-dimensional calculations, were maintained. However, the lateral distribution of the rate of influx was permitted to vary in a later set of two-dimensional areal calculations.

The best fit to the lateral variations of the mid-plane temperatures actually measured in the wells comprised a symmetrical distribution with the maximum placed in the centre. These temperatures accord satisfactorily with the steady-state temperature contours calculated with the pre-production model and a velocity plot demonstrated that the infusion of hot brine from the lower reservoir pushes much of the ingressing cold groundwater to the margins of the upper reservoir to cause lower temperatures there.

Of course, the restricted information available forced T. D. Riney and his colleagues to utilize a number of hypotheses relating to the geology, most importantly to allow for gaps in the intermittent layer of shale through which hot water from the lower reservoir migrates upwards into the upper reservoir, and to assume that the Red Hill fault is a sealing feature.[56] No doubt continuing work will clarify such matters as the real direction of groundwater flow taken as from the southeast to the northwest in the above investigation. This could be done by employing tracer tests. As the geothermal field of the Salton Sea has passed from the first stage of its development into a preliminary exploitation stage, it is expected that this and similar on-site activities will be undertaken.

By 1987 the Salton Sea Scientific Drilling Project led by Wilfred Elders of the University of California at Riverside was organized around a 3·2-km borehole drilled by the Geothermal Technology Division of the Department of Energy, support being provided by the National Science Foundation, the US Geological Survey and the Office of BES (Basic Energy Sciences). The active drilling and testing stage was completed in August 1987 when the DOE handed the well over to Kennecott Australia Ltd. Studies on the recovered cores, including those manifesting transition into zones of active greenschist metamorphism, will continue over the next few years. This well yielded the hottest fluid sample (353°C) collected in the Continental Scientific Drilling Program.

2.6.3 Olkaria, Kenya, East Africa

In Kenya there is a vast geothermal potential lying within the East African Rift Valley, where three fields have been investigated, namely Olkaria, Eburru and Lake Bogoria. However, drilling and power production have only been carried out at Olkaria and the relevant work has been discussed by J. W. Noble and S. B. Ojiambo in 1975, and by A. J. Svanbjörnsson and his fellow workers in 1983.[58,59]

In 1987 a detailed model of the Olkaria well field was designed by G. S. Bodvarsson and his associates, and this was designed to assist in the future development of the area.[60] It is three-dimensional in order to permit evaluation of vertical fluid and heat flow to be made, and a new approach was utilized so as to allow for an efficient modelling of large numbers of development wells.

Some information on the relevant region is necessary. It is related to the Olkaria volcanic complex located near the western edge of the Rift Valley and the latest eruption took place relatively recently, in fact some 300 years ago, generating the Ololbutot lava flow. The area of the field is at least 50 km^2 and it lies approximately 60 km northwest of Nairobi and immediately southwest of Lake Naivasha. There is a resistivity low. The natural heat flow through the Olkaria system has been estimated at *ca.* 400 MWt (megawatts thermal).

The actual well field is a mere 2 km^2 in extent and there are, to date, 34 wells in it. Stratigraphically, the sequence is mostly fine-grained lavas and tuffs. Basalts and rhyolites are most prominent at depths between 400 and 900 m. The hydrothermal alteration of the rocks typifies geothermal fields with temperatures exceeding 200°C. A schematic model shows an upflow zone north of the existing well field, this being implied by the substantial lateral pressure gradient in the deep boiling-water reservoir (about 11 bar/km). At between 500 and 700 m a basaltic layer caps the system. Under this is a steam zone up to 150 m thick and this possesses thermodynamic conditions like those usually present in vapour-dominated systems, the temperature and pressure being almost 240°C and 35 bar, respectively. The pressure at depth is low, namely 35 bar at 700 m, and this results from the deep water table which reaches 300 m. It is possible that this zone of steam was created by lateral flow and separation of the phases, as M. A. Grant and A. J. Whittome suggested in 1981.[61]

The thermodynamic conditions in the underlying liquid-dominated reservoir pursue a boiling point-for-depth relationship with a temperature around 300°C at a depth of 1·5 km. Over the cap rock there are shallow aquifers with cold saline fluids. Wells are usually cased down to 500 m so as to preclude the inflow of shallow and colder fluids. Most of them feed from the steam zone at between 600 and 800 m, i.e. above the 35-bar pressure contour. The reservoir fluids are sodium chloride in type, and the chloride content ranges between 200 and 500 ppm. It increases with depth and, in addition, from north to south. The lower chloride concentration at shallow depth is probably the result of steam condensing under the cap rock. Well test results demonstrated that the permeability (and transmissivity) of the rocks of the reservoir are low, the average transmissivity being given as between 1×10^{-12} and $5 \times 10^{-12}/m^3$ in the well reports issued by the Kenya Power Company Ltd, for example in 1984.[62]

As a first stage preceding the history match and prediction calculations regarding reservoir performance, all available information was analysed for input into the numerical model; among the significant parameters were rock properties and both initial and boundary conditions. Among the rock properties, one of the most important is permeability, normally assessed by pressure transient well tests, e.g. injection, fall-off and recovery tests. In respect of this parameter, the rocks comprising the Olkaria reservoir represent a dual-porosity system having both fracture and matrix porosity and permeability. A number of the above-mentioned tests have been effected upon them, but with inconsistent results. Also there appears to be no consistent relationship between permeability values so obtained and the well flow rates. Since the interpretation of pressure transient tests is based on idealized models of isothermal single-phase flow in porous media, this is explained by the fact that a non-isothermal two-phase flow regime in a fractured reservoir is being examined at Olkaria. In consequence, Bodvarsson and his co-workers used permeability as an adjustable parameter in the simulations, with the constraint that average values occur between 10^{-15} and $10^{-14}/m^2$ (1–10 mD), the most probable range, according to the pressure transient data.[60] The permeabilities in question are equivalent porous medium properties: a fractured porous medium model was not utilized because it would have been too complex.

Since, throughout most of the Olkaria reservoir, the in-situ fluid comprises both liquid water and steam, i.e. a two-phase system is in being, relative permeability is a consideration. Darcy's law is applicable using two coefficients constituting functions for this parameter. These express the fact that the presence of one phase interferes with the flow of the other, thus, in effect, lowering the permeability below the single-phase value. Correct choice of the coefficients is significant in terms of prediction of the behaviour of a geothermal reservoir, as M. L. Sorey and his colleagues indicated in 1980.[63]

Unfortunately, while there is a great deal of information available regarding gas–oil and oil–water relative permeabilities, not much exists on the subject of steam–water permeabilities. However, liquid relative permeability will most likely increase in a monotonic manner from 0 to 1 as the liquid saturation similarly rises from 0 to 1 with the vapour relative permeability having an analogous dependence on vapour saturation. Regrettably, the functional forms of the relative permeability factors not only differ between geothermal reservoirs and indeed within individual reservoirs, but they are not known exactly. This is because, although they are a material property of any particular reservoir, associated determinants such as in-situ vapour saturation are obviously extremely hard to measure.

For this reason Bodvarsson and his associates executed an analysis of relative permeability for the Krafla field in Iceland using a modified version of a method originally suggested by M. A. Grant in 1977.[60,64] The relationship that the two coefficients in question are approximately equal to unity was confirmed for four different wells over the entire range of flowing enthalpies from single-phase liquid to single-phase steam. Results at Olkaria were consistent with the trends shown at Krafla.

Turning to porosity, not much is known about this in the fracture system at Olkaria, measurements in cores showing that the matrix value varies greatly from 1 to 25%. Usually values diminish downwards. Typical fracture porosity is ca. 1% and it normally varies between 0·1 and 3% for reservoir rocks in other areas. The history matching of individual well performances seems to be more sensitive to fracture porosity than to the matrix porosity, so it is to be expected that the former will yield effective porosity values. Other rock parameters can be neglected because they play only a minor role in the exercise.

For initial and boundary conditions, the minimum requirement is the specification of temperature, pressure and vapour saturation distributions throughout the reservoir system, but other parameters should probably be examined as well, e.g. the concentration of non-condensable gases. Bodvarsson and his colleagues noted that the best approach to exploitation modelling would be to develop a detailed numerical model of the natural condition of the geothermal field.[60] Nevertheless, they adopted a simplified one, approximating the dynamic natural condition by means of a 'semi-static' model wherein the thermodynamic situation is depth-dependent with no horizontal variations. They justified it on a number of grounds, e.g. that field observations demonstrate that lateral variations in thermo-dynamic conditions over the well field area are insignificant in the natural state. Their method substituted a simple porous-medium model for the ideal dual-porosity (fracture and matrix) one which is really required. The reservoir is thick and there are considerable vertical changes in thermodynamic conditions with depth, in contrast to the laterally constant situation mentioned above. This necessitates construction of a three-dimensional model and the associated computational labour is considerable; hence the choice of the simpler model rather than the dual-porosity one.

The question arises as to the utility of the results derived from the proposed porous-medium model in reference to complex fractured reservoirs. Bodvarsson *et al.* believed that the fluid flow calculations using such a model would be accurate enough if the overall calculated pressure decline is close to that actually observed.[60] Simulations reported were effected with MULKOM, the general-purpose simulator of the Lawrence Berkeley Laboratory discussed by K. Pruess in 1983.[65] Since the salinity is low and the non-condensable gas content of the Olkaria reservoir is satisfactory, the reservoir fluid is regarded as comprising pure water, all relevant properties being derived by Bodvarsson and colleagues from steam tables provided by the International Formulation Committee in 1967.[60,66]

As regards the history match of well performance, basic data used were flow rate and enthalpy history, both compiled from information obtained from the Kenya Power Company Ltd. In appropriate simulations, calculated flow rate and enthalpy values were compared with monthly-average values of observed flow rates and enthalpies for each individual well. A multi-layered reservoir model was required because most wells possess multiple feed zones and the steam zone provides much of the fluids output. The relevant grid utilized in numerical calculations was developed on the basis of a number of factors, e.g. location of feed zones and wells. The surface locations of the wells numbers 2 to 26 were employed as nodal points for the respective grid elements. These nodal points (x,y-coordinates) were chosen so as to give elements of reasonable size and simulate recharge into the actual well field. Subsequently to their being located, interfaces were generated by using a pre-processor program (OGRE), employing perpendicular bisectors on lines connecting two elements (nodal points).

The numerical scheme used in the MULKOM code is the integral finite difference method and this facilitates construction of a complex grid.[67,68] Some reservations are necessary; for instance, the downhole location of the wells may not lie on the same x,y-coordinates as the surface sitings. On the other hand, this is not too significant because the downhole locations are not yet known. Therefore a grid based on the surface locations seems acceptable. In addition, the grid must not be too coarse to recognize the fact that the majority of the well elements cover areas equivalent to circles having radii of only about 100 m.

Prior to starting a three-dimensional simulation study, a one-dimensional radial model was utilized to find out the grid area required for the system to behave as if it was infinite throughout the history match period of six years. Three-dimensional calculations are expensive, numerical work being roughly proportional to the square of the number of grid blocks. Hence it was essential to restrict the vertical discretization. A 35-bar pressure contour was taken to represent the contact between the steam zone and its underlying liquid zone. A three-layer model incorporated a steam zone 100 m thick overlying two liquid zones, one of 250 m and the other of 500 m thickness. Thus the base of the effective reservoir is placed at 1500 m depth, the actual location of the deepest major feed zone.

Fluid flow into any well is permissible through all layers in which it has one or more feed points. A 'sandface' flow rate for each layer is not prescribed, but calculated from the productivity index (PI), the fluid mobility and the pressure in the feed zone. The approach can be expressed mathematically, as K. Pruess and his associates indicated in 1984.[69] The relevant equation is

$$q = \sum_{\beta = \text{liquid, vapour}} (k_{r\beta}/\mu_\beta)\rho_\beta \, \text{PI} \, (\rho_\beta - \rho_{\text{wb}}) \tag{2.13}$$

where $\sum$ indicates the sum of the rest of the equation, $k_{r\beta}$ is the relative permeability of the β phase, μ_β is the dynamic viscosity, ρ_β is the fluid density, PI is the productivity index, ρ_β is the pressure in the well element, and ρ_{wb} is the flowing well pressure opposite the feed zone. This equation represents Darcy's law, with the productivity index mainly representing the geometric parameters of the well feed. Clearly, in applying eqn (2.13), it is necessary to know the value of the PI of the layer involved and also the well pressure adjacent to it. Again the problem of insufficient data arises because there are none available for the Olkaria wells. Bodvarsson and his colleagues therefore adjusted the parameters in question so as to provide a fit for the observed flow rate data.[60] Other parameters cover fluid and rock properties which are calculable from the thermodynamic conditions prevailing in each layer at each time step.

Many iterations were required so as to match reasonably the observed flow rates and enthalpies of the wells; three parameters were adjusted in this procedure, namely permeabilities, porosities and PIs. Their effects are interrelated, but affect flow rates and enthalpies to differing degrees. Permeability mainly controls the chronological flow rate decline and it also affects the enthalpy transients—the lower the permeability, the higher the enthalpy. Porosity affects flow rate to a small extent only, but powerfully influences the enthalpy, particularly when the porosity is low. Rise of enthalpy in a well is directly related to $(1 - \phi)\phi$, where ϕ is the effective porosity, as G. S. Bodvarsson and his associates indicated in 1980.[70]

The approach utilized in the iterations was to assess the productivity index (PI) of wells based on initial flow rate histories, i.e. prior to any detectable well interferences. Then the raise in enthalpy was matched by adjustment of porosities, the permeabilities being adjusted until the flow rate decline was approximately represented. Obviously, any changes in the parameters concerned at any location within the well field will produce an effect in all adjacent wells. Thus numerous iterations are necessitated before a reasonable match is derivable. No information was obtainable from outside the well field involved.

The model calibration was not considered to be very sensitive to the reservoir properties and the permeabilities and porosities were periodically updated during the iterations, employing average values for the well field elements. However, over the short history-match time interval of six years, these factors are not very significant (although permeabilities and porosities of outside zones will exert a strong influence over results in making long-term performance predictions).

A best model was obtained after many iterations and the guiding philosophy was to get a good match with late time data for the wells so that future predictions would be more reliable. Generally, the flow rate decline of the wells is largely a consequence of mobility effects, i.e. due to changes in vapour saturation in the producing elements. The vapour density being lower than that of the liquid, increasing vapour saturation normally causes a flow rate decline, even though the element pressures undergo little alteration. Hence, when a well is put into service (on-line) and boiling begins near it, this starts an increase in vapour saturation, enthalpy usually rises and the flow rate falls.

In 1983, at the end of the simulation period of the exercise undertaken by Bodvarsson and his colleagues, many of the wells attained quasi-steady-state conditions with a gradual rise in enthalpy and flow rate decline.[60] The enthalpy rise in the Olkaria wells is large by comparison with other geothermal fields, probably because of low porosities and permeabilities. Much of the late time flow decline of the wells in question arose from well interference. This results from their close spacing (on average approximately 225 m); the effect manifests itself after a well has been on-line for a year or so. The inference is that the well density here is simply too high. It proved impossible to determine the porosity of the steam zone because this involved no enthalpy transients, the fluid enthalpy being that of pure vapour. A porosity of 10% was assumed for all of these elements which represent this zone. In both upper and lower liquid layers, an effective average porosity of 2% is derived, suitable for the volcanic rocks at Olkaria, and similar values were obtained for fractured basalts at Krafla by K. Pruess and others in 1984.[71] Inferred porosity shows variation over a range of 0·25–10% in the upper liquid zone and 0·5–5% in the lower one. Usually wells with the largest enthalpy were found to possess the lowest effective porosity.

Computed permeability distributions show higher values than those given by pressure transient tests, agreeing with modelling results given by S. Waruingi in 1982.[72] In the steam zone, the average permeability is 7·5 mD, within a range of 2–25 mD. Wells extracting fluids mainly from this zone need the greatest permeabilities. In the upper liquid layer, the permeabilities are considerably lower, around 4 mD, and in the lower liquid layer the average permeability is close to this also (3·5 mD).

Transmissivities calculated from calculated permeabilities in the different layers related to the wells are normally higher than those obtained from pressure transient data.

Productivity index (PI) values cannot be referred to any field information, but, interestingly, values for the parameter are usually higher for the more productive wells. Anyway a much higher PI is necessary for the steam zone than for the liquid zones. On average, the values given by Bodvarsson and his co-workers are $4·25 \times 10^{-12}$, $0·93 \times 10^{-12}$ and $0·63 \times 10^{-12}/m^3$ for the steam, upper liquid and lower liquid layers, respectively.[60]

Flow percentages from various feed points into the Olkaria wells have been estimated on the basis of geochemical information. The pressure decline of feed wells and their vapour saturation transients during the history match period vary considerably between wells and feed zones. Lower pressure decline and fluid depletion are computed for the liquid zones, as all the feeds stay in a two-phase condition. Because such two-phase fluids are highly compressible, only a relatively small area was affected by exploitation during the 6·5-year period of simulation. In order to evaluate the average pressure drop in the entire reservoir, it is necessary to consider the pressure decline in non-producing observer wells; the average calculated drawdown is roughly 0·4, 0·2 and 0·1 MPa in the steam, upper liquid and lower liquid zones, respectively. One word of caution from Bodvarsson et al.—it must be remembered that intensive exploitation began relatively recently (the time interval to which they alluded is 1981–83).[60]

In sum, therefore, Bodvarsson and his team developed a highly detailed three-dimensional well-by-well model of the East Olkaria geothermal field in Kenya, using a porous-medium description of a known fractured porous reservoir.[60] Although it is an approximation entailing simplification of both initial and boundary conditions, the model actually makes a good match with the flow rate and enthalpy data from all of the producing wells over a 6·5-year time interval from 1977 to 1983. In addition, the model concords well with the observed pressure decline in the reservoirs concerned.

Major results are summarized as follows. Firstly, there was the detection of large well interference because of small well spacing. Secondly, the determination was made that the early flow rate decline in wells is mainly the result of mobility effects (caused by an expanding boiling region around the liquid feeds). Thirdly, there is the discovery that the effective porosities of the liquid zone are low (ca. 2% on average). Fourthly, it proved possible to make an assessment of average permeabilities at Olkaria at $7·5 \times 10^{-15}$, 4×10^{-15} and $3·5 \times 10^{-15}/m^2$ for the steam, upper liquid and lower liquid layers, respectively.

Fifthly, the observation resulted that the reservoir rocks are homogeneous in the present well field (no flow barriers were required in order to obtain a reasonable match with the field data), but there is an apparent high permeability zone below 1000 m depth, extending north–south through the wells numbered 12, 15, 16 and 20.

2.6.4 Wairakei, North Island, New Zealand

The geothermal field located here is especially interesting because it is the first liquid-dominated one to be exploited for electric power. By contrast, the larger Larderello field in Italy is vapour-dominated, as also is The Geysers field in California, both being untypical of the general run of prospective geothermal resources. The system under discussion is located in the North Island of New Zealand, north of Lake Taupo and to the west of the Waikato River (see Fig. 2.10). Its area is approximately 15 km², but geophysical evidence strongly implies a much larger area enclosed by its impermeable boundaries. Drilling commenced at Wairakei in 1950 and stopped 18 years later, mass production rates peaking in the mid-1960s and slowly declining thereafter. A second reservoir at Tauhara is believed to be hydrologically interconnected with the adjacent one at Wairakei, although many earlier studies exclude Wairakei.

Exhaustive investigations of the regional geology, e.g. by G. W. Grindley in 1965 and with others in 1966, established that the hydrothermal field of

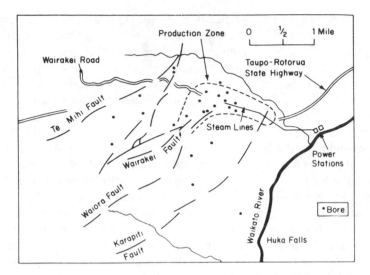

FIG. 2.10. The Wairakei (New Zealand) geothermal project region.

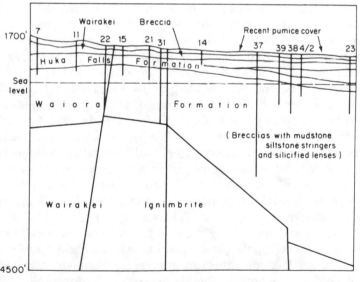

FIG. 2.11. Stratigraphic cross-section at Wairakei geothermal field in New Zealand. It is orientated northwest to southeast through the production zone and the numbers are well numbers.

Wairakei is underlain by practically horizontal Quaternary acidic volcanic rocks.[73,74] The stratigraphic sequence involved comprises, from the top downwards, a Holocene pumice cover, the Wairakei breccia (an upper aquifer), the Huka Falls Formation, the Haparangi rhyolite, the Waiora Formation (a lower aquifer), the Wairora Valley andesite, the Wairakei ignimbrites and the Ohakuri Group (which may contain a third, deepest, aquifer); see Fig. 2.11.

It is thought that most of the mass production comes from the Waiora aquifer and from the Waiora/ignimbrites interface region. As a whole the reservoir acts as a porous medium, although nearly all of the actual wells are sited in high-permeability zones associated with large-scale faulting.[75] As the low-permeability Huka Falls Formation overlies the Waiora aquifer, most of the recharge to and discharge from this aquifer has to take place through that formation. Unfortunately, the boundary between the Waiora aquifer and the Wairakei ignimbrites is ill-defined; its irregular surface and the characteristic fracture zones provide a locally high permeability for the unconformity. The Wairakei ignimbrite itself is a rock possessing low primary permeability, but it is riddled with complicated fault systems which act as conduits for hot-water recharge to the Waiora aquifer.

There have been a number of assessments of the natural heat flow at Wairakei, lying in the range 3.431×10^8–6.820×10^8 W, the best value perhaps being 4.184×10^8 W, with 285 K as a reference temperature. No measurements for mass discharge are available, but the heat flux data can be used to estimate it because most of the natural heat flow is probably associated with natural mass discharge. If a mean enthalpy of 1.025×10^6 J/kg be assumed, then a mass discharge of 440 kg/s for the time interval 1951 to 1952 is obtained, as R. G. Fisher indicated in 1964.[76]

Most probably the total natural heat flux has increased and the mass flow decreased over the life of the geothermal field. General temperature trends can only be deduced because of measurement uncertainties. Thus they rise rapidly with depth to the top of the Wairora aquifer where, in the hotter parts, a value of *ca.* 473 K is attained. The gradient is reduced in the aquifer where the maximum value is *ca.* 523 K; see Fig. 2.11.

J. W. Mercer and others in 1975 used a two-dimensional areal reservoir simulator for modelling the pre-production mass and heat transfer in the Wairakei field.[75] Excluding the Tauhara region, this model assumed that the Huka Falls Formation transmits both mass and heat vertically from the Waiora aquifer. The Wairakei ignimbrites are taken as impermeable so that heat is transmitted only through conduction. Field boundaries were also regarded as impermeable, with temperature specified as a known function of the space variables.

Because the problem is two-dimensional areally, vertically averaged temperature values were utilized. J. W. Mercer and his colleagues had taken the reservoir as single-phase before production started, so the averaged temperature values had to be reconciled with this condition and therefore the heat source at the base of the Waiora aquifer at the top of the ignimbrites was varied.[75] In the case of the Huka Falls Formation, the pressure distribution at the top was assessed by regarding the Wairakei breccia and the Holocene pumice as one aquifer with varying thickness and also by assuming that the water table reaches the ground surface. Total calculated heat and mass discharges are 9.75×10^8 W (using a reference temperature of 273 K) and 1133 kg/s, respectively. These exceed the majority of values actually recorded, probably reflecting the larger area used in the model. The single-phase model reproduced history data up to 1962 and showed practically no alteration in reservoir temperatures.

However, as J. W. Pritchett and his associates indicated in 1978, the single-phase model most probably produces too great an estimate for the recorded pressure drop.[77] There are other objections to the single-phase model. One is that no recharge from the ignimbrites was permitted in it, contrary to the geophysical evidence that they act as a hot-water source. Another is that too great a vertical permeability ($\sim 10^{-14}$ m^2) was assumed for the Huka Falls Formation.

Consequently, it is now thought that the Wairakei geothermal field is either two-phase or was almost at flashing point before production. In 1979 a new model was developed which allowed mass leakage through the ignimbrites and reduced the vertical permeability of the Huka Falls Formation to $\sim 10^{-15}$ m^2. This gave a surface mass discharge of 424 kg/s, which agreed rather well with a generally accepted value of 440 kg/s. In the new model, J. W. Mercer and C. Faust reproduced historical production data up to 1974.[57] Leakage from the ignimbrites was so calculated as to match pre-production temperatures and observed pressure drops during production.

Viable models of geothermal systems to be exploited are clearly difficult to construct. For instance, in the present case, early assumptions such as that of single-phase flow had to be altered after acquisition of later production data. Thus pre-production models must be considered as evolving ones into which information is incorporated as it becomes available, with all the modifying procedures which that entails.

REFERENCES

1. RYBACH, L., 1981. Geothermal systems, conductive heat flow, geothermal anomalies. In: *Geothermal Systems: Principles and Case Histories*, pp. 3–31, ed. L. Rybach and L. J. P. Muffler. John Wiley & Sons, Chichester, 359 pp.
2. SASS, J. H. and MUNROE, R. J., 1974. *Basic Heat-Flow Data from the United States*, US Geol. Surv., Open-file Rept, pp. 74–9.
3. CHAPMAN, D. S., POLLACK, H. N. and CERMAK, V., 1979. Global heat flow with special reference to the region of Europe. In: *Terrestrial Heat Flow in Europe*, ed. V. Cermak and L. Rybach, pp. 41–8. Springer-Verlag, Berlin.
4. FANELLI, M. and TAFFI, L., 1980. Status of geothermal research and development in the world. *Revue de l'Institut Français du Pétrole*, **XXXV**, 429–48.
5. GARG, S. K. and KASSOY, D. R., 1981. Convective heat and mass transfer in hydrothermal systems. In: *Geothermal Systems: Principles and Case Histories*, ed. L. Rybach and L. J. P. Muffler, pp. 37–76. John Wiley & Sons, Chichester, 359 pp.
6. DONALDSON, I. G. and GRANT, M. A., 1981. Heat extraction from geothermal reservoirs. In: *Geothermal Systems: Principles and Case Histories*, ed. L. Rybach and L. J. P. Muffler, pp. 145–79. John Wiley & Sons, Chichester, 359 pp.
7. WHITE, D. E., MUFFLER, J. L. P. and TRUESDELL, A. H., 1971. Vapor-dominated hydrothermal systems compared with hot water systems. *Econ. Geol.*, **66**, 75–97.
8. JESSOP, A. M. and LEWIS, T., 1978. Heat flow and heat generation in the Superior Province of the Canadian Shield. *Tectonophysics*, **50**, 55–75.
9. LUMB, J. T., 1981. Prospecting for geothermal resources. In: *Geothermal Systems: Principles and Case Histories*, ed. L. Rybach and L. J. P. Muffler, pp. 77–108. John Wiley & Sons, Chichester, 359 pp.
10. NAKAMURA, H. and SUMI, K., 1981. Exploration and development at Takinoue, Japan. In: *Geothermal Systems: Principles and Case Histories*, ed. L. Rybach and L. J. P. Muffler, pp. 247–72. John Wiley & Sons, Chichester, 359 pp.
11. STEFÁNSSON, V., 1981. The Krafla geothermal field, northeast Iceland. In: *Geothermal Systems: Principles and Case Histories*, ed. L. Rybach and L. J. P. Muffler, pp. 273–94. John Wiley & Sons, Chichester, 359 pp.
12. GARNISH, J. D., VAUX, R. and FULLER, R. W. E., 1986. *Geothermal Aquifers*, ed. R. Vaux. Energy Technology Support Unit, AERE, Harwell, Oxfordshire OX11 0RA, UK, 109 pp. and 6 appendices.
13. WALLACE, R. H., KRAEMER, T. F., TAYLOR, R. E. and WESSELMAN, J. B., 1979. Assessment of geopressurized-geothermal resources in the northern Gulf of Mexico basin. In: *Assessment of Geothermal Resources of the United States—1978*, ed. L. J. P. Muffler. US Geol. Surv. Circ., No. 790, pp. 132–55.
14. SWANBERG, C. A. and MORGAN, P., 1978. The linear relation between temperatures based on the silica content of groundwater and regional heat flow: a new heat flow map of the United States. *Pure Appl. Geophys.*, **117**, 227–41.
15. WOHLENBERG, J. and HAENEL, R., 1978. Kompilation von Temperatur-Daten für den Temperatur-Atlas der Bundesrepublik Deutschland. *Statusreport 1978—Geotechnik und Lagerstätten, Projektleitung Energieforschung*, KFA (Nuclear Research Centre), Jülich, pp. 1–12.

16. D'AMORE, F. and PANICHI, C., 1980. Evaluation of deep temperatures of hydrothermal systems by a new gas thermometer. *Geochim. Cosmochim. Acta*, **44**, 549–56.
17. BULLARD, E., 1973. Basic theories. In: *Geothermal Energy*, ed. H. C. H. Armstead. The UNESCO Press, Paris.
18. WITTKE, W., 1973. General report on the symposium 'Percolation through fissured rock'. *Bull. Internat. Assn Eng. Geol., Krefeld*, **7**, 3–28.
19. LOUIS, C., 1970. *Water Flows in Fissured Rocks and their Effects on the Stability of Rock Massifs*. Lawrence Livermore Laboratory, Livermore, CA, USA, Report UCRL-Trans-10469 (English translation of a dissertation from the University (TH) of Karlsruhe, FRG, 1967.
20. NOORISHAD, J., WITHERSPOON, P. A. and BREKKE, T. L., 1971. *A Method for Coupled Stress and Flow Analysis of Fractured Rock Masses*. University of California at Berkeley, California, USA, Report No. 71-6.
21. GRAY, W. G., O'NEILL, K. and PINDER, G. F., 1976. Simulation of heat transport in fractured, single-phase geothermal reservoirs. *Summaries Second Workshop Geothermal Reservoir Engineering, Stanford University, Stanford, California, USA*, pp. 222–8.
22. PRITCHETT, J. W., GARG, S. K., BROWNELL, D. H. JR, RICE, L. F., RICE, M. H., RINEY, T. D. and HENDRICKSON, R. R., 1976. *Geohydrological Environmental Effects of Geothermal Power Production—Phase IIA*. Systems, Science and Software, La Jolla, CA, USA, Report SSS-R-77-2998.
23. BROWNELL, D. H. JR, GARG, S. K. and PRITCHETT, J. W., 1977. Governing equations for geothermal reservoirs. *Water Resources Res.*, **13**, 929–34.
24. GARG, S. K. and PRITCHETT, J. W., 1977. On pressure-work viscous dissipation and the energy balance relation for geothermal reservoirs. *Adv. Water Resources*, **1**, 41–7.
25. RAMEY, H. J. JR, BRIGHAM, W. E., CHEN, H. K., ATKINSON, P. G. and ARIHARA, N., 1974. *Thermodynamic and Hydrodynamic Properties of Hydrothermal Systems*. Stanford Geothermal Programme Report SGP-TR-6, Stanford University, Stanford, CA, USA.
26. GARG, S. K., PRITCHETT, J. W., RICE, M. H. and RINEY, T. D., 1977. *US Gulf Coast Geopressurized Geothermal Reservoir Simulation*. Systems, Science and Software, La Jolla, CA, USA, Report SSS-R-77-3147.
27. KEENAN, J. H., KEYES, F. G., HILL, P. G. and MOORE, J. G., 1969. *Steam Tables. International Edition—Metric Units*. John Wiley & Sons, New York.
28. MORRISON, H. L., ROGERS, F. T. JR and HORTON, C. W., 1949. Convection currents in porous media—II. Observations of conditions at onset of convection. *J. Appl. Phys.*, **20**, 1027–9.
29. KASSOY, D. R. and ZEBIB, A., 1975. Variable viscosity effects on the onset of convection in porous media. *Physics of Fluids*, **18**, 1649–51.
30. WOODING, R. A., 1957. Steady-state free thermal convection of liquid in a saturated permeable medium. *J. Fluid Mech.*, **2**, 273–85.
31. STRAUS, J. M. and SCHUBERT, G., 1977. Thermal convection of water in a porous medium: effects of temperature- and pressure-dependent thermodynamic and transport properties. *J. Geophys. Res.*, **82**, 325–33.
32. MORLAND, L. W., ZEBIB, A. and KASSOY, D. R., 1977. Variable property effects on the onset of convection in an elastic porous matrix. *Physics of Fluids*, **20**, 1255–9.

33. ZEBIB, A. and KASSOY, D. R., 1976. Onset of natural convection in a box of water-saturated porous media with large temperature variation. *Physics of Fluids*, **20**, 4–9.
34. ROGERS, F. T. JR and MORRISON, H. L., 1950. Convection currents in porous media—III. Extended theory of critical gradients. *J. Appl. Phys.*, **21**, 1177–80.
35. WOODING, R. A., 1976. *Influence of Anisotropy and Variable Viscosity upon Convection in a Heated Saturated Porous Layer*. New Zealand Dept of Scientific and Industrial Research, Tech. Rept No. 55.
36. DONALDSON, I. G., 1962. Temperature gradients in the upper layers of the Earth's crust due to convective water flows. *J. Geophys. Res.*, **67**, 3449–59.
37. SOREY, M. L., 1975. *Numerical Modelling of Liquid Geothermal Systems*, US Geol. Surv., Open-file Rept 75-613.
38. RANA, R., HORNE, R. N. and CHENG, P., 1979. Natural convection in a multi-layered geothermal reservoir. *J. Heat Transfer*, **101**, 411–16.
39. RIBANDO, R. J. and TORRANCE, K. E., 1976. Natural convection in a porous medium: effects of confinement, variable permeability, and thermal boundary conditions. *J. Heat Transfer*, **98**, 42–8.
40. CALTAGIRONE, J. P., 1975. Thermoconvective instabilities in a horizontal porous layer. *J. Fluid Mech.*, **72**, 269–87.
41. STRAUS, J. M. and SCHUBERT, G., 1978. On the existence of three-dimensional convection in a rectangular box of fluid-saturated porous material. *J. Fluid Mech.*, **87**, 385–94.
42. ZEBIB, A. and KASSOY, D. R., 1978. Three-dimensional natural convection motion in a confined porous medium. *Physics of Fluids*, **21**, 1–3.
43. STRAUS, J. M. and SCHUBERT, G., 1979. Three-dimensional convection in a cubic box of fluid-saturated porous material. *J. Fluid Mech.*, **91**, 155–66.
44. EINARSSON, T., 1942. The nature of the springs of Iceland. *Rit. Visind. Isl.*, **26**, 1–92 (German text).
45. ELDER, J. W., 1966. *Heat and Mass Transfer in the Earth: Hydrothermal Systems*. New Zealand Department of Scientific and Industrial Research Bulletin No. 169.
46. DONALDSON, I. G., 1968. A possible model for hydrothermal systems and methods of studying such a model. *Proc. Third Australasian Conf. on Hydraulics and Fluid Mechanics*, pp. 200–4.
47. CHENG, P., 1978. Heat transfer in geothermal systems. In: *Advances in Heat Transfer*, ed. T. F. Irvine Jr and J. P. Hartnett, pp. 1–105. Academic Press, New York.
48. NORTON, D. and KNIGHT, J., 1977. Transport phenomena in hydrothermal systems: cooling plutons. *Am. J. Sci.*, **277**, 937–81.
49. NORTON, D., 1978. Sources, sourceregions and pathlines for fluids in hydrothermal systems related to cooling plutons. *Econ. Geol.*, **73**, 21–8.
50. CATHLES, L. M., 1977. An analysis of the cooling of intrusives by ground-water convection which includes boiling. *Econ. Geol.*, **72**, 269–387.
51. KASSOY, D. R. and ZEBIB, A., 1978. Convection fluid dynamics in a model of a fault zone in the Earth's crust. *J. Fluid Mech.*, **88**, 769–82.
52. TURCOTTE, D. L., RIBANDO, R. J. and TORRANCE, K. E., 1977. Numerical calculation of two-temperature thermal convection in a permeable layer with application to the Steamboat Springs thermal system, Nevada. In: *The Earth's*

Crust, ed. J. G. Heacock. Geophysical Monograph 20, American Geophysical Union, Washington, DC, USA, pp. 722–36.

53. GOYAL, K. P. and KASSOY, D. R., 1977. A fault-zone controlled model of the Mesa anomaly. *Proc. Third Workshop Geothermal Reservoir Engineering, Stanford University, Stanford, CA, USA*, pp. 209–13.

54. GOYAL, K. P., 1978. *Heat and Mass Transfer in a Saturated Porous Medium with Application to Geothermal Reservoirs.* PhD Thesis, Mechanical Engineering Department, University of Colorado, Boulder, CO, USA.

55. SOREY, M. L., 1976. A model of the hydrothermal system of Long Valley, Caldera, California. *Summaries Second Workshop Geothermal Reservoir Engineering, Stanford University, Stanford, CA, USA*, pp. 324–38.

56. RINEY, T. D., PRITCHETT, J. W. and GARG, S. K., 1977. Salton Sea geothermal reservoir simulations. *Proc. Third Workshop Geothermal Reservoir Engineering, Stanford University, Stanford, CA, USA*, pp. 178–84.

57. MERCER, J. W. and FAUST, C., 1979. Geothermal reservoir simulation. III: Application of liquid- and vapor-dominated hydrothermal techniques to Wairakei, New Zealand. *Water Resources Res.*, **15**, 653–71.

58. NOBLE, J. W. and OJIAMBO, S. B., 1975. Geothermal exploration in Kenya. *Proc. 2nd UN Symposium on the Development and Use of Geothermal Resources*, San Francisco, CA, USA, 20–29 May 1975.

59. SVANBJÖRNSSON, A., MATHIASSON, J., FRIMMANNSSON, H., ARNORSSON, S., BJÖRNSSON, S., STEFÁNSSON, V. and SAEMUNDSSON, K., 1983. Overview of geothermal development at Olkaria in Kenya. Presented at the Ninth Workshop on Geothermal Resource Engineering, Stanford University, Stanford, CA, USA, 13–15 December 1983.

60. BODVARSSON, G. S., PRUESS, K., STEFÁNSSON, V. and BJÖRNSSON, S., 1987. East Olkaria geothermal field, Kenya. 1: History match with production and pressure decline data. *J. Geophys. Res.*, **92**(B1), 521–39.

61. GRANT, M. A. and WHITTOME, A. J., 1981. Hydrology of Olkaria geothermal field. New Zealand Geothermal Workshop, Geothermal Institute, University of Auckland, Auckland, New Zealand.

62. KENYA POWER CO. LTD, 1984. *History Match and Performance Predictions for the Olkaria Geothermal Field, Kenya.* Report prepared by G. S. Bodvarsson and K. Pruess for Merz and McLellan and Virkir Ltd, Nairobi, Kenya.

63. SOREY, M. L., GRANT, M. A. and BRADFORD, M., 1980. Nonlinear effects in two-phase flow to wells in geothermal reservoirs. *Water Resources Res.*, **16**(4), 767–77.

64. GRANT, M. A., 1977. Permeability reduction factors at Wairakei. Presented at Heat Transfer Conference, Am. Inst. of Chem. Eng./Am. Soc. Mech. Eng., Utah, USA, 15–17 August 1977.

65. PRUESS, K., 1983. Development of the general purpose simulator MULKOM. In: *1982 Ann. Rept, Earth Sci. Div., Lawrence Berkeley Lab., Berkeley, CA, USA*, pp. 133–4.

66. INTERNATIONAL FORMULATION COMMITTEE, 1967. *The Formulation of the Thermodynamic Properties of Ordinary Water Substance.* Report, IFC Secretariat, Düsseldorf, FRG.

67. WERES, O. and SCHROEDER, R., 1977. *Documentation for Program OGRE.* Lawrence Berkeley Lab., Rept LBL-7060.

68. NARASIMHAN, T. N. and WITHERSPOON, P. A., 1976. An integrated finite difference method for analyzing fluid flow in porous media. *Water Resources Res.*, **12**(1), 57–64.

69. PRUESS, K., BODVARSSON, G. S. and STEFÁNSSON, V., 1984. Analysis of production data from the Krafla geothermal field, Iceland. *Proc. Ninth Workshop on Geothermal Resource Engineering*, Stanford University, Stanford, CA, USA, 13–15 December 1983.

70. BODVARSSON, G. S., O'SULLIVAN, M. J. and TSANG, C. F., 1980. The sensitivity of geothermal reservoir behavior to relative permeability parameters. *Proc. Sixth Workshop on Geothermal Reservoir Engineering*, Stanford University, Stanford, CA, USA.

71. PRUESS, K., BODVARSSON, G. S., STEFÁNSSON, V. and ELIASSON, E. T., 1984. The Krafla geothermal field, Iceland. 4: History match and prediction of individual well performance. *Water Resources Res.*, **20**(11), 1561–84.

72. WARUINGI, S., 1982. *A Study of the Olkaria Geothermal Reservoir when Generating Thirty Megawatts of Electricity.* Rept 82.22, Geotherm. Inst., University of Auckland, Auckland, New Zealand.

73. GRINDLEY, G. W., 1965. *The Geology, Structure and Exploitation of the Wairakei Geothermal Field, Taupo, New Zealand.* New Zealand Geol. Surv. Bulletin No. 75.

74. GRINDLEY, G. W., RISHWORTH, D. E. and WATTERS, W. A., 1966. *Geology of the Tauhara Geothermal Field, Lake Taupo.* New Zealand Geol. Surv. Geotherm. Rept No. 4.

75. MERCER, J. W., PINDER, G. F. and DONALDSON, I. G., 1975. A Galerkin finite-element analysis of the hydrothermal system at Wairakei, New Zealand. *J. Geophys. Res.*, **80**, 2608–21.

76. FISHER, R. G., 1964. Geothermal heat flow at Wairakei during 1958. *New Zealand J. Geol. Geophys.*, **7**, 172–84.

77. PRITCHETT, J. W., RICE, L. F. and GARG, S. K., 1978. *Reservoir Engineering Data: Wairakei Geothermal Field, New Zealand.* Systems, Science and Software, La Jolla, CA, USA, Report SS-R-78-3597-1.

CHAPTER 3

Geothermal Exploration

'Thou shalt be visited of the Lord of Hosts with thunder, and with earthquake, and great noise, with storm and tempest, and the flame of devouring fire.'

ISAIAH, 29:6

3.1 OBJECTIVE

The objective is to find geothermal phenomena. The appropriate geological environments are restricted mostly to regions on or near the active margins of tectonic plates at which crustal material is being either produced or destroyed and, although there may be the odd exception (e.g. Hawaii), invariably there must be a heat source. Whilst sites are determined by this factor and by the regional geology, surface phenomena or borehole manifestations of hydrological significance are recognition markers. However, they depend upon these heat sources and also upon the presence of a transport medium, normally water, responsible for the appearance of heat at the surface, together with adequate permeability in the relevant rocks.

The exploration methods employed must facilitate detection of anomalously high temperatures, subterranean fluids and permeability variations. Assessment of the first two parameters will both identify a geothermal field and give some idea of its extent. However, because hot fluids can build up in porous and not very permeable rocks over long time intervals, it will be difficult to extract them because of the paucity of permeable paths along which they can flow into boreholes. Therefore it is important to obtain a good understanding of the variations in permeability

117

so that the parts of the geothermal field capable of yielding good production can be determined prior to drilling.

Lithostatic pressure increases in tandem with increasing depth and may reduce horizontal permeability. Then almost vertical fissures in reasonably competent rocks provide the main conduits through which fluids can migrate upwards and the inference is that, in deep boreholes, inclined drilling would better the chances of transgressing such a fissure and increasing the fluid output at the surface.

Thermal waters often boil on their way up and produce less viscous steam (except inside the two-phase boiling zone). This passes more easily through lower-permeability rocks.

But the successful exploitation of a geothermal field will depend upon the production of substantial mass flow from wells and so it will be just as necessary to find high-permeability zones in vapour-dominated systems as it is in hot-water ones. Because of the high electrical resistivity and relatively greater mobility of steam compared with water, the more diffuse periphery of such a vapour-dominated field is harder to discover than is the case with a hot-water field.

Hot dry rock geothermal fields do not relate either to water or to permeability; prospection for such fields is limited to the detection of thermal anomalies. A discussion of heat extraction from hot dry crustal rock was given by M. L. Smith in 1978.[1]

Another kind of geothermal field, that found in geopressurized zones, has been mentioned in Chapter 2. It is usually located in extremely deep sedimentary basins in which fossil groundwater was entrained in sediments and, as depression occurred, heated. Maintenance of the pressure is accomplished through a cap rock which, if it is located, acts to initiate prospection. Most of the geopressurized zones so far discovered are connected with deep oil fields, for instance in the American Gulf states: in fact, they were found as a result of petroleum exploration. Along the Gulf Coast the Geothermal Technology Division research programme on such geopressurized geothermal resources reached an important point in October 1987 when the Gladys McCall well in Louisiana was shut in for pressure testing. This well had produced over $4.3 \times 10^6 \, m^3$ of brine and $19 \times 10^6 \, m^3$ (standard) of natural gas during a four-year flow test from sand zone 8. Analysis of the flow demonstrated that irreversible rock compaction was not a significant mechanism in the long-term pressure maintenance of the reservoir. The pressure build-up test may show which potential mechanism is most significant: creep, shale dewatering, cross-flow or leakage along faults.

3.2 METHODS

These may be utilized separately or in combination, depending upon the characteristics of the geothermal field which is being evaluated. Initially, all known data about an area are assembled and, ideally, basic geothermal energy resource information is incorporated into a geothermal potential map following the US Geological Survey pattern.[2] Aerial photography, infrared imagery, side-looking airborne radar (SLAR) and aeromagnetic surveys have varying degrees of value, although J. T. Lumb in 1981 had doubts about the last of these.[3]

3.2.1 Geological, Hydrological and Mineralogical Methods

Geology is basic. Its application can not only identify a possible geothermal region, but also show the economic feasibility of getting steam or hot water in it. Where several regions are involved, it aids in determining the optimum one for detailed study.

As well as ordinary geological mapping, surface geothermal features can be recorded and rough estimates made of their heat output. From their associated deposits and hydrothermal alteration, it may be feasible to guess the age of the geothermal field as well as whether it was more or less active in the past and even whether the major focus of activity has shifted. From the surface features, a minimum value for the sustainable discharge of the geothermal field may be made.

Preliminary chemical analyses of discharging fluids shed light upon the physical conditions at depth and, to some extent, upon the relevant rock types as well. A structural survey will aid in planning later work by detecting fault and joint zones, permeable beds, etc. It may prove possible to correlate faulting with maximum thermal sites.

At first, a hydrological approach entails collection of information regarding springs, rivers and rainfall. Later drilling of deep exploration wells enables a plan of the unsaturated zone and groundwater aquifer to be constructed. In addition, as drawdown takes place, the inter-relationship between shallow, normally colder water and deeper, thermal water may become apparent. If there are no surface geothermal manifestations, i.e. in a region where no recent volcanic or intrusive activity has occurred, the hydrological investigation of deep-water circulation in such deep exploration wells takes on even greater importance.

Finding adequate permeability zones is crucial and, in 1970, P. R. L. Browne showed that looking at the hydrothermal alteration in core samples from such wells can provide information leading to the identification of hotter, more permeable zones while drilling is in progress.[4] Hence it is particularly useful to know under what conditions certain minerals are in equilibrium with groundwater and surrounding rocks because these can indicate changes which happened after they formed. For instance, G. W. Grindley and P. R. L. Browne in 1976 suggested that secondary permeability may, to some extent, be associated with natural hydraulic fracturing taking place in the event that pore fluid pressures exceed the tensile strength of the rock, basing this upon petrological thin sections of drill cores.[5] Also core samples can be measured for a variety of physical parameters of differing utility, e.g. density and magnetism. The results may have a bearing upon geophysical surveys.

3.2.2 Geophysical Methods

The main application of geophysical investigations in geothermal work is the delineation of fields and the determination of appropriate drilling sites for boreholes through which deep-seated hot fluids can be extracted. Any suitable method can be used. Following the assumption that chloride waters would be characterized by anomalously low electrical resistivity, measurements of this parameter in New Zealand facilitated mapping variations in deep groundwater.[6] From this exercise it became clear that geophysics could be utilized, not merely in purely prospection activities, but also for analysis of hot fluids and to find out the factors which are most sensitive to temperature changes.

3.2.2.1 Heat Flow and the Geothermal Gradient

Obviously, accessible usable heat is the target; hence heat flow measurements and the assessment of geothermal gradients are essential aspects of any geophysical investigation.

Geothermal areas are regions with surface heat flows above the worldwide average value of *ca.* $63 \, mW/m^2$, sometimes having values exceeding this by several orders of magnitude. Usually an order of magnitude increase is entailed: an instance may be cited, namely that of the Tauhara field in New Zealand where, over 85% of the anomalous surface, the heat flow is greater than normal by a factor of 32, but over 4% it is 1000 times the normal flow.[7]

Various approaches to the measurements may be made. Conductive heat flow can be determined by temperature gradient measurements and by assessments of suitable rock samples in the laboratory. Of course, this is a laborious process because many such samples have to be examined if a representative value is to be obtained. Consequently, R. Goss and J. Combs in 1976 tried to obviate loss of time and mounting costs by prediction of thermal conductivity from standard geophysical bore logs.[8] In cases where horizontally-bedded and uniform rocks are involved, thermal conductivity may be assumed to be equally uniform, the temperature gradient then showing a distribution paralleling it. However, this is rarely possible.

In practice, almost all heat flow measurements and temperature gradient work relate to borehole data from depths up to 100 m. Shallower depths are suspect in that near-surface effects of insolation, precipitation and landscape can affect the results. This is also true of intermediate depths, at which groundwater may be a disturbing factor. Unfortunately, greater depths entail greater costs and also a diminution in the resolution capacity of the majority of geophysical methods. In addition, it must be remembered that any conclusions drawn from such measurements involve the assumption that the principle heat source is located directly underneath the borehole. It has been demonstrated that this is not invariably true, hydrothermal waters being quite capable of long-distance horizontal flow. Geothermal gradients are valuable solely for conductive heat transfer, both vertical and horizontal convection sometimes invalidating the extrapolation of temperature data, as Y. Eckstein pointed out in 1979.[9]

Heat flow measurements can be effected on the surface in areas of steaming ground and fumaroles by using a venturi meter and pitot tube, respectively. The evaporative loss from large smooth surface-water bodies can be derived from water and air temperatures, employing those relationships given in the *International Critical Tables* which are quite accurate if the wind speed is low. For boiling pools extra loss of heat is a function of the height of boiling. Heat losses through springs and seepages into rivers may be calculated from temperatures and water flows.

Since subterranean flows cannot be measured, total heat flow assessments based only upon such surface discharge data can be much too low. Making subterranean temperature measurements involves insertion of an array of thermocouples or thermistors in gradient holes; if a high-temperature gradient is anticipated, it is advisable to fill them with bentonite mud so as to obviate convection.

3.2.2.2 Remote Sensing

Accumulation of data on the surface temperatures of the Earth constitutes thermography: they can be obtained by making a record of the radiation in the intermediate and far infrared range of the electromagnetic spectrum employing optical mechanical scanners, or by making such a similar record in the microwave range using a passive microwave radiometer with a scanning antenna, as in radar.

The spectrum in question actually comprises three windows. The first is an optical one with radiation in the visible range together with small adjacent ranges, one in the near-ultraviolet and one in the near-infrared. The second is the heat window, the intermediate infrared; it depends upon the fact that a body at 300 K has a maximum radiation at a wavelength of 10 μm and, as the Earth has this temperature, its characteristics as a radiator can be observed in the thermal window. The third is the microwave window with natural radiation at a very low level so that it is suited to radar, i.e. the sending and receiving of radio signals.

From such work a thermal image can be derived which will show different tones of grey or colours based upon the temperature differences of the various surface objects. If the flying altitude is too great, a poor-resolution image results. On the other hand, poor geometric images are produced if it is too low over irregular terrain.

Several parameters can adversely affect thermographic images, e.g. air temperature, terrain inclination direction, angles of slopes in a terrain, wind velocity, moisture content of the atmosphere, soil moisture and vegetation. Nevertheless, the approach can be valuable where a particular aspect is correlated with appreciable ground temperature difference not obscured or disturbed by any or all of the above-mentioned factors. A case in point is the occurrence of heat emission losses from subterranean pipe systems and the presence of underground or surface manifestations of geothermal energy such as hot springs.

Aerial photography can provide structural information about a region and, if possible, should be coupled with an infrared imagery survey, the latter being especially valuable in hitherto unsurveyed areas.[7] Remote sensing methods employing imagery in the near or intermediate infrared region have been applied in several geothermal areas and such studies have used infrared scanners capable of detecting radiated rather than reflected energy in the 3–5 or 8–14 μm bands. They have been confined to mapping areas where the temperature exceeds ambient. Such temperatures can be related to heat flow and the principal advantage of the infrared approach is that it can spot hot regions in unsurveyed territories; see, for instance, the work of J. W. Noble and S. B. Ojiambo in 1976.[10] Also aerial infrared surveying could be useful in urban areas for monitoring alterations in the geothermal activity. Some doubt has been cast upon this idea because of the rapidity of some natural variations from which unjustified inferences regarding trends might be made.

SLAR, side-looking airborne radar, is a supplementary approach of considerable value in regions for which there are no adequate maps. It has the distinct advantage over aerial photography in that it can be carried out at any time. Sometimes an aeromagnetic survey is useful, especially in conjunction with photography.[3]

3.2.2.3 Electrical Methods

Turning now to electrical and electromagnetic methods, of which a brief mention was made early in Section 3.2.2, the resistivity approach has been especially useful because it entails the noting of variations directly connected with fluid as well as with rock properties. Consequently, a resistivity map shows the areal extent of a hot-water reservoir over the depth range through which a survey is made. Several factors influence the resistivity of a rock mass containing an electrolyte, in this case the geothermal fluid, so that the parameter is inversely related to the ionic concentration. Its value in the fluid is lower than in the host rock, varying with temperature in an inverse manner and more rapidly than in the rock. One caveat must be entered here: that when clays exist in large amounts in a rock mass, the host rock itself will have its resistivity much reduced. In regard to this, the value for the parameter at Broadlands in New Zealand could drop down to only 0·5 ohm m, which is comparable with values occurring in some geothermal fluids.[11]

Naturally, the effect is significant in tropical regions where deep weathering of rocks can produce a thick clay covering. Actually the dependence of resistivity upon temperature diminishes as the latter rises and can be disregarded above 150°C in comparison with the changes to be expected from porosity variations.

Steam shows a very high resistivity. In vapour-dominated systems, there may well be problems in interpretation of electrical resistivity data and, as a result of the fact that dry steam has an extremely high resistivity, there may be no anomaly present. Sometimes the steam zone may be overlain by a condensate layer, as for instance in the Yellowstone National Park, Wyoming, USA, where low resistivity may be due to the abundant hot water present in the condensate layer.[12] This area is shown in Fig. 3.1.

The technical aspects of electric methods are interesting. However, since resistivity is the reciprocal of conductance, some reference to the latter may perhaps be made by way of introduction. In 1927, in Alsace-Lorraine, H. G. Doll recorded the first electric log and introduced the induction logging method which overcame the problem of highly resistive drilling mud by avoiding any electrical contact with the formation under investigation. He

FIG. 3.1.   Thermal activity in Yellowstone National Park in winter. Reproduced by
courtesy of Wyoming Travel Commission.

also gave an excellent account of the electromagnetic (EM) interaction, now called geometric factor theory.

Thus surveys of electrical conductivity in boreholes were used first in the 1920s and by 1940 had become routinely applied in evaluating those drilled for hydrocarbon recovery in the USA. Until the mid-1950s the standard conductivity surveying tool was described as comprising an electrical survey (ES). In this an array of electrodes was lowered into a well and the apparent resistivity was recorded continuously as the tool moved through the hole. Usually three curves of this parameter were recorded using three electrode spacings, namely 41·1, 164·5 and 565·4 cm. The necessity for such multiple spacing measurements arose from the change in resistivity from the well bore into the undisturbed rock. Employing several spacings facilitates determination of resistivity in each of these areas separately. But no one curve gave a good representation of rock resistivity, the short ones being overly affected by the well bore and the infiltrated zone and the long ones tending to average out the resistivity over successive layers to a great extent.

From the mid-1950s to the mid-1970s, the standard choice of tool for the assessment of conductivity was the induction electrical survey (IES). Here the short electrode spacing was still used, but the other curves were replaced with an induction logging tool. In the induction log, the coupling between two coils, one a transmitter operating at approximately 20 kHz and the other a receiver, is measured. By judiciously employing several auxiliary coils, induction in the well bore and the disturbed zone are cancelled and the resulting measurement approaches a representation of the true resistivity of the rock beyond the disturbed zone.

Since the mid-1970s there has been further improvement in conductivity logging tools. A second induction device has been added, but with a shorter spacing than 102 cm, with the objective of giving a better measurement of the disturbed zone. This is termed the dual induction log (DIL). Also the electrode array log has been replaced by the spherically focused log (SFL). This is an electrode device but with auxiliary current electrodes provided in order to force the measuring current to flow into the rock rather than up and down the conductive fluid in the borehole. The SFL produces a detailed log of the electrical properties within 50–75 cm radius from the axis of the well bore.

Other electrical surveys used include, in low-conductivity terrain, a laterolog, e.g. LL-8, and, in surveying oil wells, a micro-log comprising a small-scale version of one of the logs described above. It has a pad contacting the wall of the well bore and is aimed at measuring the properties of the disturbed zone around a well bore.

Most earlier resistivity surveys were effected using direct current either in the Wenner or in the Schlumberger configuration (see also under 'Electrode array' in the Glossary). Conrad Schlumberger made experiments in the field as early as 1912 when he plotted equipotential lines on the surface of the Earth as set up by an applied electric current between two metal rods driven into the ground. Now traverses are usually made with a fixed electrode spacing in order to map variations in resistivity down to an almost constant depth below the surface, a probing depth related not merely to such spacing but also the underground distribution of resistivities. Prior determination of the latter is desirable and may be achieved by soundings, i.e. making measurements with increasing electrode spacings. As the rocks often possess parallel fissures, anisotropic resistivity is to be anticipated. In cases where very deep probing is necessary, a great deal of wire may be required for either of the survey methods mentioned.

In principle, electrodes could be a number of kilometres apart, but logistic difficulties could arise. In practice, under optimum conditions, a Schlumberger array with the current electrodes separated by a distance of 2 km can be easily arranged. Since Schlumberger-array electrode spacings are given as half the distance between the current electrodes A and B, it is possible to express this as $AB/2 = 1$ km. Unfortunately, the above-mentioned methods are incapable of probing to the depths at which geothermally productive zones are usually found.

Consequently, different techniques are applied, among which is an electrical field distribution around a dipole source. Two such configurations are common. One is the dipole–dipole, this having two similar pairs of closely spaced electrodes, of which the first carries the current and the other measures the voltage. The electrodes are traversed along profiles while they are all maintained collinear. The second is the bipole–dipole (roving dipole) in which one electrode pair (the bipole), normally with a spacing of a few hundreds of metres, is fixed and the other (with a much smaller spacing of tens of metres only) is moved around the area under investigation. An analysis of the method was given by A. Dey and H. F. Morrison in 1977.[13] From the apparent resistivities plotted, interpretation is made in terms of simple, layered, dike, vertical contact or spherical models. Ambiguity arises in the case of more complicated two-dimensional models when the detection of buried conductors is related to the choice of transmitter location. These authors compared the two situations, bearing in mind that, since apparent resistivities taken on a line collinear with the dipole are approximately equivalent to the apparent resistivities for one sounding in a dipole–dipole pseudosection, the approach is permissible.

An adaptation of the bipole–dipole approach uses a multiple (four-electrode) bipole source array and can indicate permeable zones at depth. A small rectangular pair of receiving dipoles is moved round inside the quadrilateral formed by the current electrodes and, for each receiver position, the six different current bipole combinations are used sequentially so that each measurement relates to a different current path through the ground. In this way electrical resistivity anisotropy can be inferred. An example of its value was given by G. F. Risk, who in 1976 interpreted his results as relating to fissuring created during the extrusion of a now-buried rhyolite body.[14]

Many workers have utilized the bipole–dipole method *per se*, for instance G. R. Jiracek and M. T. Gerety in 1978 in New Mexico at Los Alturas Estates.[15] In this case, however, one of the current electrodes was placed below ground in a well and the other at a distance of more than 3 km from it. The apparent resistivities derived with a roving receiving dipole using this arrangement were compared with those recorded when the down-hole electrode was replaced with one at the surface near the wellhead and, in addition, with the data obtained from dipole–dipole and Schlumberger surveys. The basic assumption was that if the electric current can be directly introduced into a sub-surface low-resistivity layer, then it becomes channelled by the latter and thus facilitates its detection. In fact, all the approaches gave comparable results, but, since an adequate three-dimensional model was unavailable and there were no deep temperature measurements for comparison with these results either, a complete assessment proved to be impossible to make.

Less unusual applications have been made by many other workers who have presented data as apparent resistivities later to be interpreted qualitatively, frequently in comparison with other, more easily interpreted, information. It is interesting that apparent resistivity maps do not necessarily visually resemble the actual electrically anomalous body and hence do not lend themselves to subjective interpretations. In fact, interpreting them is best achieved by the use of suitable models, of which perhaps the finite-element one proposed by H. M. Bibby in 1978 is the most appropriate.[16]

The collinear dipole–dipole traversing method is used often in looking for geothermal resources because of its relative simplicity and also because, if enough space is left between the dipoles, penetration to great depth becomes feasible. G. R. Jiracek and his fellow workers employed it in the Jemez Mountains of New Mexico, USA, in 1976.[17]

Dipole–dipole survey data are most often presented as a 'pseudo-section' with the apparent resistivity values plotted midway between the two dipoles and at a depth equal to half the distance between their centres. This is not necessarily the best way to indicate the results, however, because, as J. T. Lumb indicated in 1981, the true resistivity values are not used and hence the 'pseudo-section' cannot represent the real resistivity distribution along the line of the profile.[3] Consequently, such 'pseudo-sections' should be looked at merely as intermediate stages in comparing the observed with the computer-generated apparent resistivity values.

Apart from their simplicity, there is another advantage claimed for the dipole methods, namely that they appear to be somewhat insensitive to the ruggedness or otherwise of the topography. Doubt has been shed upon this by G. W. Hohmann and his associates in 1978.[18] They demonstrated by computer that a symmetrical valley having a width approximately seven times and a depth about three times the minimum dipole separation is capable of generating a 'pseudo-section' showing a low resistivity zone in it. Values would be roughly one-third of those occurring outside the valley, actually 400 ohm m compared with ≥ 1200 ohm m. A similar, but reversed, effect was observed over a ridge.

Recently, studies have been made on cross-borehole resistivity measurements for detecting anomalous bodies, especially by H. P. Ross and S. H. Ward in 1984 in regard to evaluating the application of fracture mapping in geothermal systems.[19] These could prove to be of considerable importance because F. W. Yang and S. H. Ward in 1985 indicated that such measurements constitute a more effective approach than single borehole measurements for the delineation of resistivity anomalies near a borehole.[20] Subsequently, J. X. Zhao and his co-workers in 1986 studied the effects of geological noise on such cross-borehole electrical surveys.[21]

Finally, allusion must be made to the geoelectrical self-potential (SP) method, which is dependent upon natural electric fields. Although the real source of the SP anomalies is not well known, perhaps relating to the flow of ground and other subterranean waters, the observed effects appear to vary from field to field. For instance, L. A. Anderson and G. R. Johnson in 1978 recorded high-amplitude positive anomalies exceeding 1 V but low-amplitude ones of ≤ 100 mV in Nevada.[22] They also vary with time, as was shown by K. Ernstson and H. U. Scherer in 1986.[23] In order to study the relevant parameters influencing the phenomenon, these workers made repeated measurements every two weeks over an 18-month period along fixed profiles extending over different lithological units.

An analysis of the SP recorded implies decomposition into three components of different wavelengths. Variations of the SP with wavelengths in the 0·1–1 m range and amplitudes of as much as 150 mV may have been induced by vegetation. The mean amplitudes change over a one-year time interval roughly in correlation with the temperature of the soil. A long-wavelength component (wavelengths, metres and some tens of metres; amplitudes, some 10 mV) is believed to result from changes in subterranean lithology. The long-wavelength SP and its marked variation with time are assumed to result from water movements (cf. the remark regarding the source of SP, above) in layers of differing permeabilities. There is a (disputed by others) morphology-dependent trend of SP values and this also exhibits systematic variations with time, with gradients altering in the order of 0·5 mV/m difference in elevation. There is a correlation with precipitation with a one-month delay factor. This particular effect is believed to result from depletion and recharge of perched water aquifers. Surface and shallow depth measurements established the existence of a vertical gradient of the SP showing variations with time of up to 50 mV/m. With a three-month time lag, the amplitude of the vertical gradient relates closely to the variation of the potential evapotranspiration. Groundwater recharge from gravitational water is thought to be the main cause of the vertical gradient. From these results it may be inferred that self-potentials at the near-surface must be of a streaming-potential character. Their variation with time ought not to be neglected when the SP method is applied to problems in engineering geophysics. However, since self-potentials are closely related to hydrogeological parameters, SP measurements could well become a standard hydrogeological tool.

3.2.2.4 Electromagnetic and Magnetotelluric Methods

Electromagnetic methods, too, are extremely useful in geothermal prospection. They have an advantage over d.c. methods: the penetration depth may be varied by altering the signal frequency and with no necessity to change the geometry of the field disposition. At shallow depths around 30 m, J. T. Lumb and W. J. P. Macdonald in 1970 demonstrated that the two-loop method adequately substituted for conventional d.c. resistivity and almost-surface temperature surveys, but is cheaper and quicker.[24]

Electromagnetic soundings are carried out by employing the transient, time-domain method (TDEM) in which a step-current is passed through the ground, the consequent transient magnetic field being observed and analysed. Its 'tail' results from the decay of the transient and relates to resistivities at greater depth as time increases from the onset of the signal. A sounding can be effected by one observation only, but signal-to-noise ratios are so low, especially at the end of the tail, that a number of repeat readings has to be made.

TDEM methods date back to the work of G. V. Keller in 1970 and advances during the decade are discussed by H. F. Morrison and his colleagues in 1978.[25,26] Now the method is well established as a geophysical exploration tool and many papers on its applications appear every year.

Practically all research into the behaviour of the electromagnetic field in either a two-dimensional or a three-dimensional Earth deal with Maxwell's equations applied to a secondary field in the frequency domain. This secondary field refers to the difference between the total field and the primary field, the latter being defined as the exact solution for the half-space without any inhomogeneities. The problem is so formulated in order to avoid the source description problem. Most existing schemes refer only to models where the primary field is known. Although it is possible to compute the secondary field for any two-dimensional model, the primary field can be known analytically only for a crude one-dimensional Earth. Consequently, many computed solutions contain the field behaviour and thus only lead to quantitative conclusions when applied to the real world. Y. Goldman and his co-authors in 1986 tackled the problem, presenting a numerical method for solving Maxwell's equations in the case of an arbitrary two-dimensional resistivity distribution excited by an infinite current line.[27] The concept of apparent resistivity was introduced into the TDEM method in order to facilitate intuitive interpretation, according to Y. Sheng in 1986.[28] The apparent resistivity does not contain any more data than the original field observation, but by its shape it shows more clearly the changes in Earth resistivity with depth. Thus early-time and late-time apparent resistivity approximations were used for interpretation of long-offset transient electromagnetic (LOTEM) measurements.

It is important to mention here an interesting treatment of the use and misuse of apparent resistivity in electromagnetic methods which was given by B. R. Spies and D. E. Eggers in 1986.[29] They referred to problems arising when the analogy between an apparent resistivity computed from geophysical observations and the true resistivity structure below ground is too tightly drawn, alluding also to the several definitions of the parameter of which the most commonly utilized do not always show the best behaviour. Thus many features in an apparent resistivity curve interpreted as physically significant with one definition disappear if an alternative definition is substituted. Consequently, these authors regarded it as actually misleading to compare the detection or resolution capabilities of different field systems or configurations only on the basis of the apparent resistivity curve. They pointed out that, for the in-loop transient electromagnetic (TEM) method, apparent resistivity computed from the magnetic field response shows much better behaviour than that computed from the induced voltage response.

A comparison of 'exact' and 'asymptotic' formulae for the TEM method shows that automated schemes for distinguishing early-time and late-time branches are at best tenuous and actually bound to fail for a certain type of resistivity structure, e.g. where the loop size is large compared with the layer thickness. For the magnetotelluric (MT) methods described below, apparent resistivity curves defined from the real part of the impedance exhibit better behaviour than curves based upon the conventional definition which uses the magnitude of the impedance. The results of utilizing this new definition have characteristics similar to those of apparent resistivity derived from time-domain processing.

Magnetotelluric (MT) methods employ natural magnetic fields and Earth currents. The total planetary magnetic field is complex and comprises three major 'fields'. These are the core field generated in the interior of the Earth, the lithospheric field caused by magnetization in the solid part of the outer Earth and the external field which arises from electric currents in the ionosphere–magnetosphere system caused by the interaction with the solar wind and the interplanetary magnetic field. The distinction of the 'core field' below harmonics and the 'crustal field' above harmonics became possible when the NASA MAGSAT mission (30 October 1979 to 22 June 1980) facilitated construction of a detailed spherical harmonics representation of the Earth's geomagnetic field, as N.-A. Mörner indicated in 1986.[30]

From the geothermal standpoint, there are two interesting aspects. The first is the low-frequency approach applicable to investigating large-scale features such as sources of deep heat (since hot rock is conductive) under geothermal fields. The second involves utilization of signals in the audio-frequency (AMT) range. These do not penetrate so deeply, but they can be measured quickly in reconnaissance surveying. A magnetotelluric sounding curve derived from a given location may comprise contributions from one-, two- or three-dimensional geoelectric components, either single or in combination. Modelling and interpreting such data depend upon the feasibility of assessing the degree of influence of each of these possible structural components. Methods of estimation of such geoelectric dimensions from magnetotelluric data were surveyed in 1986 by D. Beamish.[31]

Some interesting implications of magnetotelluric modelling for the deep crustal environment in the Rio Grande rift in southern USA were proposed in 1987 by G. R. Jiracek and his associates.[32] This may constitute part of the Basin and Range province and certainly is unusual in having a thinner, hotter and more highly faulted crust than typical continental crust. It extends north–south over 1000 km from Colorado through New Mexico into northern Mexico and the authors cited worked in the central part of it. Block forward modelling and smooth Backus–Gilbert linearized inversion were utilized independently so as to model two-dimensionally two east–west MT profiles. A crustal conductive zone of at least 1500 S was detected at *ca.* 10 km depth along the northern line, but along another line approximately 40 km to the south where shallow, active magma is inferred and microseismicity is maximal, no high conductivity zone was found. The conductive zone beneath this southern line is hypothesized to be present where water is trapped beneath an undisturbed ductile cap. Under the southern line, magma injection through the cap could release the water and produce a more resistive crust with no conductive, midcrustal layer. A paradox arises—where there is active magma injection into the upper crust in the Rio Grande rift, the crust is more resistive rather than more conductive. This could be relevant to geothermal work elsewhere. Incidentally, the existence of present-day magma in the study area was well attested by microseismic and reflection seismic results. Indeed, a seismic line shot by the Consortium for Continental Reflection Profiling (COCORP) crosses the northern end of the deep magma body.

It must be added that telluric current measurements alone have been employed in geothermal exploration, a three-electrode technique being applicable. This was described in 1976 by H. Beyer and his associates, who applied it in the Basin and Range valleys of Nevada in the USA.[33] This, like magnetotelluric methods, depends upon the naturally found, time-variant magnetic fields, the interpretation of collected data assuming that these fields are normal to the surface of the Earth. Obviously this is not always the case, therefore the repeatability of observations is now always optimal in regions of complex structure.

3.2.2.5 Gravity and Magnetic Methods
Gravity surveys have been used in geothermal exploration and usually yielded positive gravity residuals either in or near to geothermal areas. Often gravity anomalies are associated with background geology or buried topography, i.e. they do not have specifically geothermal significance. Thus, for instance, at the East Mesa field, a residual gravity 'high' was attributed to possible silicification and low-grade metamorphism of sediments by

C. A. Swanberg in 1976.[34] Also, in the Kurikoma field in Japan, surface manifestations of geothermal activity related to positive gravity residual anomalies, but lay on the flanks of negative Bouguer anomalies.[35]

A very detailed gravity study has been carried out in the Clear Lake region of The Geysers and was effected in conjunction with an aeromagnetic survey and described by W. F. Isherwood in 1976.[36] He constructed a pseudo-gravity anomaly map from the magnetic data which could be used to test the hypothesis that both gravity and magnetic anomalies are caused by the same structures. This was found not to be so and he inferred that part of the negative gravity anomaly results from a steam-filled zone. However, a mass deficiency has been shown to exist under the nearby Mount Hannah and this appears to be associated with an underlying body of magma there. The latter was identified from studies of recorded seismic waves traversing the magma chamber from distant earthquakes. They underwent considerable slow-down (approximately 15%) at a depth of 15–20 km, consistent with the fact that molten rock can produce this effect. It may be inferred that the zone under the volcanic area is partially molten and beneath the geothermal energy production zone there may occur a highly fractured steam reservoir overlying the magma. Such a highly fractured zone could also slow down seismic waves. From this work it has been suggested that the magma body lying more than 5 km under Mount Hannah may be the heat source of The Geysers. The magnetic survey reflects a complicated geology and indeed shows a northwest to southeast trend; see Fig. 3.2. However, when the near-surface effects are removed, the resulting anomalies still could not be exactly related to the geothermal field.

Incidentally, this had earlier proved to be the case in the Broadlands field in New Zealand, as M. P. Hochstein and T. M. Hunt indicated in 1970.[37] In 1977 the latter also showed the value of gravity monitoring in conjunction with repeated and precise levelling in assessment of the net mass loss from the nearby Wairakei field.[38] The gravity changes between surveys were found sometimes to be very small (occasionally only 0·1 μN/kg at some places), hence the need for repeated readings at the same site and accurate measurements of elevation. An extremely interesting analysis of exploitation-induced gravity changes at the Wairakei field was recorded by R. C. Allis and T. M. Hunt in 1986.[39]

Gravity changes (corrected for subsidence) of up to $-1000\,(\pm 300)\,\mu$Gal took place in the square-kilometre area of the production bore field and smaller decreases cover a 50-km^2 surrounding region. Most of these decreases occurred in the 1960s. After this the net gravity change for the

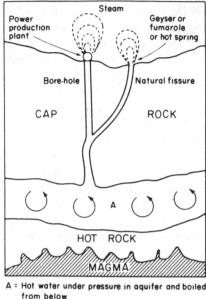

A = Hot water under pressure in aquifer and boiled from below

◯ = Circulation pattern in voids

FIG. 3.2. The Geysers, California, USA, a natural geothermal field which is hydrothermal and vapour-dominated, shown in cross-section to illustrate the subterranean geology in terms of heat and circulation patterns.

entire field has been zero, which indicates mass flow equilibrium. The main causes of gravity change have been deep liquid pressure drawdown resulting in formation of a steam zone, subsequent saturation changes in this zone, liquid temperature drop and changes in groundwater levels. Gravity models imply saturation of the steam zone was 0·7 (±0·1) in 1962, diminishing to 0·6 by 1972. Gravity increases in the northern and eastern bore field after the early 1970s were attributed to cool water invading the steam zone.

Negative magnetic anomalies have been found associated with geothermal fields in Iceland, but they are not always present. This caused G. Pálmason in 1976 to advise against making a large-scale aeromagnetic survey prior to a pilot ground truth survey.[40] He mentioned the utility of magnetic surveying in tracing dikes and faults which control the flow of low-temperature thermal water to the surface.

It may be mentioned that magnetotelluric data may be combined with differential geomagnetic sounding (DGS) surveying in order to investigate the conductivity structure of the crust and upper mantle and this was done, for instance, by M. Ritz and J. Vassal, who reported results in 1986 and 1987.[41,42] They claimed to have reconstructed the parameter in question from a traverse across the Senegal basin in West Africa and this demonstrated both strong lateral variations within the crust and regions of high conductivity, both there and at upper mantle depths. An analysis of the geomagnetic variation field identified a concentration of telluric current flow beneath the deep basin and it seems that the additional currents flowing in the strike direction of the Senegal basin are mainly controlled by sedimentary rocks of high conductivity lying at depths ≤ 5 km. Model studies showed that the local conductivity distribution can explain the currents circulating in the thick well-conducting sediments.

In recent years, techniques for carrying out gravity surveys from aircraft have advanced greatly, as J. M. Brozena indicated in 1984.[43] In 1986 D. H. Eckhardt reported that, in the following year, the Bell Aerospace/Textron Gravity Gradiometer Survey System (GGSS) would be tested by the US Air Force Physics Laboratory in an airborne survey of a 300 km × 300 km region of Oklahoma and Texas in a Lockheed Hercules C-130 cargo plane at an altitude as low as possible, optimally *ca.* 600 m.[44] Various techniques for processing the data exist and were summarized by C. Jekeli and others in 1985, Eckhardt contributing yet another in 1986.[44,45]

There have been corresponding advances in refining of instruments as well and, among these, not the least important is that described by H. D. Valliant and his colleagues in 1986.[46] They gave a convenient method for conversion of types of astatic gravity meters to electrostatically nulled Earth-tide recording instruments. The technique involves maintaining the meter at its reading line by applying an electrostatic force in opposition to changing gravity. This force is generated between the gravity meter beam and one or more fixed plates which may be either separate from or integral with the position-sensing plates. The main difficulty is in the quadratic relationship between force and voltage leading to unacceptable non-linearities in the feedback system. Earlier solutions all utilized the linearizing effects of two fixed plates, along with occasionally tedious mechanical and/or electrical adjustments to achieve linearization. The method of these workers is inherently linear and works equally well with single- and dual-force rebalance plates.

3.2.2.6 Seismic Observations and Surveys

Since geothermal exploration using seismic wave attenuation techniques started, unusually high levels of ground noise constituted a problem in interpretation of the explosive-induced records. The attenuation and source depth of seismic ground noise were estimated at Wairakei and Waiotapu by P. C. Whiteford in 1976.[47] This was done by comparing the observed amplitudes of ground particle velocity with those predicted from a theoretical equation. He obtained a maximum depth of 200 ± 100 m at Waiotapu, implying that the phenomenon could relate to a rather near-surface thermal process of some sort, possibly boiling. Another possible explanation is that the noise originates from circulation of fluids at high temperature and pressure.[48] However, seismic noise anomaly is sometimes controlled by wave amplification in an alluvial basin and this is a well-known occurrence where seismic waves traverse thick, unconsolidated strata. In such cases, it cannot be considered as indicative of a geothermal resource.

The above-mentioned noise can be categorized as one of a set of such phenomena grouped as ambient noise, specifically within ergodic 'background noise' which also includes noises due to wind, moving water and far-off, incoherently scattered, cultural noise. Other noise types comprise time-stationary noise arising from fixed machinery, noise bursts from moving sources and noise spikes from electric sources.

As well as ground noise, micro-earthquakes (those having a magnitude of 2 or less on the Richter scale) were examined at East Mesa in California; indeed they were interpreted as a means of finding an active fault and so played an important role in siting both production and injection wells.[34] However, later studies did not indicate any activity on the East Mesa fault. In fact, some have suggested that the shocks applied to defining the fault may have stemmed from the Brawley fault.[49] Clearly, there is some ambiguity about the results and from this it may be inferred that this might be the case in other geothermal areas as well. While the situation at East Mesa shows that temporary aseismicity can occur, Broadlands has been stated to be permanently in this condition.[50]

Micro-earthquakes can be distinguished from low-level and practically continuous ground noise because they are separate and much less frequent. They probably are caused by minute movements or fracturing of rocks at depth and often occur along faults. In addition, they can occur near dam sites; for example, this type of seismic activity was recorded around Thien Dam in the northwest Himalayas by S. N. Bhattacharya and his fellow workers in 1986.[51]

Another instance is that of the induced earthquakes in the Zhelin reservoir, China, discussed by D. Yuanzhang and others in 1986.[52] Here, in east China, natural seismic activity is extremely weak so that, in a 371-year period prior to impounding the reservoir mentioned, there were a mere five felt earthquakes within a range of 30 km from the lake. After impounding, a noticeable effect was observed, there being four sensory earthquakes (the largest one of M_s 3·2) and some micro-quakes over the seven years from 1972 to 1979. During the first four of these years there was a degree of correlation between the seismic activity and the reservoir levels. The events were interpreted as reservoir-induced earthquakes, although of low frequency and magnitude. Mostly they were distributed over an active normal fault with a strike-slip on the margin of the lake and this fault was the seismic structure for the felt earthquake in 1967.

In addition, micro-earthquakes can be triggered during hydraulic fracturing experiments such as those conducted during Phase II of the Los Alamos Project involving development of a reservoir in the Precambrian basement rock outside the Valles Caldera in New Mexico, USA, and described in detail in Chapter 5.

If the magnitude of the earth tremors increases above the limit of 2 on the Richter scale, then real earthquakes have taken place in the area. This happened, for instance, near Lake Bhatsa, Maharashtra, India, where events with magnitudes 4 and 4·8 occurred in August and September 1983, respectively. The region lies northwest of the Bhatsa Dam in the western part of the Deccan Volcanic Province and the two shocks followed impoundment of water in the associated reservoir a couple of months earlier. D. N. Patil and his associates discussed the subject in 1986.[53]

That variant of micro-earthquake involving extremely small-scale events (nano-earthquakes) perhaps arises from natural hydraulic fracturing taking place when pore pressures are increased through the effects of rising temperatures in regions of high thermal gradient to such an extent that they exceed the tensile strength of the rocks.

Teleseismic P-wave delays have been used in geothermal exploration and facilitated conceptual modelling of large-scale sub-surface structures showing vast zones extending from the crust into the mantle (in which seismic velocities may be reduced by as much as 25%). Although such delays have been employed in connection with a gravity-based model of crustal thinning at Imperial Valley, California, such detailed interpretations are not normally feasible because the times involved (<0.1 s) necessitate a far better knowledge of the geology than is usually available when geothermal exploration is initiated.[54] However, the approach does demonstrate large volumes of rock which can be interpreted as partially melted, therefore perhaps providing a large heat source for any geothermal field overlying it.

Attenuation of seismic waves has been investigated as a potential way of obtaining data on the large-scale structure of geothermal regions, but the results so far obtained are rather ambiguous. This has much to do with the fact that, as with P-wave delays, a more detailed knowledge of the geology than is at first offered during exploration must be accessible and normally is not. Refraction and reflection seismic surveying using explosives has been carried out in a number of geothermal areas, but has become less important since exploration emphases were switched from structural to geothermal mapping. Additionally, ground noise and signal attenuation introduce further difficulties. However, at Broadlands, it was found that refraction records were more satisfactory than reflection ones.[37]

3.2.3 Geochemical and Isotopic Methods

When the investigation of a geothermal field commences, chemical analyses of water and gas samples collected initially from hot and cold surface waters and springs, fumaroles and steaming ground will provide useful information. The connection between mercury and thermal springs is significant because unusual leakage of the metal into these springs and stream sediments can label geothermal areas. This has been noted around Monte Amiata and Larderello in Italy, and in California and Colorado in the USA as well. Details of the relevant work are given in Chapter 4 (Sections 4.1.1.2 and 4.5.1.2 under 'Italy'). High lithium contents have been suggested as a potential indicator of high sub-surface temperatures, but the method is not foolproof. This is because variations in trace metal contents can stem from factors other than high lithium, e.g. variations in lithology, in the time during which the water was in contact with rock and in the degree of pollution undergone.

In sampling surface waters, care is necessary because of the possibility of such contamination, particularly if the waters come from small fumaroles and warm, rather than hot, springs. However, mixing models and samples of cold groundwater can be used to make inferences about the conditions at depth. The samples may provide insights into possible scaling and corrosion effects of sub-surface water if development goes ahead.

In addition, the compositional range of fluid contents can be determined and this will shed light on the question of whether the fluids are economically valuable or not. The actual quantity of such surface discharge phenomena reflects the relevant heat content and area of the geothermal field itself. To decide upon its potential use in generating electricity or for some other purpose, it is necessary to assess its geothermal energy content, which can be done by drilling and by geothermometry. The latter entails the use of appropriate temperature indicators, namely thermal fluids which actually represent frozen-in equilibria of temperature-dependent reactions.

Afterwards the size of the geothermal resource and its nature, e.g. a dry steam or a steam and water mixture, must be evaluated and productive zones identified. Another significant parameter which it is desirable to know is the enthalpy of the different fluids discharged by wells in a geothermal field to be exploited. Of course, the pre-exploitation values of environmentally sensitive parameters should be determined as well. All information assembled by these procedures is applicable to future, necessary, monitoring activities. In addition, data regarding properties likely to influence adversely the development of such a geothermal field are particularly significant. These may include such factors as the possibility of subsidence or a high gas content.

It must be emphasized that the analysis of a geothermal resource does not simply involve calculation of the energy stored in a particular portion of the crust of the Earth, but rather a determination of the actual amount of earth heat which can be extracted and employed in a given region under favourable marketing conditions. Only this quantity can be compared meaningfully with the thermal energy equivalent of barrels of recoverable oil, cubic metres of gas, tons of coal or kilograms of uranium. For mankind only the recoverable geothermal energy is important and this alone constitutes the real geothermal resource. When properly determined, it gives policy makers a picture of future prospects for the energy actually producible as well as affording industrialists a reference basis for planning long-term investments and evaluating costs against the anticipated benefits. Finally, it facilitates establishment of a technical order of priority for exploration, drilling and exploitation in a prospect area.

3.2.3.1 Geothermal Fluids

Isotopic analyses, discussed in detail in various sections below, established that practically all of the water in geothermal fluids is of meteoric origin. Whether any magmatic or juvenile constituents occur is questionable. This is because elements such as boron, lithium and chlorine, once considered diagnostic of magmatic sources, are now known to be leachable in significant quantities from volcanic and sedimentary reservoir rocks of geothermal systems. Volcanic rocks are probably present in all high-temperature geothermal systems, at least at depth if not at the surface, and are definitely magmatic in origin. Hence they may comprise an intermediate reservoir for the dissolved constituents of geothermal waters. Alternatively, these waters may receive such constituents directly.

To date it is not known which of these two hypotheses is true. In attempts to elucidate the matter, comparisons have been made between the isotopic compositions of constituents of geothermal fluids and the isotopic ratios of dissolved inert gases with those of possible sources. However, such comparisons are usually unsatisfactory as a result of the fact that low-temperature surface fractionation processes yield wide ranges in isotopic compositions which include feasible magmatic values. In addition, isotopic fractionation during the crystallization and leaching processes is too small to distinguish components leached from volcanic rock from those directly contributed from magma.

Estimating the temperature of geothermal fluids by chemical geo-thermometry depends upon the temperature dependence of the concentrations of certain components, the chemical equilibria between water and minerals, gas solubilities, various chemical reactions, and isotopic distributions between the water and mineral phases. All of these approaches are applicable also to surface waters, but attempts must be made to ascertain whether these altered en route from their subterranean thermal source or not, and, if so, whether an appropriate correction factor can be introduced into the data. Normally cooling a saturated solution entails precipitation of the solute because solubility diminishes with temperature. However, some chemical compounds, e.g. silica, while retaining the effect, demonstrate it to a very small extent only. Consequently, the solution possesses a concentration only fractionally lower at relatively low temperatures than in the much hotter source. If the assumption is made that the solution was saturated from the beginning, ideally it is feasible to estimate the original temperature. In nature this becomes more difficult because of interference from other factors, e.g. cooling of the hot water in transit by mixing with cold water, which may reduce the concentration of the chemical compound, in this case silica. This leads to a low, uncorrected, silica temperature. On the other hand, where the pH is high, the silica solubility rises. These and other considerations make it mandatory to find out as much as possible regarding the progress of hot waters as they ascend from their thermal sources in order that appropriate corrections can be calculated and better temperature estimates derived.

Another approach is to use a number of constituents rather than just one, determining variations in their ratios. In this way it may be feasible to eliminate problems connected with dilution. An instance is afforded by the sodium:potassium ratio, which has been employed for many years at Wairakei.[55] This depends upon the strong temperature dependence of the partitioning of the two elements between aluminosilicates and solutions, and is related to exchange between sodium and potassium feldspars. Because of its unreliability in low-temperature and high-calcium-content waters, an empirical sodium–potassium–calcium relationship was suggested in 1973 by R. O. Fournier and A. H. Truesdell.[56] This incorporates the role of calcium in aluminosilicate reactions. Some unexpected results imply that the derived temperatures may be related to the salinity of the water and also perhaps to the nature of the bedrock with which they attained equilibrium. In addition, B. G. Weissberg and P. T. Wilson in 1977 demonstrated in the laboratory that the occurrence of montmorillonites may elevate sodium:potassium ratios and thus produce a temperature which is too low.[57] Sodium–potassium and sodium–potassium–calcium geothermometry are discussed in detail in Sections 3.2.3.2, 3.2.3.4 and 3.2.3.5.

There are alternative techniques utilizing other temperature-dependent chemical equilibria, e.g. the magnesium concentration, the ratio of molal concentration of bicarbonate to carbon dioxide, the calcium sulphate concentration, the sodium:rubidium ratio, and the equilibrium reaction between carbon dioxide and methane, this last involving gases only and the equilibrium being highly pressure-dependent. Carbon dioxide in geothermal fluids probably originated either from magmatic assimilation or from the high-temperature leaching of limestone. Either origin would be consistent with measured carbon-13 contents from localities in California, Wyoming, Larderello and Wairakei. The carbon dioxide geothermometer uses the $^{13}C/^{12}C$ ratios of the two gases as the basis of the geothermometer, assuming that the gases are in isotopic equilibrium through the exchange reaction:

$$CO_2 + 4H_2 \rightleftharpoons CH_4 + 2H_2O \qquad (3.1)$$

However, there is some ambiguity about the calculated fractionation data and, until these are verified, confidence in carbon isotopic temperatures based solely upon carbon dioxide and methane cannot be complete. Nevertheless, there has been fair agreement between this and the sulphate–water geothermometer and enthalpy–chloride relations at Yellowstone National Park. This could be substantiated by deep drilling, but this is hardly likely to take place for such a reason.

The carbon dioxide–dissolved bicarbonate geothermometer may be mentioned here. When the gas dissolves in water to form aqueous carbon dioxide, uncombined with water, and carbonic acid, in equilibrium with bicarbonate and carbonate ions, the equilibria between these species are temperature-dependent and also relate to the pH. Unfortunately, high-temperature experimental data are inconsistent.

In 1980 F. D'Amore and C. Panichi introduced another all-gas assemblage geothermometer, for instance one such as carbon dioxide, hydrogen sulphide, methane, hydrogen and nitrogen, a mixture almost invariably found in or near thermal areas.[58] The relative quantities of these gases may be formulated as a function of temperature for use as a tool in geothermal exploration. Also it seems feasible to derive a suitable geochemical model of natural hydrothermal systems in which gas–water–rock equilibrium reactions take place and from this evaluate deep geothermal temperatures from surface data on gas samples.

Earlier work, for example that by F. D'Amore and S. Nuti in 1977 and by A. H. Truesdell and N. L. Nehring in 1978/1979, shows that both hydrogen and hydrogen-sulphide partial pressures are controlled by mineral buffers; i.e. both, together with iron sulphides and sulphate ions plus iron oxides and silicates, may be in some chemical equilibrium in the geothermal environment.[59,60]

It may be mentioned also that the Fischer–Tropsch reaction (eqn (3.1)) has been successfully applied to geothermal localities in order to explain the origin of methane.[59] But this reaction cannot cover all the chemical concentrations and isotopic characteristics of the methane occurring in geothermal manifestations; see, for instance, the report by W. M. Sackett and H. Moses Chung in 1979.[61] In fact, some methane may be the result of a reaction synthesis between carbon-bearing matter and molecular hydrogen, as C. Panichi and his fellow-workers proposed in 1978.[62]

F. D'Amore and C. Panichi made chemical analyses of the gas compositions for 34 thermal systems from all over the world and developed an empirical relationship between the relative concentrations of hydrogen sulphide, hydrogen, methane and carbon dioxide, and the reservoir temperature.[58] Evaluated temperatures were expressed by

$$t\,(^\circ\mathrm{C}) = \frac{24\,775}{\alpha + \beta + 36 \cdot 05} - 273 \tag{3.2}$$

where

$$\alpha = 2 \log \frac{[\mathrm{CH_4}]}{[\mathrm{CO_2}]} - 6 \log \frac{[\mathrm{H_2}]}{[\mathrm{CO_2}]} - 3 \log \frac{[\mathrm{H_2S}]}{[\mathrm{CO_2}]} \tag{3.3}$$

(concentrations in % by volume and $\beta = -7 \log P_{\mathrm{CO_2}}$). Some interesting results of applying the above to geothermal regions discussed earlier are given in Table 3.1.[58]

Another chemical geothermometer involves the solid solution of aluminium in quartz crystals; this was tested at Broadlands yielding temperatures ranging from 177 to 645°C, showing unreliability.[63,64]

TABLE 3.1
Temperatures calculated using eqn (3.2)[58]

Locality	α	Gas concentrations (% by vol.)	β	t from eqn (3.2) ($^\circ C$)	t of reservoir ($^\circ C$)
The Geysers	4·2	$CO_2 \leq 75$	7	240	247
Larderello	11·2	$CO_2 \geq 75$; $CH_4 \leq 2H_2$; $H_2S \leq 2H_2$	0	251	250
Broadlands	17·0	$CO_2 \geq 75$; $CH_4 \geq 2H_2$; $H_2S \geq 2H_2$	−7	265	275

The temperature dependence of isotopic fractionation of nuclides of several relatively light elements constitutes another set of geothermometers. The most widely used are the following:

$$^{34}S/^{32}S \qquad ^{13}C/^{12}C \qquad ^{18}O/^{16}O \qquad ^{2}H/^{1}H$$

The last of these is often written as D/H.

As with the chemical geothermometers, using isotopes is dependent upon equilibria established at depth being maintained while fluid ascends to the surface of the Earth. Sad to say, the time necessary to re-establish equilibrium is extremely long in certain cases and so it may be that the observed equilibria relate to conditions deeper in the crust than are interesting in geothermal exploration. Thus the temperatures calculated will be anomalously high. There may be two exchange reactions with relatively short and hence acceptable half-times, as J. R. Hulston indicated in 1977.[65] One is the oxygen isotope exchange between bisulphate and water which has a temperature- and pH-dependent half-time in Wairakei water of some four months. This provides a temperature range of 250–300°C for Wairakei. The second is hydrogen isotope exchange with water and it has an extremely short half-time of approximately a week. Using this approach gave a temperature for Wairakei water of *ca.* 260°C, similar to the actual measurements.

An added complication is that of surface oxidation, which is especially important in relation to the discharges of hot springs.[55] It may even be that a kinetic rather than an equilibrium isotopic fractionation is involved. However, it could be that equilibration does occur. All this shows the complexity of the problem and explains why it is not yet fully understood.

3.2.3.2 Hydrothermal Reactions
In geothermal reservoirs with relatively high permeabilities and rather long residence times for fluid water (up to years), water and rock should attain

chemical equilibrium, particularly where the temperatures exceed 200°C. At equilibrium the ratios of cations in solution are controlled by temperature-dependent exchange reactions of the following types, as indicated in 1981 by R. O. Fournier:[66]

$$NaAlSi_3O_8 + K^+ \rightleftharpoons KAlSi_3O_8 + Na^+ \qquad (3.4)$$

Albite   K-feldspar

$$K_{eq} = (Na^+)/(K^+) \qquad (3.5)$$

and

$$CaAl_2Si_4O_{12} \cdot 2H_2O + 2SiO_2 + 2Na^+ = 2NaAlSi_3O_8 + Ca^{2+} + 2H_2O$$

Wairakite   Quartz   Albite

$$(3.6)$$

$$K_{eq} = [Ca^{2+}][H_2O]^2/[Na^+]^2 \qquad (3.7)$$

The pH is controlled by such hydrolysis reactions as, for example,

$$3KAlSi_3O_8 + 2H^+ = KAl_3Si_3O_{10}(OH)_2 + 6SiO_2 + 2K^+ \qquad (3.8)$$

K-feldspar   K-mica

$$K_{eq} = [K^+]/[H^+] \qquad (3.9)$$

The factor K_{eq} in the above equations is the equilibrium constant for the stated reaction, assuming unit activities of the solid phases, the brackets indicating the activities of the dissolved species.

Testing the water–rock equilibration is effected by collecting and analysing representative samples of both water and rock. The former derive from fluids collected after drilling mud and make-up water have been flushed out of the well, and the latter from cutting and core material obtained during drilling. Chemical analyses of the water samples facilitate calculation of its former thermodynamic state and for this computer programs may be utilized.[67] Alternatively, the equilibrium may be determined by plotting the activity ratios of aqueous species, calculated by computer, as theoretical activity diagrams. This was done by A. J. Ellis and W. A. J. Mahon in 1977,[55] among others.

As noted earlier, many chemical and isotopic reactions can be used to estimate reservoirs, and among the most widely used of these possible geothermometers are the silica, the sodium–potassium, the sodium–potassium–calcium and the sulphate–oxygen, $\delta^{18}O(SO_4^{2-}-H_2O)$, the last using the fractionation of oxygen isotopes between water and dissolved sulphate; see W. F. McKenzie and A. H. Truesdell (1977).[68] Table 3.2 lists

TABLE 3.2

Equations showing the temperature dependence of the geothermometers listed

Geothermometer	Equation[a]	Restrictions
1. Quartz		
(i) No steam loss	$t\,(°C) = \dfrac{1309}{5·19 - \log Co} - 273·15$	$t = 0-250°C$
(ii) Maximum steam loss	$t\,(°C) = \dfrac{1522}{5·75 - \log Co} - 273·15$	$t = 0-250°C$
2. Chalcedony	$t\,(°C) = \dfrac{1032}{4·69 - \log Co} - 273·15$	$t = 0-250°C$
3. α-Cristobalite	$t\,(°C) = \dfrac{1000}{4·78 - \log Co} - 273·15$	$t = 0-250°C$
4. β-Cristobalite	$t\,(°C) = \dfrac{781}{4·51 - \log Co} - 273·15$	$t = 0-250°C$
5. Amorphous silica	$t\,(°C) = \dfrac{731}{4·52 - \log Co} - 273·15$	$t = 0-250°C$
6. Na/K (Fournier)	$t\,(°C) = \dfrac{1217}{\log(Na/K) + 1·483} - 273·15$	$t \geq 150°C$
7. Na/K (Truesdell)	$t\,(°C) = \dfrac{855·6}{\log(Na/K) + 0·8573} - 273·15$	$t \geq 150°C$
8. Na–K–Ca	$t\,(°C) = \dfrac{1647}{\log(Na/K) + \beta[\log(\sqrt{Ca}/Na) + 2·06] + 2·47} - 273·15$	$t \leq 100°C,\ \beta = 4/3$
		$t \geq 100°C,\ \beta = 1/3$
9. $\delta^{18}O(SO_4^{2-}-H_2O)^{b}$	$1000 \ln \alpha = 2·88(10^6 T^{-2}) - 4·1$	

[a] Co = concentration of dissolved silica; all concentrations are in mg/kg.

[b] $\alpha = [1000 + \delta^{18}O(HSO_4^-)]/[1000 + \delta^{18}O(HSO_4^-)]$ and T is absolute temperature (K).

the temperature dependence of the geothermometers mentioned and is taken from R. O. Fournier (1981).[66]

The following observations may be made on these geothermometers.

3.2.3.3 Silica

Originally proposed by G. Bodvarsson in 1960, this was at first purely empirical and qualitative, but later experimental work on the solubility of quartz in water at the vapour pressure of the solution, together with higher-temperature data, produced a theoretical basis for the geothermometer.[69] Subsequently, it was demonstrated that the silica concentration in geothermal waters in New Zealand is controlled by the solubility of quartz and a method was described for using the parameter in hot-spring and well waters in order to make quantitative assessments of reservoir temperatures.[70] This geothermometer is optimum at sub-surface temperatures exceeding 150°C, but, while functioning satisfactorily for hot-spring waters, it may give erroneous results if applied without adequate care.

Various factors must be borne in mind when the silica geothermometer is to be used. Firstly, the correct temperature range must be available (cf. Table 3.2). Secondly, the effects of steam separation must be considered. As water boils the quantity of silica remaining in the residual liquid rises in proportion to the amount of steam separating. Consequently, where boiling springs are being examined, it is necessary to correct for the quantity of steam being formed or the proportion of adiabatic as against conductive cooling.[70] Vigorously boiling springs possessing mass flow rates exceeding 130 kg/min are taken as having cooled mainly adiabatically. Thirdly, it must be remembered that silica may precipitate from a solution before the relevant sample is collected. The geothermometer functions on the basis that the rate of quartz precipitation drops rapidly as the temperature declines. In the range 200–250°C, water attains equilibrium with quartz within a few days at most, but, below 100°C, solutions can stay supersaturated with regard to quartz for many years. Where waters reach the surface from reservoirs having temperatures less than about 225°C and cool fast, i.e. within a few hours, quartz is very likely to precipitate. On the other hand, if the temperature of a reservoir is higher, some quartz may well precipitate in the deep and hot part of the system as cooling of the solution proceeds. Because amorphous silica precipitates at low temperatures and quartz at high ones, the quartz geothermometer when applied to the waters of hot springs almost invariably fails to show temperatures exceeding 250°C, even when such higher reservoir temperatures are known to exist.

Precipitation of silica after sample collection depends upon the following relations. Amorphous silica has a solubility at 25°C of *ca.* 115 mg/kg. The solubility of quartz is 115 mg/kg at *ca.* 145°C. Hence water coming from a reservoir possessing a temperature exceeding 145°C becomes supersaturated with regard to amorphous silica when cooled to 25°C. At high temperatures, all of the dissolved silica in equilibrium with quartz is in the monomeric form. Now, when a cooling solution becomes saturated as regards amorphous silica, some of the normally monomeric silica is converted into highly polymerized dissolved silica species with or without accompanying precipitation of amorphous silica. Fourthly, polymerization may take place either before or after sampling. Fifthly, it is advisable to assess the effect of pH on quartz solubility; the two rise together. This can be calculated from the work of, for example, R. H. Busey and R. E. Mesmer in 1977.[71]

Finally, there is the possibility that hot water may become diluted with cold water before it reaches the surface. Where subterranean mixing of hot and cold waters has taken place, a common event, the dissolved silica concentration of the resultant mixed water can be employed to determine the temperature of the hot-water constituent, the simplest method of calculation involving a plot of dissolved silica against enthalpy of liquid water. Since the combined heat contents of two waters at different temperatures are conserved when they mix, apart from small heat-of-dilution effects, the combined temperatures are not. Thus, although temperature is a measured parameter and enthalpy is derived from steam tables, if the temperature, pressure and salinity are known, enthalpy is preferred as a coordinate rather than temperature. Usually the solutions involved are so dilute that enthalpies of pure water can be utilized in constructing such plots.

3.2.3.4 Sodium–Potassium

That the variation of sodium and potassium in natural geothermal waters is temperature-dependent was observed by many workers.[55,56] The resultant geothermometer is especially applicable to waters arising from high-temperature environments in which $t > 180$ or 200°C. Its principal advantage is that this is less influenced by dilution and steam separation than other geothermometers, providing that there is little Na^+ and K^+ in the diluting water as compared with the water of the reservoir. It is also unaffected by subterranean boiling.

3.2.3.5 Sodium–Potassium–Calcium

This geothermometer was developed in order to handle calcium-rich waters yielding anomalously high calculated temperatures by the sodium–potassium method. The relevant empirical equation showing the variation of temperature with sodium, potassium and calcium is given in Table 3.2, item 8. In employing it the temperature is initially calculated using a value of $\beta = 4/3$ and cation concentrations given as either mg/kg or ppm. When the result is $\leq 100°C$ and $[\log(\sqrt{Ca/Na}) + 2·06]$ is positive, it is not possible to proceed. However, if the $\beta = 4/3$ calculated temperature $\geq 100°C$ or if $[\log(\sqrt{Ca/Na}) + 2·06]$ is negative, the factor $\beta = 1/3$ is applicable and may be utilized to calculate the temperature.

The changes in concentration caused by boiling and also through mixing with cold dilute water affect this geothermometer. The former produce loss of carbon dioxide which can lead to the precipitation of calcium carbonate. The loss of aqueous Ca^{2+} usually causes Na–K–Ca-calculated temperatures to be too high.

Anomalously high results can result also when the geothermometer is applied to waters rich in Mg^{2+}, as R. O. Fournier and R. W. Potter II indicated in 1979.[72] Temperature corrections can be calculated using equations derived by these authors. If $R = [Mg/(Mg + Ca + K)] \times 100$, with concentrations expressed in equivalents, t_{Mg} (°C) is the temperature correction to be subtracted from the Na–K–Ca-calculated temperature and T (K) is the Na–K–Ca-calculated temperature, then, for $5 \leq R \leq 50$,

$$t_{Mg} = 10·66 - 4·7415R + 325·87(\log R)^2 - 1·032 \times 10(\log R)^2/T$$
$$- 1·968 \times 10(\log R)^2/T^2 + 1·605 \times 10(\log R)/T^2 \qquad (3.10)$$

and, for $R \leq 5$,

$$t_{Mg} = -1·03 + 59·971 \log R + 145·05(\log R)^2 - 36711(\log R)^2/T$$
$$- 1·67 \times 10^7 \log R/T^2 \qquad (3.11)$$

In some circumstances, the eqns (3.10) and (3.11) can give negative values for t_{Mg}, in which case no magnesium correction need be applied to the Na–K–Ca geothermometer.

From the foregoing it is apparent that care is needed in applying the Na–K–Ca geothermometer where magnesium-rich waters occur. For instance, in regard to seawater–basalt interactions and genesis of coastal thermal waters in Maharashtra, India, calcium increases during these at elevated temperatures would appear to produce base temperatures computed using this geothermometer which are much lower than those actually prevailing in the relevant geothermal system. This matter is discussed in detail in Chapter 4, Section 4.5.1.2, on the geothermal resources of India.

3.2.3.6 Sulphate–Water $[\delta^{18}O(SO_4^{2-}-H_2O)]$

This geothermometer was originated in 1968 by R. M. Lloyd, who measured the exchange of oxygen-16 and oxygen-18 between sulphate and water at 350°C, and in 1969 and 1972 by Y. Mituzani and T. A. Rafter, who measured the exchange of oxygen-16 and oxygen-18 between water and HSO_4^- at 100–200°C.[73–75]

The agreement found between the 100–200°C HSO_4^- data and the 350°C SO_4^{2-} data implies that no fractionation of oxygen-16 and oxygen-18 between SO_4^{2-} and HSO_4^- occurs.

With the majority of natural waters the pH value is usually such that SO_4^{2-} rather than HSO_4^- is present.

The rate of exchange is strongly dependent upon pH, with

$$\log t_{1/2} = 2500/(T+b)$$

where $t_{1/2}$ is the half-time of the exchange in hours, T is the absolute temperature, and b is 0.28 at pH 9, -1.17 at pH 7 and -2.07 at pH 3.8. Most deep geothermal waters have pH values from 6 to 7 and the time for 90% exchange at 250°C is approximately one year.

The equilibration rate was measured by R. M. Lloyd in 1968.[73] In the majority of pH situations with natural waters, the rates of the sulphate-oxygen isotopic exchange reactions are extremely slow compared with the solubility of silica and cation-exchange reactions; this is advantageous because, after reaching equilibrium subsequent to a long residence time in a reservoir at high temperature, there will be very little re-equilibration of the oxygen isotopes of sulphate as the water cools down in its ascent to the surface, unless this process is extremely slow. However, if there is steam separation during cooling, there will be a change in the oxygen isotope composition. Fractionation of the oxygen-16 and oxygen-18 between liquid water and steam is temperature-dependent and attains equilibrium practically at once when the temperature drops to 100°C. Hence, where a water cools adiabatically from a higher temperature to this temperature, that liquid water residual when boiling ends possesses a different isotopic composition depending upon whether steam remained in contact with it throughout, separating only at the final temperature (100°C), or was continuously removed over a range of temperature.

From this argument it is clear that boiling complicates matters, but it does not prevent the sulphate–oxygen geothermometer from being employed. It has been shown that temperatures derived from it can be calculated from several end-member models, namely conductive cooling, one-step steam loss at any specified temperature and continuous steam loss.[68] Obviously, in the case where water is obtained from a well and steam separated at a known temperature or pressure, the correct model to utilize can be selected.

The effects of subterranean boiling are significant in that they entail partitioning of dissolved elements between the steam and the residual liquid, dissolved gases and other volatiles concentrating in the former and the non-volatiles in the latter. For the non-volatile constituents remaining with residual liquid as steam separates, the final concentration, C_f, after single-stage steam separation at a given temperature, t_f, is given by the expression

$$C_f = [(H_s - H_f)/(H_s - H_i)]C_i \qquad (3.12)$$

where C_i is the initial concentration prior to boiling, H_i is the enthalpy of the initial liquid prior to boiling, and H_f and H_s are the enthalpies of the final liquid and steam at t_f. For solutions with salinities under *ca.* 10 000 mg/kg, enthalpies of pure water tabulated in steam tables can be employed to solve eqn (3.12).

The result of subterranean boiling due to decreasing hydrostatic head on different geothermometers may be summarized. The sodium–potassium one is unaffected, but the sodium–potassium–calcium one is the case where loss of carbon dioxide produces precipitation of calcium carbonate and causes calculated temperatures to be too high. The $\delta^{18}O(SO_4^{2-}-H_2O)$ geothermometer must be corrected also for steam loss because of the temperature-dependent fractionation of oxygen isotopes between liquid water and steam.

One disadvantage of the sulphate–water geothermometer is that temperatures calculated using it can be invalidated by mixing of hot or cold waters. However, corrections can be made for the changes in the isotopic composition of both the sulphate and the water resulting from such mixing. Even when no sulphate occurs within the cold constituent of the mixture, the calculated sulphate–oxygen isotopic temperature will be erroneous unless the isotopic composition of the water is corrected back to the composition of the water in the hot constituent before mixing took place. Some instances of calculated corrections were given by McKenzie and Truesdell in 1977 and by Fournier and his colleagues in 1979 for waters from Yellowstone National Park and Long Valley in California.[68,76]

3.2.3.7 Other Isotope Geothermometers
Originally, it was assumed that the heat and most of the water entered geothermal systems as magmatic steam, an hypothesis which could not be

tested until the development of isotopic methods allowed comparison to be made between the hydrogen and oxygen isotope compositions of thermal waters and possible source waters. Variations in isotope composition of surface waters result from fractionation effects. Fractionation of oxygen-18 and deuterium into condensed phases occurs at low temperatures; clouds arising from the evaporation of tropical seawater progressively deplete in the heavier stable isotopes with precipitation during movement towards the poles, inland over land masses and to high elevations. The last two are known as the continental and altitude effects. Freshwater compositions lie along the well-known meteoric water line of H. Craig (1961), described by

$$\delta D\,(\permil) = 8\delta^{18}O + d \qquad (3.13)$$

where d is the deuterium excess (normally 10 but higher in some coastal areas and lower in dryer inland areas).[77]

It should be noted that the 'per mil' (parts per thousand, $\permil$) notation is used, differences from isotopic standards being expressed as $\delta\,(\permil) = [(R_{Sa}/R_S) - 1)] \times 1000$, where R_{Sa} = oxygen-18/oxygen-16, D/H, etc., in the unknown, and R_S represents the same ratios in a standard. The standard for water is standard mean ocean water (SMOW), for carbon the Peedee belemnite (PDB), and for sulphur the Canyon Diablo troilite.

(a) Hydrogen isotope geothermometers. A minimum of five gases containing hydrogen are found in large quantities in geothermal fluids: these are water, hydrogen sulphide, hydrogen, methane and ammonia. It is probable that four independent isotopic geothermometers based upon hydrogen–deuterium fractionation are available.

The relevant fractionation factors were given by Y. Bottinga in 1969 and by P. Richet and his associates in 1977.[78,79] Those for H_2–H_2S and H_2–H_2O are in good agreement with the restricted experimental data. The agreement for the system H_2–CH_4 is not so good and there is practically no experimental information available yet for ammonia. The D/H ratios for hydrogen sulphide and ammonia have not been adequately tested as possible geothermometers, but some measurements have been made of D/H ratios for coexistent hydrogen, water and methane in geothermal fluids. Of course, in geothermometers including water, the isotope fractionation consequent upon boiling must be considered and the water composition at the point of equilibrium either calculated or measured. Measurement of both the amounts and isotope compositions of steam and water from a separator from hot-water systems is feasible, and it is also possible to calculate the composition of the hot water which supplies the drill hole.

The hydrogen–water geothermometer is based upon the assumption that hydrogen equilibrates with liquid water. It was applied in Iceland by B. Arnason in 1977 and he obtained good agreement with silica geothermometric measurements, also with the measured downhole temperatures in most areas.[80] However, at Reykjanes, the isotopic temperature exceeds other temperatures by 70–90°C, probably because of incoming contributions from a deeper and higher temperature source. Other applications of the geothermometer seem to point to variable degrees of re-equilibration between the sampling point and the reservoir. Indicated temperatures include 115 and 122°C for fumaroles in the Yellowstone National Park in Wyoming, USA.[81] The Salton Sea wells gave temperatures of 140 and 220°C, according to H. Craig (unpublished data of 1975), cited by A. H. Truesdell and J. R. Hulston in 1980.[82] In both of these cases, the values are 120–240°C lower than the measured or estimated aquifer temperatures.

The water–steam geothermometer depends upon the fractionation of both hydrogen and oxygen isotopes between liquid water and steam, and is a valuable instrument in those rather rare geothermal systems in which both steam and water can be separately sampled and their isotopic differences compared with experimental data. Even in such systems, however, there may be incomplete or multi-stage separation of the two; hence the differences in oxygen-18 or deuterium content of the two phases may be greater or smaller than the single-stage enrichment factors and divergent temperatures may be given by the two isotopes. The approach has been utilized at The Geysers, where reservoir steam δD and $\delta^{18}O$ separately gave temperatures of 240 and 260°C, respectively. The temperature derived from $\delta D/\delta^{18}O$ was 247°C, exactly agreeing with the observed well-bottom temperature. Of course, The Geysers is a rather extreme instance because at this site only steam is produced since the reservoir water is immobile and vaporizes totally in response to production-induced pressure decreases.

The methane–hydrogen geothermometer is another potential tool. Isotope exchange between the two components in the temperature range 200–500°C was obtained in the laboratory using a catalyst. It is believed that, in geothermal systems, the hydrogen–water isotopic exchange is probably faster. Since the equilibrium constants for both of these reactions are similar, it is to be expected that derived temperatures will be similar. They are indeed so: Yellowstone National Park fumaroles yielding values around 70°C and wells at the Salton Sea giving 125 and 255°C. The isotopic exchange between methane and water could provide a useful geothermometer, but not enough experimental work has been done on it as yet and the theoretical calculations are ambiguous.[79]

(b) Oxygen isotope geothermometer. Both hydrogen and oxygen constitute the water molecule, which, therefore, contains a large reserve of both elements. Boiling during ascent of geothermal fluids promotes isotopic fractionation. The oxygen-18 is more highly fractionated than deuterium by steam separation taking place above 200°C and is not much affected by reaction with rocks including abundant oxygen and relatively little hydrogen. Despite these problems, a number of oxygen isotope geothermometers appears to be rather promising. In fact, oxygen-18 has been given a particular prominence among isotope tracers for many years.

As regards oxygen isotopic fractionation among minerals, appropriate fractionation equations were given by Y. Bottinga and M. Javoy in 1973, 1975 and 1977.[83-85] In 1987 these authors stated that oxygen isotope exchange between quartz and water under hydrothermal conditions is at least a two-step process, but the transition between these processes appears not to take place simultaneously for oxygen-17, oxygen-16 and oxygen-18/oxygen-16 exchange.[86] Of all naturally occurring minerals, the oxides of iron seem to be the most depleted in heavy oxygen isotopes and hence present potentially very sensitive oxygen isotope geothermometers when paired with oxygen-18-enriched minerals such as quartz. While laboratory data on the oxygen isotope fractionation between magnetite and water are available only for temperatures ≥ 300°C, a calculated curve exists for the entire geological temperature range. This shows an isotopic cross-over at ≈ -20°C. At Wairakei magnetite grows in the pipelines of the geothermal power station, a serious corrosion process taking place under well-known conditions. The millimetre-deep surface layers of the mineral gave the following oxygen isotope calibration points of magnetite and water below the previously available experimental range: (175°C) -7.9 ± 0.5 and (112°C) -3.7 ± 0.5. This information was given by P. Blattner and his colleagues in 1983 and supports the earliest proposed magnetite–water fractionation curve, with an isotopic cross-over near 50°C.[87] It could indicate a relatively low-temperature origin for some Precambrian iron formations.

(c) Carbon dioxide–water geothermometer. This is based upon the exchange of oxygen isotopes between carbon dioxide and water vapour. It has been applied to show reservoir temperatures at Larderello in Italy by C.

Panichi and others in 1977.[88] The exchange is thought to occur through the activated complex HCO_3^- which exists solely in the presence of liquid water. Consequently, exchange would take place only in the reservoir and not in the ascending column of steam. The temperatures given by CO_2–$H_2O(g)$ fractionation data are equal to or exceed the wellhead steam temperatures. It is possible that the use of fractionation data for CO_2–$H_2O(g)$ may be incorrect since oxygen isotope exchange takes place between carbon dioxide and liquid water in the reservoir and not between carbon dioxide and water vapour in the steam. The employment of fractionation data for CO_2–$H_2O(l)$ eliminates a good deal of the apparent difference between the indicated and the wellhead temperatures.

3.2.3.8 Isotopic Dating of Geothermal Waters

It has been shown from deuterium and oxygen isotope analyses that the water of most geothermal systems originates as local precipitation which obtains heat through deep circulation. However, such stable isotopes are unable to indicate the duration of this phenomenon. This would be desirable in assessment of the total energy content of the systems. In order to date them, appropriate radioisotopes must be used, namely tritium and carbon-14.

Tritium is the radioactive isotope of hydrogen, ^3H (often written simply as T), and it has a half-life of approximately 12·3 years. It is continuously produced in the upper atmosphere by cosmic ray neutrons interacting with nitrogen-14, thus

$$(^{14}_{7}N, ^{1}_{0}n)(^{3}_{1}H, ^{12}_{6}C)$$

Its natural abundance is from 5 to 10 atoms per 10^{18} atoms of hydrogen, i.e. 5 to 10 tritium units (TU). Atmospheric testing of thermonuclear weapons during the period 1954 to 1965 injected pulses of the radioisotope and the levels rose by up to three orders of magnitude in the northern hemisphere and five-fold in the southern hemisphere, remaining somewhat elevated to this day. Once formed tritium is rapidly incorporated into the water molecule and is removed from the atmosphere in meteoric precipitation. The residence time of tritiated water in the lower stratosphere is between one and ten years, but once it has reached the lower troposphere tritiated water rains out in five to 20 days.

Tritium has labelled sub-surface waters of which the circulation times do not exceed approximately 40 years. Unfortunately, most geothermal waters have much longer circulation times than that, and so contain such small quantities of the radioisotope that these cannot be measured, being less than the detection limit of *ca*. 0·1 TU. Such a tritium level merely indicates that the water containing it was out of contact with the atmosphere for more than half a century. However, tritium analyses proved valuable at Larderello, where recharge occurs through permeable reservoir rock outcrops south and east of the geothermal field. Wells near the infiltration areas are usually unproductive because of surface water flooding. Wells somewhat further away yield steam and this has been found to contain 40 TU. By contrast, steam from wells in the centre of the field contains under 0·5 TU, as R. Celati and his colleagues indicated in 1973.[89] The pattern of tritium occurrence implies lateral recharge, but contributions of the radioisotope from above the reservoir must be taken into account as well because the tritium-containing wells are among the oldest and shallowest in this geothermal field. Tritium also occurs in hot springs and can indicate the mixing of deep, usually tritium-poor, thermal waters with shallow waters.

The value of T, indicating the extent of mixing, relates to the fact that hot springs giving large flows may stem from mixing of hot and cold waters or from conductive heating during the process of deep circulation followed by little heat loss during ascent. If only the latter mechanism is involved, then drilling will yield water which is only a little hotter than that emerging at the surface. On the other hand, if the temperature of the hot spring is the result of mixing, then such drilling may well give high-temperature fluids suitable for the production of power. Tritium investigations have been made on hot springs in the eastern USA by W. A. Hobba and his fellow-workers, as they reported in 1978, and in the Dead Sea Rift in Israel by E. Mazor, who described his work there in 1977.[90,91]

Carbon-14 (radiocarbon) is produced in the upper atmosphere by a variety of nuclear reactions based upon the interaction of cosmic ray-produced neutrons with the stable isotopes of nitrogen, oxygen and carbon. The most important of these is that between slow cosmic-ray neutrons and the nuclei of stable nitrogen-14. This can be written as follows:

$$({}^{14}_{7}N, {}^{1}_{0}n)({}^{14}_{6}C, {}^{1}_{1}H)$$

Afterwards the atoms of radiocarbon are incorporated into carbon dioxide molecules by reactions with oxygen or by exchange reactions with stable carbon isotopes in molecules of carbon monoxide or carbon dioxide. The molecules of carbon dioxide containing carbon-14 are mixed rapidly throughout the atmosphere and the hydrosphere and attain constant levels of concentration representing a steady-state equilibrium. When the carbon dioxide combines with water to form carbonic acid, this may enter the groundwater system or a geothermal system. Thereafter the radiocarbon decays by emission of a negative β-particle and forms stable nitrogen-14, thus

$$^{14}_{6}C \rightarrow \, ^{14}_{7}N + \beta^- + \gamma + Q$$

and the end-point energy (Q) is 0·156 MeV. The decay is directly to the ground state of nitrogen-14 and no γ-ray is emitted.

Radiocarbon, carbon-14, has a half-life of 5730 years and therefore could be valuable in dating geothermal fields were it not for the fact that it becomes very diluted through the entry of old radiocarbon-free carbon dioxide into the geothermal fluid and carbonate within the reservoir. In such systems, this radiocarbon-free carbon dioxide may arise from the thermal metamorphism of limestones or perhaps even directly from magma; either way it can easily dilute the relatively small concentrations of carbon dioxide in the recharge waters. Also a good deal of the metamorphic carbon dioxide deposits in the reservoir as hydrothermal calcite, which is then available for isotopic exchange with radiocarbon-bearing carbon dioxide. At temperatures exceeding 200°C, the isotopic exchange between calcite and dissolved carbon dioxide takes place rapidly and the introduced radiocarbon mixes with this large reservoir of old carbon as well. In view of all this, it is hardly surprising that deep geothermal fluids have been found to possess very little radiocarbon. Thus H. Craig in 1963 stated that borehole fluids from The Geysers, Steamboat Springs and the Salton Sea contained under 0·5% modern carbon (pmc), and a similar level was reported previously from a deep borehole at Wairakei by G. J. Fergusson and F. B. Knox in 1959.[92,93]

Possibly other radioisotopes could be utilized in dating geothermal fluids. One such is cosmogenic argon-39, which is cosmic-ray-produced in the upper atmosphere and has a half-life of 269 years. Its concentration in the atmosphere is 0·112 ± 0·010 dpm/litre of argon and of this about 5% was injected by thermonuclear weapons testing. It is chemically inert and its concentration in groundwater and geothermal waters depends upon the time elapsed since such waters were last in contact with the atmosphere, its main source (because it does not interact with solids, being a noble gas radionuclide). Analyses entail collecting very large samples of these waters; also there is always the possibility that excess argon-39 may be produced by the following reaction in rocks which contain appreciable quantities of uranium and thorium:

$$^{39}K(n, p)^{39}Ar$$

Even if the age of a geothermal water could be assessed, this would be difficult to relate to the volume of the reservoir, which is usually too vast to correspond with any simplistic hydrological model. Other noble gases have been proposed for such dating work in regard to geothermal waters. For instance, F. D'Amore in 1975 made a radon-222 survey at Larderello, H. Craig and J. E. Lupton in 1976 examined primordial neon, helium and hydrogen in oceanic basalts, and E. Mazor in 1976 studied the potential applications of atmospheric and radiogenic noble gases in geothermal waters to prospection and steam production.[94-96] The common isotopes of neon, argon, krypton, xenon and usually nitrogen dissolved in geothermal waters originate almost entirely from the atmospheric gas content of infiltrating recharge waters, as E. Mazor indicated in 1976.[96] In those geothermal waters which have not boiled below ground, i.e. which have behaved as closed systems to inert gases, the actual contents of such gases will depend only upon the temperature of equilibration of the water with the atmosphere at the point of recharge. In the case where boiling took place, the residual water is depleted in these gases and the depletion pattern depends upon the temperature and the extent of boiling.

Some of the isotopes of the noble gases, e.g. helium-4, argon-40, xenon-136 and radon-222, are produced by the radioactive decay of uranium, thorium and potassium-40 in rocks and indicate the contents of radioactive elements in a geothermal system and also the water:rock ratio. The average content of radiogenic argon in the steam at Larderello decreased from 23 to 12% during the period 1951 to 1963, according to G. Ferrara and others in 1963.[97] This implies that the atmospheric argon is introduced with increased recharge caused by pressure drawdown.

Radon-222 has a very short half-life of 3·8 days and has been tested as a radiotracer in geothermal studies on the reservoirs of vapour-dominated systems, for instance by A. K. Stoker and P. Kruger in 1976.[98] Radium-226, a daughter of uranium-238, decays to produce radon, but is not itself soluble in steam; therefore the radon concentration must be related to the areas of rock and water surface in contact with steam during the last few days or so prior to its reaching the surface. P. Kruger and his associates in 1977 noticed that, when the flow of a well at The Geysers was lowered from 90 to 45 tons/h, the radon concentration decreased from an average of 16 to 8 nCi/kg.[99] This could be an indication that, during its passage to the well, the steam moves through large fractures which have a low surface:volume ratio for time intervals similar to the half-life of radon. At Larderello, F. D'Amore demonstrated that the ratio of radon to total dry gas is greatest at the peripheries of the geothermal field.[94] Such rather large radon:total gas ratios could well indicate heavily produced zones which contain steam boiling from gas-depleted water into huge, dried volumes in which radon emanates from the rock surfaces. In vapour-dominated geothermal reservoirs, radon could prove to be a valuable tool. Since uranium-238 produces radon-222 through radium-226 and resembles calcium chemically, the two may co-precipitate in spots where carbon dioxide is lost at the surface or in zones of subterranean boiling. Consequently, radon in hot-water systems is associated with silicic rocks of rather high uranium content and with travertine-depositing springs. However, there are almost no systematic data regarding the distribution of radon in hot-water reservoirs or its concentration in boreholes as a function of flow.

Helium-3 is not formed through radioactive decay of rock material, apart from minute quantities resulting from tritium decay, and is found in the atmosphere in quantities much smaller than that of helium-4, the ratio usually being $^3He/^4He \approx 1.4 \times 10^{-6}$. It occurs more abundantly in geothermal systems and in ocean waters near spreading centres, in the latter case probably arising from the outgassing of primordial mantle. The ratio in Lassen, the Salton Sea and Hawaii ranges from 3 to 16 times that of the atmosphere.[94] High ratios have also been recorded in geothermal fluids in Iceland by V. I. Kononov and B. G. Polak in 1976.[100] The presence of such anomalous helium-3/helium-4 ratios in volcanic rocks in Iceland and East Africa as well as in oceanic basalt implies that the observed helium-3/helium-4 ratios of geothermal fluids may just as well have been caused by rock leaching as by direct contributions from the mantle of the Earth.

3.2.4 Drilling

The investigatory methods described above contribute to the correct siting of exploratory boreholes and the drilling of the first well facilitates geographical delimitation of the geothermal field and also shows whether predictions about its capacity were accurate or not. Since this work is costly, an attempt is usually made to locate it at a place from which a demonstration can be made that sufficient steam or water can be produced to justify exploitation. Should this not be obtained, some hard thought is necessary. However, if it is successive wells are drilled in order to confirm the findings of the first one and to discover the most likely places for getting results.

The approach may be simply to treat the first well as a potential future production well. This could be more economic, because it was shown in 1976 by S. Arnósson and his fellow-workers that narrow boreholes deliberately orientated towards exploration can only cause problems.[101] For instance, the duration of drilling was longer since the penetration rates at the top of the hole were slower because the shorter and lighter drill stem did not permit the drill bit to be fully loaded, and there were also losses of the drilling fluid. In addition, the operation was not as cheap as had been anticipated.

In the course of exploratory drilling, information may be derived from three factors, namely (1) the cuttings and cores collected, (2) the temperature of the returned lubricating mud, and (3) mud flow rates and the circulation losses from which permeability can be assessed. Coring is often a routine practice, carried out at regular intervals and followed by mineralogical analysis. When drilling has ceased and mud has been flushed out, the permeability is tested by simply measuring the pressure, either at the wellhead or at a fixed depth while cold water is introduced over a broad range of flow rates. If pressure rises rapidly as the flow rate increases, then poor permeability is evident. If it remains constant, this indicates high permeability.

Finding out where the permeable zone is can be achieved from down-hole temperatures recorded in the course of the cold-water injection. Another approach is to measure temperatures and pressures in the well over a month or so while it is heating after the injection test in a static condition. Such measurements will demonstrate, too, whether the well has penetrated a zone of boiling or single-phase water or if both steam and gas occur in any appreciable quantities.[54] Of course, the down-hole pressures may not represent conditions which were obtained before drilling commenced, since the well may well provide an additional link between aquifers and circulation may occur within a static well. When such a phenomenon takes place, the effect is manifested in a constant temperature over the interval where flow occurs or it may be observed in down-hole flow measurements.

The output of a well after it is heated can be made by means of mass flow and heat measurements over various discharging pressures. In the case of those wells which discharge only a single phase such as steam, these can be made without difficulty. The problem is complicated when two-phase mixtures are involved. There are appropriate techniques available such as the lip-pressure method or calorimetry.[102,103] The measurement of pressure drawdown while fluids are being discharged from a well can indicate permeability near it and sometimes even distinguish fissure from porous bed permeability. When drilling is undertaken, evaluation of geophysical and geochemical results is feasible and often these can be refined. Obviously, knowledge of the down-hole temperatures facilitates a reappraisal of both the surface heat flow and the temperature gradient.

Once discharge has taken place and drilling mud and cold water from completion tests have been removed, chemical analyses of waters which have not been contaminated by shallow groundwater become possible, samples being collected at the surface if a well is flowing or at chosen depths by means of a down-hole sampling bottle. However, it is not yet possible to utilize the familiar well-logging methods such as electric, neutron–gamma, density and sonic techniques without cooling down the well.

When enough data have been accumulated, a model of the geothermal field may be constructed. This matter has been discussed in detail in Chapter 2, in which various types such as pipe models and large-scale convection cell models were described. Here, therefore, it is necessary to make only a brief allusion to the subject. In any particular geothermal field, the first model will be simple and probably based upon analogy with others which are better understood. However, it will become more complex as geophysical, geochemical, geothermometric and isotope surveys increase knowledge of the geothermal field. Drilling will be succeeded by output testing. Thereafter predictive modelling may facilitate the locating of future wells as well as estimating the capacity of the geothermal field. Well measurements of both temperature and pressure can be combined with drawdown, transient and interference tests to supply information from which models showing reservoir characteristics can be derived. Over a longer time interval, the rundown features of wells can be assessed and occasionally the interaction between geothermal waters and cold groundwater can be evaluated in addition.

3.2.5 Geothermal Energy Recovery using Oil Exploration Data

The potential for geothermal energy recovery using petroleum exploration data has been discussed by H.-L. Lam and F. W. Jones in 1986 in regard to the Calgary area of southern Alberta, Canada.[104] The sedimentary basin which underlies most of this province possesses large quantities of low-grade geothermal energy contained in hot water of porous and permeable rocks. So far its existence has been somewhat neglected because of the sparcity of population, but Calgary, as a city of over half a million, constitutes a potentially large market for geothermal energy. This is despite the fact that it is located in an area of normal geothermal gradient of approximately 24°C/km. However, it lies on the flank of a deep part of the Alberta Basin where sediments reach a depth of over 4000 m. This implies that a large temperature range is present. There had been earlier studies of petroleum exploration data for geothermal purposes in the western Canadian sedimentary basin, e.g. by Sproule Associates Limited in 1977.[105] This one followed the same approach and represents a continuation of an earlier investigation of geothermal potential in the Hinton–Edson area of west–central Alberta by H.-L. Lam and F. W. Jones, of which details were published in 1984 and 1985.[106,107] Petroleum exploration data were used as a database and included bottom-hole temperatures, lithological logs, drill stem tests and water analyses.

Good aquifers were shown to exist in the carbonate rocks of the Elkton Formation of the Mississippian and the Wabamun Formation of the Late Devonian. The water temperatures range from 60 to 90°C, with salinities of 70 000–100 000 mg/litre, two to three times greater than that of average seawater. High water flow rates of up to 700 m³/h from the Elkton Formation at moderate depths are obtainable in areas northwest and south of the city. Although the flow rates for the Wabamun are lower, it and the deeper Leduc Formations could supplement the Elkton as the main target for geothermal purposes. During the work, available petroleum exploration data were synthesized in order to identify these water-bearing formations, their permeabilities and both the water chemistry and temperatures in the Calgary area. The availability of hot water of reasonable quality at reasonable depth offers the city useful recoverable geothermal energy.

REFERENCES

1. SMITH, M. L., 1978. Heat extraction from hot, dry crustal rock. *Pure Appl. Geophys.*, **117**, 290–6.
2. MUFFLER, L. J. P. (ed.), 1979. *Assessment of Geothermal Resources of the United States—1978*, US Geol. Surv. Circ. No. 790, 163 pp.
3. LUMB, J. T., 1981. Prospecting for geothermal resources. In: *Geothermal Systems: Principles and Case Histories*, ed. L. Rybach and L. J. P. Muffler, pp. 77–103. John Wiley & Sons, Chichester, 359 pp.
4. BROWNE, P. R. L., 1970. Hydrothermal alteration as an aid in investigating geothermal fields. *Geothermics*, Sp. issue **2**(2), 564–70.
5. GRINDLEY, G. W. and BROWNE, P. R. L., 1976. Structural and hydrologic factors controlling the permeabilities of some hot-water geothermal fields. *Proc. 2nd UN Symp.*, Part 1, pp. 377–86. US Govt Printing Office, Washington, DC, USA.
6. HATHERTON, T., MACDONALD, W. J. P. and THOMPSON, G. E. K., 1966. Geophysical methods in geothermal prospecting in New Zealand. *Bull. Volcanologique*, **29**, 485–98.
7. DICKINSON, D. J., 1976. An airborne infrared survey of the Tauhara geothermal field, New Zealand. *Proc. 2nd UN Symp.*, Part 2, pp. 955–61. US Govt Printing Office, Washington, DC, USA.
8. GOSS, R. and COMBS, J., 1976. Thermal conductivity measurements and

prediction from geophysical well-log parameters with borehole application. *Proc. 2nd UN Symp.*, Part 2, pp. 1019–27. US Govt Printing Office, Washington, DC, USA.

9. ECKSTEIN, Y., 1979. Heat flow and the hydrologic cycle: example from Israel. In: *Terrestrial Heat Flow in Europe*, ed. V. Cermak and L. Rybach. Springer-Verlag, Berlin.

10. NOBLE, J. W. and OJIAMBO, S. B., 1976. Geothermal exploration in Kenya. *Proc. 2nd UN Symp.*, Part 1, pp. 189–204. US Govt Printing Office, Washington, DC, USA.

11. MEIDAV, T., 1970. Application of electrical resistivity and gravimetry in deep geothermal exploration. *Geothermics*, Sp. issue **2**(2), 203–310.

12. ZOHDY, A. A. R., ANDERSON, L. A. and MUFFLER, L. J. P., 1973. Resistivity, self-potential, and induced polarization surveys of a vapor-dominated geothermal system. *Geophysics*, **38**, 1130–44.

13. DEY, A. and MORRISON, H. F., 1977. An analysis of the bipole–dipole method of resistivity surveying. *Geothermics*, **6**, 47–81.

14. RISK, G. F., 1976. Detection of buried zones of fissured rock in geothermal fields using resistivity anisotropy measurements. *Proc. 2nd UN Symp.*, Part 2, pp. 1191–8. US Govt Printing Office, Washington, DC, USA.

15. JIRACEK, G. R. and GERETY, M. T., 1978. Comparison of surface and downhole resistivity mapping of geothermal reservoirs in New Mexico. *Geothermal Resources Council Trans.*, **2**, 335–6.

16. BIBBY, H. M., 1978. Direct current resistivity modelling for axially symmetric bodies using the finite element method. *Geophysics*, **43**, 550–62.

17. JIRACEK, G. R., SMITH, C. and DORN, C. A., 1976. Deep geothermal exploration in New Mexico using electrical resistivity. *Proc. 2nd UN Symp.*, Part 2, pp. 1095–102. US Govt Printing Office, Washington, DC, USA.

18. HOHMANN, G. W., FOX, R. C. and RIJO, L., 1978. Topographic effects in resistivity surveys. *Geothermal Resources Council Trans.*, **2**, 287–90.

19. ROSS, H. P. and WARD, S. H., 1984. *Borehole Electrical Geophysical Methods:* a review of the state-of-the-art and preliminary evaluation of the application of fracture mapping in geothermal systems. University of Utah Research Institute, Earth Sciences Lab. Rept DOE/SAN/12196-2.

20. YANG, F. W. and WARD, S. H., 1985. Single-borehole and cross-borehole resistivity anomalies of thin ellipsoids and spheroids. *Geophysics*, **50**, 637–55.

21. ZHAO, J. X., RIJO, L. and WARD, S. H., 1986. Effects of geologic noise on cross-borehole electrical surveys. *Geophysics*, **51**(10), 1978–91.

22. ANDERSON, L. A. and JOHNSON, G. R., 1978. Some observations of the self-potential effect in geothermal areas in Hawaii and Nevada. *Geothermal Resources Council Trans.*, **2**, 9–12.

23. ERNSTSON, K. and SCHERER, H. U., 1986. Self-potential variations with time and their relation to hydrogeologic and meteorological parameters. *Geophysics*, **51**(10), 1967–77.

24. LUMB, J. T. and MACDONALD, W. J. P., 1970. Near-surface resistivity surveys of geothermal areas using the electromagnetic method. *Geothermics*, Sp. issue **2**(2), 311–17.

25. KELLER, G. V., 1970. Induction methods in prospecting for hot water. *Geothermics*, Sp. issue **2**(2), 318–32.

26. MORRISON, H. F., GOLDSTEIN, N. E., HOVERSTEN, M., OPPLIGER, G. and RIVEROS, C., 1978. Controlled-source electromagnetic system. In: *Geothermal Exploration Technology: Annual Rept 1978, Lawrence Berkeley Laboratory, University of California*, pp. 9–12.

27. GOLDMAN, Y., HUBANS, C., NICOLETIS, S. and SPITZ, S., 1986. A finite-element solution for the transient electromagnetic response of an arbitrary two-dimensional resistivity distribution. *Geophysics*, **51**(7), 1450–61.

28. SHENG, Y., 1986. A single apparent resistivity expression for long-offset transient electromagnetics. *Geophysics*, **51**(6), 1291–7.

29. SPIES, B. R. and EGGERS, D. E., 1986. The use and misuse of apparent resistivity in electromagnetic methods. *Geophysics*, **51**(7), 1462–71.

30. MÖRNER, N.-A., 1986. The lithospheric geomagnetic field: origin and dynamics of long wavelength anomalies. *Physics of the Earth and Planetary Interiors*, **44**, 36–72.

31. BEAMISH, D., 1986. Geoelectric structural dimensions from magnetotelluric data: methods of estimation, old and new. *Geophysics*, **51**(6), 1298–309.

32. JIRACEK, G. R., RODI, W. L. and VANYAN, L. L., 1987. Implications of magnetotelluric modeling for the deep crustal environment in the Rio Grande Rift. *Physics of the Earth and Planetary Interiors*, **45**, 179–92.

33. BEYER, H., MORRISON, H. F. and DEY, A., 1976. Electrical exploration of geothermal systems in the Basin and Range valleys of Nevada. *Proc. 2nd UN Symp.*, Part 2, pp. 889–94. US Govt Printing Office, Washington, DC, USA.

34. SWANBERG, C. A., 1976. The Mesa geothermal anomaly, Imperial Valley, California; a comparison and evaluation of results obtained from surface geophysics and deep drilling. *Proc. 2nd UN Symp.*, Part 2, pp. 1217–29. US Govt Printing Office, Washington, DC, USA.

35. BABA, K., 1976. Gravimetric survey of geothermal areas in Kurikoma and elsewhere in Japan. *Proc. 2nd UN Symp.*, Part 2, pp. 865–70. US Govt Printing Office, Washington, DC, USA.

36. ISHERWOOD, W. F., 1976. Gravity and magnetic studies of the The Geysers–Clear Lake geothermal region, California, USA. *Proc. 2nd UN Symp.*, Part 2, pp. 1065–73. US Govt Printing Office, Washington, DC, USA.

37. HOCHSTEIN, M. P. and HUNT, T. M., 1970. Seismic, gravity and magnetic studies, Broadlands geothermal field. *Geothermics*, Sp. issue **2**(2), 333–46.

38. HUNT, T. M., 1977. Recharge of water in Wairakei geothermal field determined from repeat gravity measurements. *NZ J. Geol. Geophys.*, **20**, 303–17.

39. ALLIS, R. G. and HUNT, T. M., 1986. Analysis of exploitation-induced gravity changes at Wairakei geothermal field. *Geophysics*, **51**(8), 1647–60.

40. PÁLMASON, G., 1976. Geophysical methods in geothermal exploration. *Proc. 2nd UN Symp.*, Part 2, pp. 1175–84. US Govt Printing Office, Washington, DC, USA.

41. RITZ, M. and VASSAL, J., 1986. Geoelectrical structure of the northern part of the Senegal basin from joint interpretation of magnetotelluric and geomagnetic data. *J. Geophys. Res.*, **91**(B10), 10443–56.

42. RITZ, M. and VASSAL, J., 1987. Geoelectromagnetic measurements across the southern Senegal basin (West Africa). *Physics of the Earth and Planetary Interiors*, **45**, 75–84.

43. BROZENA, J. M., 1984. A preliminary analysis of the NRL airborne gravity system. *Geophysics*, **49**, 1060–9.
44. ECKHARDT, D. H., 1986. Isomorphic geodetic and electrical networks: an application to the analysis of airborne gravity gradiometer survey data. *Geophysics*, **51**(11), 2145–55.
45. JEKELI, C., WHITE, J. V. and GOLDSTEIN, J. D., 1985. A review of data processing in gravity gradiometry. Paper presented at 3rd Int. Symp. on Inertial Techniques for Surveying and Geodesy, Banff.
46. VALLIANT, H. D., GAGNON, C. and HALFPENNY, J. F., 1986. An inherently linear electrostatic feedback method for gravity meters. *J. Geophys. Res.*, **91**(B10), 10463–9.
47. WHITEFORD, P. C., 1976. Studies of the propagation and source location of geothermal seismic noise. *Proc. 2nd UN Symp.*, Part 2, pp. 1263–71. US Govt Printing Office, Washington, DC, USA.
48. IYER, H. M. and HITCHCOCK, T., 1976. Seismic noise measurements in Yellowstone National Park. *Geophysics*, **39**, 389–400.
49. MCEVILLY, T. V., SCHLECHTER, B. and MAYER, E. L., 1978. East Mesa seismic study. In: *Geothermal Exploration Technology: Annual Rept 1978, Lawrence Berkeley Laboratory, University of California, USA*, pp. 23–5.
50. EVISON, F. F., ROBINSON, R. and ARABASZ, W. J., 1976. Microearthquakes, geothermal activity and structure, central North Island, New Zealand. *NZ J. Geol. Geophys.*, **19**, 625–37.
51. BHATTACHARYA, S. N., PARKASH, C. and SRIVASTAVA, H. N., 1986. Microearthquake observations around Thien Dam in northwest Himalayas. *Physics of the Earth and Planetary Interiors*, **44**, 169–78.
52. YUANZHANG, D., ANYU, X. and YIMING, C., 1986. Induced earthquakes in Zhelin reservoir, China. *Physics of the Earth and Planetary Interiors*, **44**, 107–14.
53. PATIL, D. N., BHOSALE, V. N., GUHA, S. K. and POWAR, K. B., 1986. Reservoir induced seismicity in the vicinity of Lake Bhatsa, Maharashtra, India. *Physics of the Earth and Planetary Interiors*, **44**, 73–81.
54. STEEPLES, D. W. and IYER, H. M., 1976. Teleseismic P-wave delays in geothermal exploration. *Proc. 2nd UN Symp.*, Part 2, pp. 1199–206. US Govt Printing Office, Washington, DC, USA.
55. ELLIS, A. J. and MAHON, W. A. J., 1977. *Chemistry and Geothermal Systems*. Academic Press, New York, 392 pp.
56. FOURNIER, R. O. and TRUESDELL, A. H., 1973. An empirical Na–K–Ca geothermometer for natural waters. *Geochim. Cosmochim. Acta*, **37**, 1255–75.
57. WEISSBERG, B. G. and WILSON, P. T., 1977. Montmorillonites and the Na/K geothermometer. In: *Geochemistry 1977*, compiled by A. J. Ellis. New Zealand Dept Sci. Indust. Res. Bull. No. 218, pp. 31–4.
58. D'AMORE, F. and PANICHI, C., 1980. Evaluation of deep temperatures of hydrothermal systems by a new gas geothermometer. *Geochim. Cosmochim. Acta*, **44**, 549–56.
59. D'AMORE, F. and NUTI, S., 1977. Notes on the chemistry of geothermal gases. *Geothermics*, **6**, 39–45.
60. TRUESDELL, A. H. and NEHRING, N. L., 1978/1979. Gases and water isotopes in a geochemical section across the Larderello, Italy, geothermal field. *Pageoph.*, **117**, 276–89.

61. SACKETT, W. M. and MOSES CHUNG, H., 1979. Experimental confirmation of the lack of carbon isotope exchange between methane and carbon oxides at high temperature. *Geochim. Cosmochim. Acta*, **43**, 273–6.

62. PANICHI, C., NUTI, S. and NOTO, P., 1978. Remarks on the use of the isotope geothermometers in the Larderello geothermal field. *Proc. Int. Atomic Energy Agency Symp. on Nuclear Techniques in Hydrology, München, FRG.*

63. DENNEN, W. H., BLACKBURN, W. H. and QUESADA, A., 1970. Aluminium in quartz as a geothermometer. *Cont. to Mineral. and Petrol.*, **27**, 332–42.

64. BROWNE, P. R. L. and WODZICKI, A., 1977. The aluminium-in-quartz geothermometer: a field test. In: *Geochemistry 1977*, compiled by A. J. Ellis. New Zealand Dept Sci. Indust. Res. Bull. No. 218, pp. 35–6.

65. HULSTON, J. R., 1977. Isotope work applied to geothermal systems at the Institute of Nuclear Sciences, New Zealand. *Geothermics*, **5**, 89–96.

66. FOURNIER, R. O., 1981. 4. Application of water geochemistry to geothermal exploration and reservoir engineering. In: *Geothermal Systems: Principles and Case Histories*, ed. L. Rybach and L. J. P. Muffler, pp. 109–43. John Wiley & Sons, Chichester, 359 pp.

67. KHARAKA, Y. K. and BARNES, I., 1973. *SOLMNEQ: Solution–Mineral Equilibrium Computations.* US Geol. Surv. Computer Center, US Dept of Commerce, National Technical Information Service, Springfield, Virginia 22151, USA, Rept PB-215899, 82 pp.

68. MCKENZIE, W. F. and TRUESDELL, A. H., 1977. Geothermal reservoir temperatures estimated from the oxygen isotope compositions of dissolved sulfate and water from hot springs and shallow drillholes. *Geothermics*, **5**, 51–61.

69. BODVARSSON, G., 1960. Exploration and exploitation of natural heat in Iceland. *Bull. Volcanol.*, **23**, 241–50.

70. FOURNIER, R. O. and ROWE, J. J., 1966. Estimation of underground temperatures from the silica content of water from hot springs and wet-steam wells. *Am. J. Sci.*, **264**, 685–97.

71. BUSEY, R. H. and MESMER, R. E., 1977. Ionization equilibria of silicic acid and polysilicate formation in aqueous sodium chloride solutions to 300°C. *Inorg. Chem.*, **16**, 2444–50.

72. FOURNIER, R. O. and POTTER, R. W. II, 1979. Magnesium correction to the Na–K–Ca chemical geothermometer. *Geochim. Cosmochim. Acta*, **43**, 1543–50.

73. LLOYD, R. M., 1968. Oxygen isotope behavior in the sulfate–water system. *J. Geophys. Res.*, **73**, 6099–110.

74. MIZUTANI, Y. and RAFTER, T. A., 1969. Oxygen isotopic composition of sulfates—Part 3: Oxygen isotopic fractionation in the bisulfate ion–water system. *New Zealand J. Sci.*, **12**(1), 54–9.

75. MIZUTANI, Y., 1972. Isotopic composition and underground temperature of the Otake geothermal water, Kyushu, Japan. *Geochem. J., Japan*, **6**(2), 67–73.

76. FOURNIER, R. O., SOREY, M. L., MARINER, R. H. and TRUESDELL, A. H., 1979. Chemical and isotopic prediction of aquifer temperatures in the geothermal system at Long Valley, California. *J. Volcanol. Geotherm. Res.*, **5**, 17–34.

77. CRAIG, H., 1961. Isotopic variations in meteoric waters. *Science*, **133**, 1702.

78. BOTTINGA, Y., 1969. Calculated fractionation factors for carbon and hydrogen isotope exchange in the system calcite–carbon dioxide–graphite–methane–hydrogen–water vapor. *Geochim. Cosmochim. Acta*, **33**, 49–64.

79. RICHET, P., BOTTINGA, Y. and JAVOY, M., 1977. A review of hydrogen, carbon, nitrogen, oxygen, sulfur, and chlorine stable isotope fractionation among gaseous molecules. *Ann. Rev. Earth Planet. Sci.*, **5**, 65–110.
80. ARNASON, B., 1977. The hydrogen and water isotope thermometer applied to geothermal areas in Iceland. *Geothermics*, **5**, 75–80.
81. GUNTER, B. D. and MUSGRAVE, B. C., 1971. New evidence on the origin of methane in hydrothermal gases. *Geochim. Cosmochim. Acta*, **35**, 113–18.
82. TRUESDELL, A. H. and HULSTON, J. R., 1980. Isotopic evidence on environments of geothermal systems. In: *Handbook of Environmental Isotope Geochemistry*, ed. P. Fritz and J. Ch. Fontes, pp. 179–226. Elsevier Scientific Publishing Company, Amsterdam, 545 pp.
83. BOTTINGA, Y. and JAVOY, M., 1973. Comments on oxygen isotope geothermometry. *Earth Planet. Sci. Lett.*, **20**, 250–65.
84. BOTTINGA, Y. and JAVOY, M., 1975. Oxygen isotope partitioning among minerals in igneous and metamorphic rocks. *Rev. Geophys. Space Phys.*, **13**, 401–18.
85. JAVOY, M., 1977. Stable isotopes and geothermometry. *J. Geol. Soc., London*, **133**, 609–36.
86. BOTTINGA, Y. and JAVOY, M., 1987. Comments on stable isotope geothermometry: the system quartz–water. *Earth Planet. Sci. Lett.*, **84**, 406–14.
87. BLATTNER, P., BRAITHWAITE, W. R. and GLOVER, R. B., 1983. New evidence on magnetite oxygen isotope geothermometers at 175° and 112° in Wairakei steam pipelines (New Zealand). *Isotope Geosci.*, **1**, 195–204.
88. PANICHI, C., FERRARA, G. C. and GONFIANTINI, R., 1977. Isotope thermometry in the Larderello (Italy) geothermal field. *Geothermics*, **5**, 81–8.
89. CELATI, R., NOTO, P., PANICHI, C., SQUARCI, P. and TAFFI, L., 1973. Interactions between the steam reservoir and surrounding aquifers in the Larderello geothermal field. *Geothermics*, **2**, 174–85.
90. HOBBA, W. A., FISHER, D. W., PEARSON, F. J. JR and CHEMERYS, J. C., 1978. *Hydrology and Geochemistry of Thermal Springs of the Appalachians, II.* Geochemistry US Geol. Surv. Paper 1044-E.
91. MAZOR, E., LEVITTE, D., TRUESDELL, A. H., HEALY, J., GAT, J. and NISSENBAUM, A., 1977. *Mixing Models and Geothermometers Applied to the Warm (up to 60°C) Springs of the Jordan Rift Valley, Israel.* Weizmann Institute (Israel), Int. Rept, 32 pp.
92. CRAIG, H., 1963. The isotopic geochemistry of water and carbon in geothermal areas. In: *Nuclear Geology in Geothermal Areas, Spoleto*, ed. E. Tongiorgi. Consiglio Nazionale delle Ricerche, Laboratorio di Geologia Nucleare, Pisa, Italy, pp. 17–53.
93. FERGUSSON, G. J. and KNOX, F. B., 1959. The possibilities of natural radiocarbon as a ground water tracer in thermal areas. *New Zealand J. Sci.*, **2**, 431–41.
94. D'AMORE, F., 1975. Radon-222 survey in Larderello geothermal field, Italy, 1. *Geothermics*, **4**, 96–108.
95. CRAIG, H. and LUPTON, J. E., 1976. Primordial neon, helium, and hydrogen in oceanic basalts. *Earth Sci. Planet. Lett.*, **31**, 369–85.
96. MAZOR, E., 1976. Atmospheric and radiogenic noble gases in thermal waters: their potential application to prospecting and steam production studies. In:

Proc. 2nd UN Symp., Part 1, pp. 793–802. US Govt Printing Office, Washington, DC, USA.

97. FERRARA, G., GONFIANTINI, R. and PISTOIA, P., 1963. Isotopic composition of argon from steam jets of Tuscany (Italy). In: *Nuclear Geology in Geothermal Areas, Spoleto,* ed. E. Tongiorgi. Consiglio Nazionale delle Ricerche, Laboratorio di Geologia Nucleare, Pisa, pp. 267–75.

98. STOKER, A. K. and KRUGER, P., 1976. Radon in geothermal reservoirs. In: *Proc. 2nd UN Symp.,* pp. 1797–803. US Govt Printing Office, Washington, DC, USA.

99. KRUGER, P., STOKER, A. K. and UMANA, A., 1977. Radon in geothermal reservoir engineering. *Geothermics,* **5,** 13–20.

100. KONONOV, V. I. and POLAK, B. G., 1976. Indicators of abyssal heat recharge of recent hydrothermal phenomena. In: *Proc. 2nd UN Symp.,* pp. 767–73. US Govt Printing Office, Washington, DC, USA.

101. ARNÓSSON, S., BJÓRNSSON, A., GISLASON, G. and GUDMUNDSSON, G., 1976. Systematic exploration of the Krisuvik high temperature area, Reykjanes Peninsula, Iceland. *Proc. 2nd UN Symp.,* Part 2, pp. 853–64. US Govt Printing Office, Washington, DC, USA.

102. JAMES, R., 1970. Factors controlling borehole performance. *Geothermics,* Sp. issue **2**(2), 1502–15.

103. WAINWRIGHT, D. K., 1970. Subsurface and output measurements on geothermal bores in New Zealand. *Geothermics,* Sp. issue **2**(2), 746–67.

104. LAM, H.-L. and JONES, F. W., 1986. An investigation of the potential for geothermal energy recovery in the Calgary area in southern Alberta, Canada, using petroleum exploration data. *Geophysics,* **51**(8), 1661–70.

105. SPROULE ASSOCIATES LTD, 1977. Report on study of geothermal resources in Western Canada sedimentary basins from existing data, phase two: The Sproule Report, Earth Physics Branch, open-file 77-14.

106. LAM, H.-L. and JONES, F. W., 1984. An assessment of the low grade geothermal potential in a Foothills area off west–central Alberta. In: *Energy Developments: New Forms, Renewables, Conservation,* ed. F. Curtis, pp. 273–8. Pergamon Press, Oxford.

107. LAM, H.-L. and JONES, F. W., 1985. Geothermal energy potential in the Hinton–Edson area of west–central Alberta. *Can. J. Earth Sci.,* **22,** 369–83.

CHAPTER 4

Geothermal Resource Assessment

'The world is a very ancient battered tome, in which its tale is writ: Alas! the first and the concluding pages have fallen out of it.'

MIRZA MUHAMMAD ALI, 1605–77 ('Saib')

4.1 GENERAL CONSIDERATIONS

Several considerations must be taken into account before a geothermal assessment of a field can be made.

Firstly, there are the physico-geological factors such as the temperature distribution in, and specific heat of, the relevant rock, its total and effective porosity and permeability, the fluid circulation patterns, the distribution of the fluid phases and the depth of the reservoir. The terrain over which a survey has to be made may limit the places where observations, sample collection and measurements are possible as well as influencing the selection of geophysical methods to be employed. In regard to the logistics of fieldwork, the available manpower has to be taken into account and this will be affected by the amount of money available. Thus, where labour is not a problem, it is feasible to transport equipment into locations difficult to access. If there is a shortage of manpower, then steep inclines or thick forests could obviate the utilization of Schlumberger or Wenner resistivity surveys. In the majority of geophysical methods, it is possible to correct for the effects of terrain on data interpretation, but in rough country the requisite corrections for resistivity methods are hard to compute. Also tall trees in a region characterized by high winds can limit the applicability of seismic methods. Obviously, adverse climatic factors act to reduce the number of working days available because of heavy rain or snowfalls and ılso to impose the times of year during which fieldwork is feasible. Also it must not be forgotten that high humidity and other climatic conditions can affect the operation of some geophysical instruments.

Secondly, engineering matters such as the optimum drilling technology, he method of extraction of the geothermal energy either by means of natural fluids or by employing thermohydraulic loops, the conversion of hermal energy to electric energy, the various uses of extracted fluids and he disposal of residual gases or water must be examined.

Thirdly, there are the inevitable economic constraints involved in the overall capital outlay with its associated costing of the many elements of he utilization plant. This aspect entails careful valuation of the geothermal energy which can be used either directly or for the production of electricity.

Finally, there are the associated questions of law, national energy policy, social and ecological restraints.

Much also depends upon the end-usage of the geothermal development project, i.e. whether it is to generate electricity or for other purposes such as space-heating and agriculture (especially greenhouses). In the latter case, only low-grade heating is required and so the requirement of high-enthalpy fluid is less important.

The main aim of an exploration programme is to find out whether the geothermal field is a vapour- or water-dominated hydrothermal one. Since he nature of the field plays a vital part in designing an exploration programme, where this is not known beforehand a flexible approach is mandatory. However, in most cases, the preliminary geochemical reconnaissance will suffice to settle the issue prior to its commencement.

4.2 METHODS

Proceeding with the above considerations in mind, it is important briefly to review appropriate methodologies for the assessment of geothermal resources before examining firstly their main disadvantage and then the nomenclature utilized within this field of inquiry.

There are four significant approaches, namely surface thermal flux, volume, planar fracture and magmatic heat budget.

4.2.1 Surface Heat Flux

The surface heat flux method is based upon the calculation of thermal power, P, which is released from the soil through conductive heat flow, P_1, and thermal effluents of hot springs, fumaroles and so on, P_2, in the area which is under investigation. Clearly, $P = P_1 + P_2$. If the parameter P is known, it becomes possible to calculate the total energy, H, stored below ground, assuming that all the energy is dissipated in a given geological time interval, t, say of 10 000 or 100 000 years, without any contemporaneous resupply of heat from below the crust. The energy *in situ* is given by

$$H = (P_1 + P_2) \times t \qquad (4.1)$$

It is possible to estimate the recoverable fraction of this, i.e. the resource H_r, by application of a suitable recovery factor, R_g, which theoretically could range from 0 to 100%. In practice, it rarely exceeds 25%.

Knowledge so derived from existing and developed geothermal fields may then be applied to relate the energy loss rate to the rate at which thermal energy might be produced through boreholes; see the work of Kenzo Baba published in 1975.[1]

Conceptually, this is a very simple idea, but it suffers from the serious drawback that the geological time interval t, i.e. the number of years over which the natural discharge endures, is practically impossible to estimate. Hence it is an approach which can be only semi-quantitative and, unless a rather long time interval of say 10 million to 100 million years is considered, the value obtained for the *in-situ* energy will be greatly underestimated.

4.2.2 Volume

The volume method requires the calculation of that thermal energy which is present in any given volume of rock and water, an estimate then being made of the amount of this energy which is recoverable. All calculations are referred to a mean annual temperature, T_0. The area under examination is divided into a set of geologically homogeneous rock volumes, V_i, each of which has a mean temperature, T_i. The energy *in situ*, H_i, in each volume is calculated and $\sum H_i$ = accessible resource base in the area. It is then feasible to determine the resources and the reserves by using an appropriate recovery factor, R_g, which again could attain a value of 100% but usually varies between 0·5 and 1%, rarely exceeding a maximum of 25%. As far as the determination of reserves is concerned, R. Cataldi and P. Squarci in 1978 estimated the geothermal potential by using a reserve recovery factor derived from the resource recovery factor through a reductive factor entailing the incidence of the costs of drilling in extraction of the geothermal energy from reservoirs located at different depths down to 3 km.[2] They assumed that the reserve recovery factor varies between 0 at 3 km and R_g at the ground surface.

There are two main problems in the method, apart from the difficulties inherent in defining the geological and physical parameters necessary for calculation of the accessible resource base.

One is that the resource recovery factor represents the fraction of extractable energy without reference to the time needed for the actual extraction to be accomplished. Obviously, this is of critical importance for industrial development of a geothermal resource. L. J. P. Muffler and R. Cataldi in 1978 believed that it is a function of the fluid production model (intergranular vaporization, intergranular or planar flow) as well as of temperature, effective porosity and depth within the said model.[3] This assumption of a relationship between resource recovery factor and effective porosity is one approach to solving the problem of determining the rate of extraction. Another approach involves consideration of reservoir permeability and possible flow. Despite its importance, attempts have been made to evade estimating the extraction rate. For instance, in 1975 M. Nathenson and L. J. P. Muffler presented the hypothesis that, for water-dominated reservoirs, an average of only half of the entire volume of any given reservoir is porous and permeable, the rest being almost impermeable.[4] From this it is apparent that there is, as yet, no satisfactory solution to the problem of determining the recovery factor and establishing the rate of extraction in a geothermal reservoir.

The second problem is that the volume method regards the *in situ* energy only and takes no stock of possible resupply of heat from greater depth and/or from rock volumes adjacent to the geothermal reservoir.

An alternative approach is to calculate the thermal energy roughly as the product of the volume, temperature and an assumed volumetric specific heat, as for instance C. A. Brook and his associates did in 1979.[5]

4.2.3 Planar Fracture

The planar fracture method derives from the work of G. Bodvarsson in the early 1970s and involves a model in which thermal energy is extracted from an impermeable rock through water flowing along planar fractures.[6,7]

If, as reference model, a horizontal planar fracture penetrating an impermeable rock mass is considered, then heat propagation in it is achieved through conduction, heat transfer along the fracture taking place by means of flowing water. Taking T_0 as the initial rock temperature, and the temperature of the recharge water entering the fracture as T_r, then the temperature of the water flowing out will drop from T_0 at the commencement of fluid extraction to a minimum value, T_m, after a production period t_0. The heat which can be extracted theoretically from a unit fracture area can be calculated as a function of T_0 and the 'end-temperature ratio' which is expressed by

$$r = (T_m - T_r)/(T_0 - T_r) \qquad (4.2)$$

and applies to production periods of 25 and 50 years. Of course, the single fracture model can be extended to more complicated situations involving a series of parallel planar fractures so long as the distance between individual fractures is sufficiently great to prevent thermal interaction.

The method does not require any knowledge of the accessible resource base or any explicit estimation of the recovery factor and can be utilized in order to calculate the energy which may be extracted from a geothermal system within a given time interval, but it is applicable only in certain terrains, e.g. the unfolded flood basalts of Iceland. In the rather complicated, three-dimensional fracture systems which characterize the majority of hydrothermal systems, it is of little use.

4.2.4 Magmatic Heat Budget

The fourth method necessitates calculation of the thermal energy which remains in young igneous intrusions and adjacent country rock as a function of emplacement temperature, size and age, cooling mechanism and so on. It is useful in providing some idea of the order of magnitude of geothermal energy to be anticipated in young volcanic terrains, but the estimates are not directly convertible into geothermal resources. Also the method is restricted to active or recently active volcanic regions. In Japan, T. Noguchi in 1970 assumed that a volcano develops every 500 years, each one being associated with an intrusive body of approximately $400\,km^3$ situated at a depth of *ca.* 10 km.[8] If the time of cooling of each intrusion entails a temperature decline from 1200 to 900°C in 61 500 years with the accompanying liberation of 5% of its weight in steam and 124 such intrusions exist in Japan of ages 0, 500, 1000, 1500 ... 61 500 years, then it may be calculated that the total energy recoverable from release of magmatic steam there is $1 \times 10^{22}\,J$.

A rather different approach was employed in the United States, where R. L. Smith and H. R. Shaw in 1975 estimated the volume of silicic magma chambers probably emplaced within the upper 10 km of the crust.[9] If these be assumed to be of the same age as the younger silicic volcanics which outcrop in the vicinity of the said intrusions, the quantity of residual energy in both the latter and the adjacent country rock can be estimated. The aims of both approaches are to calculate that energy present in a magma chamber when it is emplaced and to estimate the heat liberated from the magma and spreading towards the surface in a given time interval.

4.3 RESUPPLY OF HEAT

The disadvantage with all of the above-mentioned methods is that they avoid the question of a possible resupply of thermal energy to the reservoir while it is under exploitation. For instance, the geothermal resource might benefit from a heat flux arising from deeper levels, the movement of water acting to concentrate it. Additionally, the exploitation of a geothermal field may accelerate fluid, and hence heat, flow from surrounding hot rock into it. L. J. P. Muffler and R. Cataldi in 1978 tried to estimate to what extent this resupply of heat could meaningfully enlarge a geothermal resource, using simple analytical models selected so as to represent three feasible mechanisms through which thermal energy could be resupplied to a hydrothermal reservoir.[3] One of these was by heat conduction from a subjacent intrusion, a second involved the transfer of regional heat flow into the reservoir by means of the horizontal flow of water and the third entailed thermal transfer by water flow out of rock surrounding the reservoir. In terms of significance, Muffler and Cataldi stipulated a resupply of heat of at least 10% of the thermal energy extractable from the reservoir.[3]

Reverting to the three mechanisms discussed above, such a quantity would be impossible to achieve by means of conduction from an intrusion of area equal to that of an overlying reservoir unless it was very close to the latter, say within a few hundreds of metres. Where horizontal flow is concerned, useful resupply of heat is attainable only when the volumes and areas from which the heat is derived far exceed those of the reservoir; clearly, therefore, this mechanism is restricted to relatively small reservoirs. Consequently, resupply is most likely to occur by the transfer of heat extracted by the fluid from the adjacent and subjacent rocks.

The question of the rate of this resupply plays a role in the geothermal resource assessment only when some information regarding it can be obtained prior to the exploitation of a reservoir. Both H. J. Ramey Jr in 1970 and M. Nathenson in 1975 assumed that the rate of resupply corresponds to the total surface heat output from both conductive flow and natural fluid discharge through geysers and so on; M. Nathenson also analysed the situation regarding the resupply of heat in a number of geothermal fields throughout the world.[10,11] Some interesting implications result from this work. If reservoirs very near to magmatic bodies be excluded, then conductive resupply of heat may be neglected. Also negligible is convective resupply of heat in vapour-dominated hydro-thermal systems since a high water influx at great depth is incompatible with these systems. While there may be significant water recharge in some exploited steam fields, this normally takes place through relatively shallow permeable layers in which pressure equilibria and/or fluid discharge from the reservoir were initially possible, as R. Celati and his colleagues indicated in 1973, as also did C. Petracco and P. Squarci in 1975.[12,13] Of course, convective resupply of heat may become important in liquid-dominated hydrothermal systems and indeed this may be so at Wairakei. It was estimated that the original natural heat discharge there was of the order of 20% of subsequent field production.[11] Nevertheless, there are many such hydrothermal systems having a negligible natural heat discharge.

4.4 NOMENCLATURE

It is advisable to provide some background information regarding the terminology of geothermal energy assessment. In 1977 T. Leardini proposed the term 'geothermal potential' for the thermal energy of the Earth, without regard to matters of classification and resource assess-ment.[14] In 1978, proceeding from a definition given by S. H. Schurr and B. C. Netschert in 1960 that the resource base of any given material is all of it within the crust of the Earth, whether its existence is known or not, regardless of considerations of cost,[15] L. J. P. Muffler and R. Cataldi stated that the geothermal resource base comprises all of the geothermal energy in the planetary crust under a specific region, measured from local mean annual temperature.[3] Clearly, this term refers to an instant in time and neglects the transfer of heat from the mantle as well as not taking into account whether it would ever be possible to recover such energy, either technically or economically.

As mentioned already (see Section 3.1), the recoverable geothermal resource is the only part of the thermal energy in the ground (referenced to mean annual temperature) of interest to society. The ratio between this and the thermal energy remaining in place is termed the recovery factor. It can range up to 1 for some metallic deposits where practically all of the material concerned can be mined and brought up to the surface. This is hardly likely to be the case with geothermal energy, however, and therefore its value will almost invariably be much lower than unity in this case. Resource itself may be defined as that portion of the resource base (together with its reserves) which may become available under certain technological and economic circumstances.[16]

Finally, reserve was defined by P. T. Flawn in 1966 as that quantity of minerals which may be reasonably assumed to exist and which is producible under current economic conditions and using existing technology.[17]

Needless to say, the above definitions are not restricted to geothermal energy, but apply to any mineral or fuel. Ideally, the terminology of the latter should be compatible with that utilized for the other energy sources such as petroleum.

Subsequently, the components of geothermal energy may be incorporated in a McKelvey diagram.[3,18] This is a rectangular diagram which depicts the degree of economic feasibility on the vertical axis and the degree of geological assurance on the horizontal one. It is widely used by the US Bureau of Mines and the US Geological Survey and encompasses total resources, the main vertical subdivisions being economic and uneconomic, and the main horizontal subdivisions being identified and undiscovered.

It should be mentioned that J. J. Schanz in 1975 noted that the McKelvey diagram of the US Geological Survey is open-ended and hence does not cover all of a given material.[19] He added that, in using it, it is necessary to remember the physical existence of materials and energy beyond resources, meaning those materials and energy which may be used only in the far-distant future. He gave an excellent instance, namely aluminium. This metal occurs in almost all crustal rocks, but aluminium resources as such are not considered as including these, but rather refer solely to specific and quite restricted rock types in which aluminium is found in such a form and in such concentration, say as bauxite, that it is extractable under favourable economic and technological circumstances.

Another approach commences with the resource base; accessible reserves are divided from inaccessible by depth, both referring to an instant in time.[13] Thereafter the accessible reserves are divided into useful and residual, these being distinguished by economics in the future.

As regards this latter point, R. Cataldi and R. Celati in 1983 proposed that, in order to make comparison between estimates from different countries, it might be established by convention that short-term (< 5 years), medium-term (10 years?) and long-term (20–25 years?) projections could be referred to, each giving a different degree of reliability of the 'economics at some future time'.[20]

One objection to the McKelvey diagram for classifying geothermal resources is that hot dry rock cannot easily be fitted into it and hence the EEC in 1980 proposed that this should be considered as a sub-economic resource undiscovered in known regions.[21]

Finally, the useful are divided into economic and sub-economic, the division between the two being determined by the economics prevailing at the time of the categorization. Both resources and reserves are calculated as heat-producible at wellhead prior to the losses entailed by transportation, direct utilization or conversion of the thermal energy produced. As R. Cataldi and R. Celati pointed out in 1983, in order to account for the use which can be made of the wellhead-produced energy, a temperature limit must be defined below which it is economically inadvisable to use processes of conversion to electric power.[20] This means reference being made to a local mean annual temperature.

These authors also mentioned the problems of defining the borderline between categories of identified resources and undiscovered resources.[20] 'Identified' relates to specific concentrations of geothermal energy known and characterized by drilling or geological, geochemical and geophysical evidence. By contrast, 'undiscovered' refers to surmised and unspecified concentrations. In situations involving such limited information, a distinction between the relatively lower level of assessment and reliability attainable and that implicit in others having a great deal of data available may be advisable and, in 1977, an attempt to do so was made by I. G. Donaldson and M. A. Grant in New Zealand.[22] They noted that the temperatures of the majority of identified geothermal fields there are bracketed by Wairakei and Broadlands (260 and 300°C maximum, respectively). In 1977 and 1978 they concluded that the electrical capacity of these two fields per unit area should be bracketed by their values (Wairakei, 13–14 MW/km^2; Broadlands, 10–11 MW/km^2).[22,23] If adequate permeability is assumed, they argued that the power potential of each identified, but so far undeveloped, field can be estimated from the relevant area (obtained from resistivity and other geophysical surveys) and the sub-surface temperatures (either from direct measurement or by means of geothermometry). Their estimate for proven, inferred and speculative electrical capacities for 13 geothermal fields in the North Island is approximately 1100–2500 MW, a resource likely to last for a rather long time. However, their terminology is inapplicable outside New Zealand.

R. Cataldi and R. Celati in 1983 indicated that, in a small country like Italy, it is permissible to regard geothermal resources as all belonging to the identified category, although this is an exaggeration when the specific thermal knowledge available on Tuscany, Latium and Campania is compared with that existing for other regions there.[20] They also discussed the difficulty of fixing the depth limit between the inaccessible and the accessible resource base. This reflects the fact that at the moment there are insufficient physico-geological data on depths below 2 or 3 km to do so, and any theoretical assumptions about the lower levels could prove to be disastrously incorrect. Then, too, there is the financial side of the drilling, the costs of which comprise the main part of a geothermal budget in every country. This renders attempts to fix the limit in question at more than 5 km quite futile. The suggestion was made that, to avoid ambiguities and facilitate comparison of results from various areas, reference should be made to a standard depth, say 2 km or 2 and 3 km. Heat resupply is involved in the separation between accessible and inaccessible resource base, the possibility existing that fluid may rise from considerable depth entirely or partially to replace fluid extracted by wells drilled within this 2 or 3 km limit.

In view of the fact that it is at present extremely difficult to make very short-term appraisals, the utility of the above considerations is limited and it is even hard to discriminate between sub-economic and economically useful accessible resource bases.

In Italy, the economic problems were attacked by establishing a mean of 3 km as the depth below which, at least now and until the mid-1990s, it is unprofitable to extract geothermal energy. All rock·volumes possessing temperatures less than 60°C were ruled out and the resources were regarded as all the thermal energy which can be extracted from rocks at > 60°C, but referred to 15°C. The resource for the generation of electricity was restricted to thermal energy which can be recovered from reservoirs having temperatures > 130°C.

4.5 GLOBAL GEOTHERMAL RESOURCES

Geothermal resources are exploited in many countries for two main purposes, namely to generate electricity and for non-electrical uses such as space heating and agricultural uses, mostly heating greenhouses. A look at past and present electricity-generating capacity is instructive. In 1979 it was estimated that the total throughout the world was 2063 MWe (megawatts electrical energy) distributed between the countries which are most advanced as regards geothermal energy as follows: USA, 908 MWe, i.e. 44% or almost half; Italy, 421 MWe (approximately 20%); New Zealand, 202 MWe (just under 10%); and Japan, 171 MWe (about 8%). For 1985, the forecast given by M. Fanelli and L. Taffi in 1980 was *ca.* 6500 MWe.[24] T. Meidav, S. Sanyal and G. Facca in 1977 went even further, estimating that installed geothermoelectric capacity could reach *ca.* 12 000 MWe by

1985.[25] However, both views turned out to be over-optimistic, as can be seen from Table 4.1.

It is interesting to compare the information in this table with overall world electric power, which in 1981 was a staggering 1 910 000 MW. It is apparent that geothermally generated electric power then accounted for around 0·2% of this total and there is no reason to think that the percentage has changed. Certainly, installed geothermal power almost doubled over the following three or four years, but the conventional electric power-generating capacity has increased in step with this. It could be inferred that geothermal energy has only a very small role to play in the energy scene of

TABLE 4.1
Electricity-generating capacity from geothermal resources

Country	Capacity (MWe)				
	Installed (1979)[a]	Estimated (1985)[a]	Installed (1982)[b]	Installed (1985)[b]	Estimated (1990)[b]
Azores (Portugal)	—	—	3	3	—
Chile	—	—	—	—	15
China	3	10	4	14	30
El Salvador	60	100	95	95	150
Ethiopia	—	—	—	—	5
Greece	—	—	—	2	10
Guatemala	—	—	—	—	15
Iceland	63	63	41	41	71
India	—	—	—	—	5
Indonesia	—	43	30	32	144
Italy	420·6	480	440	520	700
Japan	171·5	500	215	215	325
Kenya	—	30	30	45	120
Mexico	112·5	310	180	645	2 440
New Zealand	202·6	352	202	167	302
Nicaragua	—	30	35	35	180
Philippines	110	1 320	570	894	1 042
Turkey	5	5	0·5	20	130
USA	908	3 100	936	2 022	4 370
USSR	5·75	18	11	11	241
West Indies (French)	—	—	—	4	—
Totals	2 061·95	6 394	2 792·5	4 766	10 300

[a] Data from M. Fanelli and L. Taffi (1980).[24]
[b] Data from E. Barbier and M. Fanelli in 1983,[26] R. Di Pippo in 1985[27] and E. Barbier in 1986.[28]

the world. However, the picture alters if the industrialized countries are considered separately from the under-developed ones. The former possess installed electric power reaching hundreds of thousands of megawatts, hence geothermal energy is unlikely to make a contribution exceeding 1% or so to their needs over the next decade or two. On the other hand, the latter have much more restricted electric power consumption associated with good geothermal prospects which could make a significant contribution to their needs; for example, in 1986 this contribution was 17·5% in the Philippines (see Table 4.2).

TABLE 4.2
Total electric power compared with that derived from geothermal sources in selected developed and under-developed countries
Data from the *UN Statistical Yearbook 1981* (1983)[29] and the Geothermal Research Council (1985)[30]

	Total electric power (MW)	Geothermal electric power (1985)	
		(MW)	(% of total power)
Industrialized countries			
Italy	56 000	520	1·0
Japan	150 000	215	0·1
USA	652 000	2 022	0·3
Under-developed countries			
El Salvador	502	95	18·9
Kenya	541	45	8·3
Mexico	19 895	645	3·2
Nicaragua	370	35	9·5
Philippines	5 099	894	17·5

The significance of the geothermal contribution, actual or potential, in the developing countries is emphasized by the data in Table 4.3, from which it is clear that in 1962 the OECD countries, the USSR and eastern Europe consumed 81% of the total energy and the developing countries 19%. In 1972 this rose to 82% as opposed to 18%, but in 1982 it fell to 77% as against 23%. On a population basis, the Third World, in which 68% of the peoples of the world live, in 1982 consumed 23% of the energy of the world.[29] How this was employed is indicated in Table 4.4.

This information dramatically indicates that the demand for domestic energy is relatively much smaller in developing countries. Admittedly, they usually have more favourable climatic conditions, but even so only 13% of

TABLE 4.3

World energy consumption (MW) for various selected years and groups of countries[28]

	1962	1972	1982
OECD countries[a]	2 105	3 445	3 550
USSR and eastern Europe	550	1 060	1 625
Developing countries			
OPEC	40 ⎱ 615	100 ⎱ 960	225 ⎱ 1 525
Other	575 ⎰	860 ⎰	1 300 ⎰
Total	3 720	5 465	6 700

[a] EEC, Austria, Finland, Iceland, Norway, Portugal, Spain, Sweden, Switzerland, Australia, Canada, Japan, New Zealand, Turkey, USA, Yugoslavia.

their total energy requirements are utilized for domestic purposes. As this factor is basic in assessing living standards, the inference is obvious.

As regards geothermal power plants, 188 were operational throughout the world in 1985, each comprising one turbine and an electric genetator: 71% of these were under 31 MW in size and the biggest, with a capacity of 133 MW, is installed at The Geysers in California, USA. Of the 188 units, 73 (39%) were dry-steam, 61 were single-flash (32%), 23 were dual-flash (12%) and 31 were binary-cycle (16%). Table 4.5 lists capital costs per kilowatt installed of geothermal electric power plants compared with some other types.

The average world cost for geothermal energy can be estimated as being between US $50 and $140 (1987 values) for the energy of 1 OEBBL (barrel of oil equivalent on a thermal basis, with 1 barrel = 137 kg × 10 000

TABLE 4.4

The principal energy consumption (MW) of various countries in 1983[28]

	USA	OECD	Japan	Developing countries
Domestic	365	305	65	115
Industry energy use and loss	510	325	95	285
Industry	255	255	90	225
Transport	435	230	65	210
Total	1 565	1 115	315	835

kcal/kg = 1 370 000 kcal), assuming the transfer of heat to geothermally derived electric power. The higher value approximately corresponds to 8·8 ¢/kW h, the actual cost of geothermal power production in the Japanese power stations in 1983, as M. Kaneko indicated.[31]

The World Bank has taken a great interest in the development of geothermal energy resources. Geothermal power is generated in at least nine developing countries, representing almost 40% of the geothermal electric power generating capacity of the world. In six of these, the World Bank has aided in financing geothermal investment. Its lending complements grants from other sources and was confined initially to funding the power-plant components of geothermal projects. Now it has expanded its

TABLE 4.5
Economics of generating electric power in terms of 1984 US dollars[28]

Power plant type	Capital cost ($/kW)	Bus-bar cost (¢/kW h)				
		Italy	USA	Japan	Guadaloupe (France) (4 MW)	Djibouti (20 MW)
Geothermal						
Dry-steam flash	1 000	2·2	3·9	8·8	8·4–11·8	9·6–12·6
Binary	3 600	—	5·1–16·5	—	—	—
HDR (flash)	2 750	—	4·9	—	—	—
Nuclear	1 500	2·2	3·6	—	—	—
Oil	725	4·2	6·3	8·8	—	20
Coal	1 100	2·9	3·4	—	—	—

role to include assisting with exploration and development as well. Beginning with the Cerro Prieto geothermal field in Mexico and the Ahuachapán geothermal field in El Salvador in the late 1960s and early 1970s, a total of 18 projects were assisted up to the middle of 1985 (Table 4.6).

The efforts of the World Bank have been concentrated in two areas, according to W. Bertelsmeier and J. A. Koch in 1986, namely geothermal power projects and geothermal exploration projects.[32] As regards the first, since 1968, and following Mexico (75 MW) and El Salvador (65 MW), Kenya (45 MW) and Indonesia (110 MW) have recently come on stream to bring the total for these four countries alone to 295 MW of electric power-generating capacity. In the second area, the World Bank began lending for exploration purposes in 1982 in the Philippines and thereafter in the Yemen

TABLE 4.6
World Bank Group financial involvement in geothermal exploration and development[32]

Country	Date	Geothermal loan: credit amount (US$ million)	Geothermal project: component cost (US$ million)
Djibouti	1984	6·0	16·6
El Salvador	1973	23·4	25·8
El Salvador	1976	39·0	72·8
Ethiopia	1983	2·1	2·2
Honduras	1982	0·9	1·1
Indonesia	1982	120·3	136·7
Indonesia	1983	4·0	4·8
Kenya	1980	40·0	89·0
Kenya	1983	12·0	41·6
Kenya	1984	24·5	34·3
Mexico	1968/1970/1972	a	a
Nicaragua	1977	4·1	4·9
Philippines	1982	36·0	71·5
Yemen Arab Republic	1980	0·1	0·1
Yemen Arab Republic	1984	13·0	15·3
Yugoslavia	1981	0·7	0·7

[a] The three bank loans to Mexico financed a part of that country's power investment programme during the period 1968 to 1974 and this included Cerro Prieto.

Arab Republic, Kenya and Djibouti. These projects together financed more than 60 deep exploration-appraisal wells. In addition, the World Bank supported smaller geothermal components in the form of technical assistance and studies in several countries, including construction of some wells in connection with non-power uses of geothermal energy, for instance in the Philippines and Yugoslavia. With geothermal power capacity installed and being constructed in developing countries presently standing at approximately 2000 MW, World Bank loans and credits account for some 15% of this. Also World Bank funding comprises approximately 8% of the estimated US$4 billion total investment in geothermal projects in the Third World.

4.5.1 Overview
This comprises a summary of the existing situation and the short-term prospects for geothermal exploration and exploitation throughout the world. Where relevant, more details on some of the existing geothermal

fields are given in others parts of this book; hence this section is intended to provide a bird's eye view of the present state of affairs.

Nobody can predict the future with certainty, although it is safe to say that increasing dependence upon fossil fuels, non-renewable reserves of energy like oil and coal, which are rapidly diminishing as both populations and expectations inexorably rise, *pace* AIDS, makes the search for alternative sources ever more urgent. This is also necessitated by the mounting damage to the environment resulting from such factors as oil spillage, strip mining, sulphur emission, aerosols and the accumulation of solid chemical and radioactive wastes which threaten all Mankind. The total geothermal energy stored in the uppermost 10 km of the planetary crust in places where the temperature exceeds 100°C is greater by orders of magnitude than all the thermal energy available from such fossil fuels sources and nuclear power combined.

From a human perspective, geothermal reserves are theoretically so vast as to be practically inexhaustible, self-renewing and non-polluting. Although not yet fully located and exploited, their significance is so great and their potential so valuable that the long-term scenario certainly should, but most probably will not, involve the assignment of much more investment to research and development.

4.5.1.1 The New World

Almost 60% of the installed electricity-generating capacity from geothermal resources in the world in 1985 came from geothermal resources in the New World, mainly from the USA (cf. Table 4.1);[26-28] hence it is logical to commence the overview in this region.

(a) United States of America. Researches into exploitation of geothermal energy in the USA date back to immediately after the Great War of 1914–18, the first well having been drilled in The Geysers field in California in 1921. However, only within the last couple of decades has exploration really progressed and received large-scale financial backing from the Federal Government. The US Department of Energy assessed the position, dividing the nation into four regions.

Possibly the most productive is the Pacific region, comprising California, Oregon, Washington, Alaska and Hawaii, where most of the geothermal resources are of convective hydrothermal type. Although hot-water resources predominate, The Geysers constitute a vast vapour-dominated resource of which the installed capacity was around 600 MW in 1979, rising steadily thereafter, now to exceed 2000 MW. They are located on the north

bank of Big Sulfur Creek about 36 km northeast of Geyserville and 27 km
east of Cloverdale near the Californian coast west of Sacramento (see Figs
4.1 and 4.2). Existing geothermal facilities are situated approximately
160 km north of San Francisco on the steep slopes of a canyon near an
extinct volcano called Cobb Mountain. The first recorded sighting of it was
by William Bell Elliott, an explorer and surveyor who was looking for
grizzly bears in 1947 and stumbled upon steam emerging from a canyon for
about half a kilometre of its length.

FIG. 4.1.   The Geysers, California, USA: general view of operations. Reproduced
by courtesy of the Pacific Gas and Electric Company, California, USA.

The geology of the area is shown in Fig. 3.2. It consists of highly
permeable fractured shales and basalts of Jurassic to Cretaceous age. Fault
and shear zones developed during the seismic events of the early Cenozoic
also contribute to the permeability and permit a large geothermal reservoir
to exist. Steam and water are found in this reservoir and temperatures range
between 260 and 290°C, the shut-in pressures of wells deeper than 600 m
lying between 450 and 480 psig (3·1–3·3 MPa). Wells up to almost 3000 m
have been drilled, many far removed from the natural steam emissions, and
they produce steam flows reaching almost 175 tonnes/h. Steam wells were
also drilled near steam vents and this was done on 60–300 m centres and to

depths of between 120 and 300 m to produce steam flows in the range 18–36 tonnes/h.

It is believed that a molten magma mass is still cooling underneath The Geysers. As noted in Chapter 3, a mass deficiency was identified under Mount Hannah and this is associated with such a magma body by means of a slowing-down of seismic waves from distant earthquakes at 15–20 km depth, implying a molten mass of this type. Under the geothermal energy production zone, there may be a highly fractured steam reservoir overlying

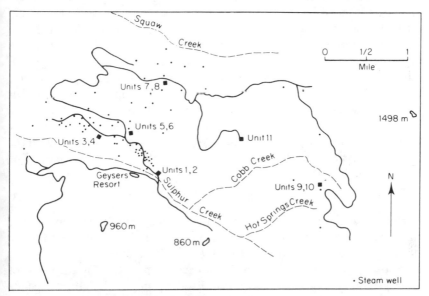

FIG. 4.2.   Map of The Geysers, California, USA, showing the layout of the various units and locations of steam wells.

the magma, this also being capable of slowing down seismic waves. The magma body is thought to be located approximately 5 km under Mount Hannah and may be the heat source of The Geysers.

In 1967 the Union Oil Company of California entered into partnership with the original developers of The Geysers, the Magma Power Corporation and Thermal Power Company, steam being produced being sold to the Pacific Gas and Electric Company for use in condensing turbines driving electric power generators. As the reservoir pressures reduce at the wellhead, the turbines are designed to operate at 80–100 psig (0·5–0·7 MPa) intake pressure and need a constant throughput of steam.

Some curious phenomena were observed; for instance, the known gravity anomaly covers about 250 km² and within it there are some wells drilled to considerable depth which produce no steam. This must relate to some subterranean inadequacy of permeability.

The actual producing zone is graywacke sandstone (the Franciscan Formation), this being highly metamorphosed and hard to penetrate. Mud drilling causes lost circulation to permeable zones with fluid pressures under hydrostatic. Air drilling improves penetration rates and eliminates the lost circulation factor, but can be effected only below segments where significant water-bearing beds have been cased (otherwise water would enter the borehole and promote cave-in of the formation). There is heavy wear on equipment due to abrasion and heating. When steam is encountered, air circulation pressures rise and annular velocities approach sonic values, which causes tool joints to take the wear of high-speed particle impact. The production rate obtainable from a completed steam well is highly sensitive to the size of the borehole itself and also of the casing installed in it. Very high mass flow rates may be achieved and the only limitations relate to these dimensions. At The Geysers, conformity with the classic expression for the performance of a gass well is attained. The relevant equation is

$$W = C(P_s^2 - P_f^2)^n \qquad (4.3)$$

in which W is the steam flow rate (lb/h); C is a factor which is a function of time, reservoir matrix, fluid properties, flowrate and well bore conditions; P_s is the static reservoir pressure (psia); P_f is the well bore pressure at steam entry (psia); and n is an exponent, usually constant with W in the range of usual production rates ($0.5 \leq n \leq 1.0$). Production, W, therefore depends upon the bottom-hole pressure, P_f, i.e. the sum of the wellhead flowing pressure, the total frictional resistance pressure and the column weight of the fluid. As might be expected, the cost of drilling sharply increases with depth; hence the possibility of locating an economic steam-producing well lessens with increasing depth because the length of the steam-flow conduit increases as well. As there is no way in which the productivity of a well can be predicted before it is drilled, the values for C and n also cannot accurately be foreseen.

As at Larderello, a constant production rate cannot always be maintained at The Geysers, individual wells declining and extra wells becoming necessary in order to maintain the steam supplied to the generating units. However, two wells at The Geysers, part of a block-of-power producing steam for generating unit 3 since 1967, had an almost constant wellhead flowing pressure throughout their productive lives. Pressure build-up tests on them demonstrated that no nearby well bore alteration in permeability took place; hence neither the wells nor the formation suffered plugging. The implication is that production decline probably results, at least partly, from pressure depletion in the geothermal reservoir itself. A system was developed at The Geysers to deliver 2×10^6 lb/h of dry superheated steam through two separate paths to two units, each of which generated 110 MW of electricity. Near each wellhead there is a meter continuously recording the production rate and the producing pressure of the well. Also routine measurements of both enthalpy and steam quality are effected. The steam at The Geysers actually contains approximately 1% of non-condensable gases. It also contains dust which may accumulate on the inside of the turbine blade shrouds and result in failure. This problem was resolved by installing heavy-duty replacement blades and shrouds in units constructed earlier.

Studies were carried out in order to determine the suitability of various materials for the mechanical equipment and piping before the detailed design of the first unit at the site. The steam emerging from the wells with a small amount of superheat proved to be relatively non-corrosive and here carbon steel is feasible for piping. In addition, the turbines do not require any special corrosion-resistant materials. But, as the steam condenses, the non-condensable gases become ever more concentrated and hydrogen sulphide partly oxidizes to form sulphuric acid. Thus steam and condensate become more and more corrosive with time and carbon steel becomes inadequate (as also do copper-based alloys, cadmium and zinc). Austenitic stainless steels or aluminium or epoxy-fibre glass are quite suitable. The hydrogen sulphide may corrode electrical equipment as well and, initially, lack of appropriate components in unit 1 caused difficulties with electrical contacts and other metal parts. Aluminium offers satisfactory resistance to corrosion in this manner and so does stainless steel and some precious metals such as platinum. Protective relays are especially liable to corrosion: therefore special ones made out of corrosion-resistant materials were supplied for units 2, 3 and 4. From unit 5 on, relays, communications equipment, 480V switchgear and generator excitation cubicles were placed in a clean environment consisting of three rooms on three levels maintained at a slight positive pressure using clean air from activated carbon filters.

At first the sufficiency of the steam source had to be established and it had to be confirmed that there is a sufficiently large flow from existing wells to supply the first turbine generator unit. The steam was found to have a constant enthalpy of 1200–1500 Btu/lb. So as to exploit the steam energy fully, it was decided to use condensing steam turbines exhausting below atmospheric pressure. There was no source of condenser cooling water in the region and therefore cooling towers were incorporated into the cycle as a heat sink. These cooling towers are of the induced-draught type and their structural supports are of redwood. The tower siding is transite. Polyvinyl chloride was used as fill because of its fire-retarding qualities. Cooling-tower basins are made of reinforced concrete coated with coal-tar epoxy so as to arrest deterioration of the concrete. Cooling towers are designed to cool condensate from 48 to 27°C at 18°C wet-bulb temperature. The turbines are manufactured for low-pressure and low-temperature service. Usually, the blades and nozzles are of 11–13% chrome steel and carbon steel was used for the casings. Austenitic stainless-steel inserts are provided in casings opposite the rotating blades in order to prevent moisture erosion of the casings. Table 4.7 embodies relevant data.

TABLE 4.7
Steam inlet conditions for the turbine generator units

Unit no.	Rating (kW)	Steam flow (lb/h)	Steam pressure (psig)	Temperature (°C)
1	12 500	240 000	93·9	175·5
2	13 750	255 475	78·9	172·2
3, 4	27 500	509 600	78·0	172·2
5–10	55 000	907 530	113·7	179·4
11	110 000	1 808 000	113·7	179·4

In the case of earlier units, lower steam pressures were employed because the steam involved originated from the shallow, low-pressure reservoir. Later units are supplied from the deep, high-pressure reservoir. The electric generators are typical of similarly sized units used elsewhere, the largest being hydrogen-cooled and automatically purged under certain circumstances.

Outdoor generator oil circuit breakers are utilized. The potential transformers are out-of-doors as well and their potential taps to the bus are supplied by the cable-bus manufacturer. Four aluminium cables employed per phase for each generator give a rating of 3000 amperes (A). Cables from each generator oil circuit breaker are connected directly to the single main transformer terminals and are shielded with cross-linked polyethylene insulation. On units 1–4, three different types of excitation system exist.

Thereafter static excitation is used with power potential transformers and saturable current transformers.

The elimination of commutators is desirable in the environment of The Geysers. As for the transformers, each of units 1 to 4 has its own step-up one, an essential in these early units which contributed to the local power supply. Subsequent increase of the size unit to 55 000 kW was accompanied by the utilization of one main three-phase transformer of 132 000 VA for each two units. This was an economy measure aimed at reducing costs. The transmission voltages ranged from 60 000 to 230 000 V. For units 3 to 10, dual-voltage step-up transformers were acquired, but, from unit 11 onwards, single-voltage 230 000 V transformers are feasible because the combined plant output necessitates operation at this level.

The steam purchase contract stipulated payment for power delivered to transmission. Units 1 and 2, and units 3 and 4, have transmission voltage metering sets of 60 000 and 115 000 V, respectively. As 230 000 V metering equipment is expensive, units 5 and 6 and subsequent units are metered at the low-voltage side of the main transformer. The transformer losses are determined from the ampere-squared–hour and volt-squared–hour meters and subtracted.

The power cycle involves the introduction of steam from wells into the turbines; this exhausts the direct-contact condensers below them and the combined condensed steam and cooling water are pumped back by two condensate pumps to the cooling water.

The turbine back-pressure on all units is approximately 100 mm Hg absolute. Cooled water from the tower basin is returned to the condenser under gravity and the vacuum head developed by the latter. The evaporation rate of the cooling tower is less than the turbine steam flow and therefore an excess of water develops. The flow depends upon the dry-bulb temperature and the relative humidity, but, under all operational conditions, an excess persists. Over years this has been returned to the steam suppliers for re-injection through wells into the steam reservoir. At the beginning, it was thought that this might quench the producing steam wells, but no deleterious effects seem to have taken place. Actually, the re-injection may well prolong the productive life of the steam reservoirs because there is a possibility that there is more heat in the reservoir than there is vapour available to extract it. Two-stage steam-jet injectors are used to purge non-condensable gases from the turbine condensers. The condensers for the ejectors are again of direct-contact design. Figure 4.3 illustrates the relevant vapour-turbine cycle and shows isobutane as the fluid used, but other fluids may be substituted; for example, Freon has been employed in the USSR. However, isobutane has demonstrated optimum economy for developing power from water at a temperature of approximately 180°C.

Finally, as to costs, these are favourable in comparison with running other types of electricity-generating power plants, the fixed charges of a conventional fossil fuel operation being similar to those for a geothermal

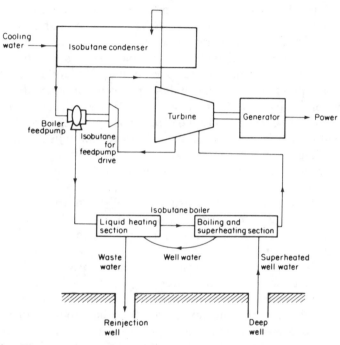

FIG. 4.3.   The vapour–turbine cycle at The Geysers geothermal power plant. (Although isobutane is shown as the fluid utilized, it is possible to use others, for instance Freon.)

one. Either of them costs less than a hydropower plant or a nuclear plant. Maintenance expenses are minimal; see Table 4.8.

An advantage of geothermal power plants is that they constitute an extremely low pollution hazard. Two factors are significant in geothermal costing. One is the need for increased profit so as to promote new exploration and field development as well as expansion of existing facilities. The second is the desirability of increasing efficiency by using a multiple operation in which all aspects of a geothermal system can be exploited.

Unfortunately, little or no control can be exerted over some parameters such as the pressure, temperature and fluid characteristics of a geothermal reservoir, factors developed over millions of years.

Lastly, allusion may be made once again to the existence of a number of active fault systems near The Geysers and Clear Lake, the San Andreas lying only some 50 km to the southwest. Continuous induced seismicity entailing micro-earthquakes occurs, these latter taking place on fractures which are less than 1 km long. Although benign, the seismic activity relates to the environment and results directly from production-related work. Four power units were added since March 1979 and power production has increased 70% through these units. Near three of them seismic activity has developed in areas formerly aseismic. However, the inducing mechanism is

TABLE 4.8
Average geothermal costs at The Geysers from 1961 to 1972
(capacity factor = 77·5%)

Expenditure item	Power cost (mills/net kWh)	Annual cost (US$/kW)
Fixed charges	2·25	14·08
Maintenance	0·387	2·43
Operation	0·208	1·30
Steam	2·587	—
Total	5·432	17·81

not fully understood. The matter is discussed in Chapter 6, which covers environmental aspects.

Incidentally, it is interesting to note that there may be another geothermal system of the same type as The Geysers at Mount Lassen. Other geothermal developments in California include a 7-MW binary-cycle geothermal project at Mammoth Lake which is being operated by the Ben Holt Company (see Fig. 4.4).

Hot dry rock resources certainly exist in Hawaii. The Puhimau thermal area of the Volcanoes National Park there is interesting because it has an anomalously high heat flow and the heat source may be a shallow magma body. Much geophysical information was accumulated and the self-potential data imply a steeply northwardly dipping intrusion. In fact, a controlled-source audiofrequency magnetotelluric electromagnetic survey gave results according with a dike-like feature containing a good conductor

Fig. 4.4. Mammoth Lakes, California, USA: 7-MW binary-cycle geothermal project designed and operated by the Ben Holt company. Reproduced by courtesy of the Southern California Edison Company, California, USA.

Fig. 4.5. Brawley, California, USA: 10-MW single-flash geothermal demonstration facility in operation. This was dismantled after the successful demonstration. Reproduced by courtesy of the Southern California Edison Company, California, USA.

approximately 200 m deep, according to L. C. Bartel and R. D. Jacobson in 1987.[33] However, this may not be magma, but rather hot mineralized water. Above it is a 40-fold less conductive zone probably representing a region having less than 100% saturation. The 'dike' connects to a conducting basal layer at a depth of 350 m.

In addition, there is direct-heat use of geothermal fluids at several places, for instance at Klamath Falls in Oregon as well as at Susanville in California. Unocal and Magma–Dow have major plant construction activities and drilling projects in the Salton Sea geothermal field. This and the Imperial Valley are located in the Salton trough, an active rift between the Pacific and North American plates. Also Ormat and Geothermal Resources International are developing the East Mesa field. This, too, lies in the Imperial Valley region, which is regarded by many as the birthplace of the San Andreas fault system in the sense that the Imperial fault seems to be the structure along which most of the plate boundary strain is transferred northwards into two of the major fault branches of the San Andreas system, namely the San Jacinto and the San Andreas (Banning–Mission Creek) faults. The San Jacinto and San Andreas faults are demonstrably linked in creep and strain behaviour to the Imperial fault. In addition, Oxbow is developing the Dixie Valley field in Nevada and building a 340-km transmission line to Bishop, California. Several hundred megawatts of power will be added when the geothermal plants are completed. In November 1987 the California Energy Company completed the 32-MW power plant at Coso Hot Springs, California. In addition, mention may be made of a 10-MW single-fluid geothermal demonstration facility at Brawley which is owned by Southern California Edison Company and is now dismantled after a successful run (see Fig. 4.5).

There can be little doubt that the geothermal energy of the Pacific Region could probably supply a good proportion of the total energy requirement of the five states comprising it.

The Rocky Mountain Basin and Range Region includes ten states, namely Arizona, Colorado, Idaho, Montana, Nevada, New Mexico, North and South Dakota, Utah and Wyoming. Its geothermal resources comprise both high-temperature and many low-to-moderate-temperature convective hydrothermal ones. One dry hot rock resource has been located in the Jemez Mountains near the Valles Caldera. At least two areas are optimal for potentially generating electric power from geothermal sources and these are Roosevelt Hot Springs in Utah and the Valles Caldera in New Mexico.

There is an Office of Basic Energy Sciences of the Department of Energy project in the Valles Caldera for which two boreholes were completed by 1987, with the aim of investigating the physics, chemistry and evolution of the mature hydrothermal system of the caldera. Fraser Goff and Jamie Gardner of Los Alamos National Laboratory and Jeff Hulen of the University of Utah Research Institute are the chief scientists for this investigation. Flow tests from selected horizons with fluid sampling are continuing at present. A hot dry rock experiment is being carried out near the Valles Caldera, at Fenton Hill, and a new reservoir has been developed in Precambrian basement there. Details of this work are given in Section 5.1 in Chapter 5.

The Geothermal Technology Division of the Department of Energy financed research by the University of North Dakota, which demonstrated the coincidence of a large geothermal anomaly with a -50 milligal gravity low in western South Dakota. An area of $10\,000\,km^2$ has a heat flow of over $100\,mW/m^2$ and seven permeable aquifers providing a geothermal resource base of 11×10^{21} J.

In Nevada, laboratory experiments effected by the University of Nevada, Las Vegas, in Reno indicated that gold recovery could be increased, by 17%, to 40% through the use of geothermal water to heat the cyanide heap-leaching of ore. Such heat could permit the continuation of such operations throughout the winter as well. Other uses for geothermal resources within the region include the application of geothermal fluids in space heating and agriculture.

The Gulf Coast Region is made up of Louisiana and Texas, and possesses geopressurized geothermal resources comprising thermal, dissolved-methane and hydraulic-pressure energies. They occur both on- and off-shore, and estimates of recoverable geothermal energy extend to as much as 38 000 MW-centuries. It is expected that electric generating capacity from the geothermal resources could reach anything from 500 to 2500 MW by AD 2000.[24]

The Eastern US Region includes all of the country east of the Rocky Mountains, apart from the geopressurized resources of Louisiana and Texas mentioned above. There are restricted temperatures in the region and its resources could be applied most beneficially to direct heating of industrial and agricultural processing, and for space heating rather than for the generation of electric power.

The direct use of geothermal energy is steadily increasing in the USA. Many direct-use developments are paid for by state and local government agencies, and private investment is common in the agricultural utilization of geothermal energy. District heating systems now comprise the greatest use of low-temperature ($\leq 100°C$) geothermal energy and facilities were recently expanded at Susanville, California; at Boise, Idaho; and at Klamath Falls, Oregon. Other innovative applications include sewage treatment at San Bernardo in California and the cultivation of roses at Animas, New Mexico, and Salt Lake City in Utah.

(b) Canada. A research programme in the enormous country of Canada was initiated in 1975, specifically to assess the geothermal resources of British Columbia and the Yukon, the likeliest areas in which to find useful ones. A semi-geothermal application was organized in Whitehorse, Yukon, whereby 7600 litres/min of 7°C water is mixed with the municipal water supply during the winter in order to combat problems of freezing.[34,35]

A low-grade geothermal potential survey of western Canada, including a reconnaissance survey of petroleum exploration data, was made in the 1970s and, in 1986, an investigation of the potential for geothermal energy recovery in the Calgary area, southern Alberta, Canada, was described. These matters were mentioned in Chapter 3; see Refs 104 and 105 at the end of that chapter.

(c) Central America. There are several countries of interest in Central and South America. In Central America, Mexico, Guatemala, El Salvador, Nicaragua and Costa Rica merit attention.

(i) Mexico. Almost all of the geothermal energy is situated along a belt of rather recent volcanism running from the Pacific Ocean across the Gulf of Mexico, and nine suitable areas have been defined. These are Ixtlan de Los Hervores, Los Negritos and Los Azufres in Michoacan State; San Marcos, Los Hervores de la Vega, La Primavera and La Soledad in Jalisco State; Pathé in Hidalgo State; and Los Humeros in Pueblo State. At Los Azufres a 55-MW power plant is under construction, and several new 5-MW wellhead generators have been installed there and also at La Primavera geothermal field. The sole geothermal field outside the belt in question is Cerro Prieto. Exploratory wells have been drilled at Ixtlan de Los Hervores and Los Negritos, and produce a mixture of water and steam. A small generating power plant at Pathé was installed in 1959, operated for 13 years and produced 3·5 MW. The exploration at Cerro Prieto commenced in 1959 and the geothermal field began to generate electric power in 1973 with an installed capacity in 1980 of 112·5 MW in three units of 37·5 MW each. The geological position of the hot-water zone is shown in Fig. 4.6. An additional 37·5-MW unit is added and output should rise to about 500 MW by the end of the century.[24]

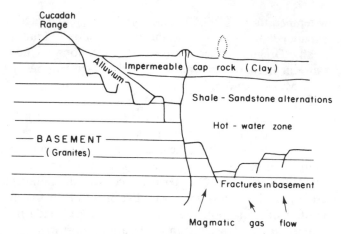

FIG. 4.6.   Geological cross-section of the Cerro Prieto geothermal field in Mexico.
It has been adapted from the brochure issued by the Comision Federal de
Electridad, Mexico, in 1971.

(ii) Guatemala. Geothermal exploration commenced in 1971 and
showed that there are at least three geothermal areas in Guatemala, namely
Moyuta, Amititlan and Zunil, but exploratory drilling has been undertaken
only at the last site, at which a 15–20-MW electric power-generating unit is
planned.

(iii) El Salvador. El Salvador has been explored for geothermal
resources since 1966 and the Ahuachapán geothermal field identified. The
first of two geothermoelectric plants commenced operating there in May
1975 and the second began in June 1976. As each of these medium-pressure
units has a capacity of 30 MW, the installed capacity then became 60 MW.
That this is no small quantity is clear when it is realized that the amount in
question represents 32·3% of the total electricity output of the entire
country. A third unit, capacity 35 MW, was planned and the search for
alternative areas extended to nearby Salitre and, about 125 km east-
southeast to Berlin, where good preliminary results were obtained,
Chinameca, Lempa and San Vicente, as was recorded by the United
Nations University in 1979.[36]

(iv) Nicaragua. Exploration in Nicaragua began in 1966 and two
geothermal fields were located, one at Momotombo and the other at San
Jacinto, both with a reservoir temperature of 230°C or so and with an
estimated potential of 140 and 70 MW, respectively. Wells drilled at
Momotombo during 1975 and 1976 established that a reservoir exists there

at depths of between 300 and 400 m. It is capable of supplying a 30-MW unit.[35,37]

(v) Costa Rica. Geothermal exploration has gone on sporadically in Costa Rica ever since 1959 and a significant geothermal anomaly has been located in the La Union–Las Hornillas–La Fortuna area below the Miravalles volcano.[24] Fluid circulation exists within this region and temperatures up to 200°C were recorded at depths up to 300 m. A geothermoelectric plant was planned.

(d) South America. The most interesting country from the geothermal energy standpoint is Chile.

(i) Chile. Here the fluids of El Tatio were evaluated with a view to possible utilization as a geothermal resource as early as 1908. Wells were drilled at this geothermal field in 1921, but only within the past quarter of a century has systematic research been effected there. From this has come ample substantiation of the economic significance of the field in question. Also other potential geothermal areas in the Andean Cordillera at an average elevation of 4250 m above sea-level have been identified. Perhaps the most promising are at Puchuldiza and Suriri with potential outputs of 180 and 50 MW, respectively, but El Tatio is by far the most advanced with an estimated potential exceeding 200 MW. Existing producing wells contain geothermal fluids at about 250°C and can supply a 50-MW electric power-generating plant. For this about 500 tonnes/h of steam are necessary. The first 30-MW unit was planned to commence work in this decade.

(ii) The rest of South America. Investigations are proceeding into the geothermal energy potentials, if any, in a number of other countries in Latin America, including Argentina in the Capahue zone of Neuquen Province; Bolivia at Salar de Empexa and Laguna Colorada; Brazil: Colombia in the Ruiz volcanic massif; Dominica; Ecuador; Honduras; Panama at Cerro Pando, Chiriqui Province; Peru; and Venezuela in the El Pilar–Casanay area.[24,38]

4.5.1.2 The Old World

Although accounting for less geothermal-energy-generated electricity output from installed power plants in the late 1980s, the region contains some big producers such as the Philippines, Italy and Japan as well as the most richly endowed area in comparison with its size and population, namely Iceland.

(a) Europe. Taking into account its small size and population, Iceland is the most fortunate country in the world from a geothermal energy standpoint and is discussed first. Afterwards the relevant countries are examined, passing from the Azores through the United Kingdom, France, Italy, Hungary, Greece and the USSR to the rest of this continent.

(i) Iceland. Iceland is perhaps the most fascinating part of the Old World and possibly the most fruitful geothermal region on Earth, its thermal gradient exceeding 150°C in the first kilometre and representing an active volcanic belt transgressing the country from southwest to northeast

FIG. 4.7. Map of Iceland showing the main high-temperature geothermal fields lying roughly along a southwest-to-northeast belt of relatively young neovolcanics.

(see Fig. 4.7). There is a relatively low-temperature zone in the west and south which flanks this high-temperature one. The potential of the main thermal belt, ignoring exploitation problems arising from the inaccessibility of many of the resources, is thought to be 6000 MW for at least half a century.

First employed in 1921 in order to heat some houses, geothermal energy is utilized widely today and space heating is still its main use. In fact, about two-thirds of the entire population obtain home heating from geothermal fluids and the total capacity of the Reykjavik District Heating System ten years ago was about 540 MW (Fig. 4.8). However, geothermal energy is applied also to agriculture in regard to greenhouse heating and smaller-

FIG. 4.8.   Blue Lagoon, Iceland.

FIG. 4.9.   Stóra Víti, Mt Krafla, Iceland.

scale industrial uses include the famous diatomite factory at Lake Mývatn (from which the material comes) and a factor for drying seaweed at Karsley

Hydroelectric power is abundant in Iceland and so geotherma development was not as rapid as it might otherwise have been. However now it is widely used and, by 1981, the total installed geothermal capacity o Iceland was some 600 MW thermal (MWt).[39] Of this about 500 MW represented the net geothermal consumption at that time and most of i came from rather low-temperature fields, permitting direct usage of the 60–120°C water, for instance in greenhouse cultivation.

A proposal to erect an hydroelectric power plant in northeast Iceland was abandoned in 1973 because of possible adverse effects on the environment. Therefore it was decided in 1975 to construct a geotherma power plant at Krafla, a high-temperature field approximately 16 km north of Námafjall, and to do so concomitantly with drilling for steam (Fig. 4.9) This would supplement a small, 3-MWt geothermoelectric generating plan at Námafjall which was installed in 1969.[24,39,40] At the same time, the Svartsengi high-temperature field was developed in southwest Iceland and a 30-MWt heat-exchange plant now operates there. A 235°C saline geothermal brine is applied to heating fresh water for space heating and domestic usage and a 1-MW electric power generator is employed for the internal power supply of the plant.[41]

Exploration at Krafla had begun in 1970 under the Programme for the Exploration of High-Temperature Areas in Iceland and, as work went on, i became apparent that the reservoir is complex, comprising two geotherma zones. These are an upper, water-dominated one having temperatures around 205°C, and a lower-boiling zone with higher temperatures of from 300 to 350°C. This was not anticipated and influenced the plans for the power plant so that, on completion in 1977, the steam available could produce only 7 MW of electricity, whereas the plant was designed to yield 60 MW. In December of 1975 a volcanic eruption took place about 2 km from the Krafla power plant, actually at Leirhnjúkur, and initiated a rifting event in the fissure swarm which intersects the Krafla Caldera Subsequently, this volcanism continued and affected the production characteristics of the geothermal field.

It is interesting to note that, in February 1988, talks were proceeding between the British Government, in the person of Energy Secretary Cecil Parkinson, and the Government of Iceland as regards the possible construction of a power cable between the two countries in order that Icelandic electric power could contribute to the needs of the United Kingdom.

(ii) Azores. Not too far from Iceland lie the Azores, a Portuguese geothermal resource, wells on one of the islands, Sao Miguel, having produced geothermal fluid with a temperature of 200°C at some 900 m depth. It was planned to utilize these in a small 3-MW electric power-generating plant and, in addition, directly to apply some of them to such activities as space and greenhouse heating, fresh-water production and food processing.[35]

(iii) United Kingdom. Energy Paper 9, published in 1976, assessed the prospects for exploiting geothermal energy in the UK, emphasizing the paucity of relevant data on deep sub-surface geology and hydrogeology. Actually, heat flow information was scanty as well, but showed conditions similar to those in France, i.e. typical of continental land masses removed from active volcanic areas. Since deep aquifers were exploited in France as heat providers, for instance at Villeneuve-la-Garenne, the British Geological Survey was co-opted as the main investigator of geothermal aquifers which might be able to supply 4×10^6 tonnes of coal equivalent (Mtce) by AD 2000. It was proposed to locate and characterize these by means of regional studies of the main sedimentary basins accompanied by supportive deep drilling, and also to define how identified resources may be used and to establish under what conditions their usage would be economic.[42]

Six sedimentary basins in the UK comprising Permo–Triassic sandstones, namely Wessex, East Yorkshire and Lincolnshire, Worcester, west Lancashire and Cheshire, Carlisle and Northern Ireland, were examined by the British Geological Survey. These are optimum areas and the work led to the drilling of the first exploratory borehole at Marchwood in 1979/1980. At the commencement of the 1980/1984 phase, assessment was made of basins outside the Wessex one where the first borehole is located. Drilling at intermediate depths to investigate the 40–60°C resource was excluded.

The following aquifers containing waters above 60°C were identified in 1980.

East Yorkshire and Lincolnshire Basin: the Early Permian Sandstone was believed to be relatively thin and perhaps discontinuous. The Sherwood Sandstone was estimated to be at 40–50°C.

Cheshire Basin: the 60°C isotherm located centrally near Crewe was assessed at 2300–2500 m, with the lowermost 500 m of the Sherwood Sandstone and the Permain Collyhurst Sandstone exploitable. However, the low temperature gradient, *ca.* 23°C/km, is a disadvantage.

Northern Ireland Basin: at Larne, it may be that the 60°C isotherm would be in the Sherwood Sandstone, but 700 m of this formation would be at a higher temperature together with all the Permian Sandstone.

Carlisle Basin: it was anticipated that the Penrith Sandstone would be above 60°C, perhaps reaching 90°C, but this is based only on a single borehole.

West Lancashire Basin: an earlier heat flow measurement demonstrated an unusually high heat flow, but whether this is caused by a localized anomaly or is general throughout the area is uncertain. Additional heat flow measurements were recommended before deep drilling is undertaken.

It is doubtful whether other basins, specifically the older Palaeozoic ones, could be exploited. For instance, in the South Wales, East Midlands and West Pennines Basins, there are no formations with high enough cumulative transmissivities (except in unusual and highly localized areas).

Summarizing the current situation as regards geothermal aquifers above 60°C, initial results are rather disappointing and the economic analysis showed that geothermal heat is not economic compared with current gas prices. A significant real increase in fossil-fuel prices, especially of natural gas, would be essential to justify the cost of deep aquifer schemes. Also geothermal schemes are capital-intensive and entail a major outlay for drilling costs and rig hiring. However, the possible rewards of successful hydrocarbon exploration are so much greater than those for hot water that such services reflect the latter market. Nevertheless, were there to be extensive geothermal activity, a separate market for rig hire might well result and have its own pricing policy. There is even some evidence to indicate that this has occurred in France, where the costs of geothermal drilling are significantly less than in Great Britain. As regards geothermal aquifers below 60°C, these far exceed the 60°C category in areal extent, but the economics of using them are not attractive enough at current fossil-fuel prices to interest local authorities in changing from local heating of apartments and houses to group heating schemes.

(iv) France. In France, low-enthalpy fluids are recovered and used in heating systems, this having gone on since the 1960s. Subsequent development produced at least 20 000 housing units which were so heated by 1978, some in conjunction with other heating methods. Of these, 1704 were at Villeneuve-la-Garenne. However, there were 4000 at Creil, 4000 at Melun-Sénart and 3000 at Melun Almont, all of these being within the Paris region. In addition, there were others at Mont-de-Marsan in the

Aquitaine Basin. At the beginning of the present decade, many more homes were so heated, not only in the Paris area and the Aquitaine Basin but also in eastern France.

The Bureau de Recherches Géologiques et Minières, Paris and Orléans, is effecting a programme of investigations into thermal waters and, in 1986, recorded some interesting observations from the thermal station at Balaruc-les-Bains (Hérault), on anomalous radon levels in the soils as well as characterization of geothermal waters in the Massif Central at Cauterets.[43-45]

Low- and high-energy geothermal resources were under investigation also: the former aimed at developing a second generation of exploitation procedures. Special attention was given to studying the Dogger in the Paris Basin, from which it was inferred that the contained geothermal fluids represent meteoric water which has infiltrated and mixed with other waters as well as exchanging isotopically through water–rock reactions, as C. Fouillac and his associates indicated in 1986.[46] Work on high-energy resources entailed a joint effort regarding the island of Milos, a site approved by the EEC for erection of a laboratory charged with testing and comparing various geophysical exploration methods. In addition, a Franco–German hot dry rock programme initiated in 1977 was continued at Soultz-sous-Forêts (Alsace).[43]

Other overseas work being effected by ORSTOM related to quantitative evaluation of aquifer recharge in an arid environment, specifically Burkino-Faso, for which the physical aspects of the unsaturated zone at the selected site were incorporated in an adapted algorithm of the model suitable for allowing calculation of recharge over a period of several years. A similar model, GARDENSOL, has been tested in France at several locations. The work on site in Burkino-Faso was scheduled to commence in 1987 or 1988.

(v) Italy. Whilst geothermal energy was used industrially in Italy in the last century (Figs 4.10 and 4.11), the Larderello dry-steam geothermal field in the Etruscan Swell has been exploited since 1904 and, in 1905, a 20-kW power plant was installed, this gradually expanding until it yielded 250 kW by 1913, 2·75 MW a year later and 126 MW by 1940.

The Larderello–Travale field (Figs 4.12–4.16) possesses an installed capacity of about 400 MW, but Tuscany has a second geothermal field at Monte Amiata, detected in 1959 about 80 km south of Larderello and with a capacity of over 22 MW. Both of these fields are vapour-dominated and produce superheated steam. At Monte Amiata, there is an excellent cap rock sealing the subterranean permeable complex associated with an adequate supply of water and a geothermal anomaly. However, for

FIG. 4.10. The inside of the crater at Vulcano.

FIG. 4.11. A pumice quarry at Vulcano.

FIG. 4.12. San Martino power station, Larderello.

FIG. 4.13. Le Rancia power station, Travale-Radicondoli.

FIG. 4.14.   Le Pianacce power station, Travale-Radicondoli.

FIG. 4.15.   Steam pipeline feeding Larderello 3 at Casuonova power plant.

comparable initial pressures (20–40 kg/cm^2), temperatures are much lower than at Larderello.

This and other factors have induced some authors to take these geothermal fields as separate sub-types of vapour-dominated systems. One characteristic which they both have in common is a clear halo of mercury around each of them, identified from stream sediments by M. Dall'Aglio and his fellow-workers in 1966.[47] Mercury was known to be connected with thermal springs in California as long ago as 1935 and so this Italian

FIG. 4.16. Steam pipeline feeding Larderello 3 at Casuonova power plant.

instance is valuable in demonstrating its occurrence elsewhere. Later occurrences at Glenwood Springs in Colorado were recorded by R. W. Klusman and others in 1977.[48] They examined concentrations of the metal in soils of six Colorado geothermal areas utilizing regression analysis. This incorporated the secondary effects of pH and the organic carbon, iron and manganese concentrations on mercury in these soils and the technique applied to evaluation of a geothermal anomaly. The results showed that secondary influences on mercury in soils of geothermal areas are important in such work because they can show anomalous mercury leakage.

At Monte Amiata, steam is initially a minor component and, after production and decompression, the initial vapour of high steam content is flushed out of the reservoir and replaced by relatively low-pressure steam of lower gas content resulting from the boiling of water at moderate temperatures.

Another characteristic was noted by R. Cataldi in 1967, namely a trend in fluids produced from the Bagnore field of the Monte Amiata district to go from dry vapour to vapour and liquid water.[49]

According to M. Guglielminetti and C. Sommaruga in 1979, 50 000 m^2 of

FIG. 4.17.   Alfina 1 drilling rig, north Latium.

geothermally heated greenhouses have been proposed for the Monte Amiata area, where a portable seismic station used by E. Del Pezzo and his colleagues in 1975 measured seismic noise.[50,51]

In northern Latium (Fig. 4.17), the geothermal fields at Alfino and Cesano are liquid- and vapour-dominated, respectively. At the former, the water temperature is 150°C and a 10–15-MW power plant is envisaged. At the latter, temperatures up to 300°C have been recorded down to 3000 m depth and the fluids have a high saline content, actually reaching 300 g/litre, which constitutes a problem in regard to exploitation. Yet another geothermal system exists at Lago di Patria in the Phlegraen Fields caldera

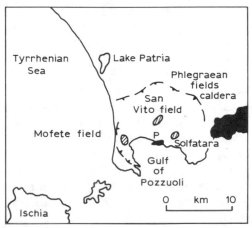

FIG. 4.18. Phlegraean Field, Italy. P represents Pozzuoli; the large black area is Naples.

region of Campania (see Fig. 4.18). The associated fluids have temperatures of 250–300°C, a power plant of 20 MW capacity being foreseen.[24] Natural electric field vectors have been measured here by recording the potential differences between two orthogonal electrode pairs with an electrode spacing of 20–50 m. The 106 stations involved were distributed fairly regularly throughout the region.

Also in this region is the Mofete geothermal field lying some 15 km west of Naples. Here exploration commenced in 1977 and hot waters were located in specific areas. In addition, a geoelectric survey identified low resistivity between Lakes Fusaro and Averno, and a gravity survey demonstrated that a high-density body exists along the caldera. The presence of the latter was attributed to several factors, including dense masses of lava and hydrothermally altered rocks as well as deeper metamorphosed rocks connected to the actual magma chamber. A magnetic survey detected a low-susceptibility region between the two lakes, this again being considered to result from rocks being hydrothermally altered through the circulation of hot fluids. Seven wells were drilled, three vertical and four deviated to intercept fractures. The results were four productive wells, two injection wells and a dry well. However, the final disposition of these will be decided only after data from interference tests in progress have been interpreted.[52] Wells were usually finished with casings and slotted liners. Permeability was found to increase with depth and three aquifers were identified, all in volcanic-sedimentary tuffites and at the following depths: 500–1000 m, 1800–2000 m and 2500–2700 m. Production came from the first two only, the deepest aquifer tapped by a single well rapidly ceasing to contribute due to blockage of the hole by salt deposits. Injection tests to assess reservoir transmissivity, the locations of the permeable horizons, temperature and pressure profiles, and skin effects were succeeded by production tests lasting three days and entailing discharge into artificial ponds with capacities up to 5000 m^3. These were interconnected with pipes and pumps.

As M. Guglielminetti indicated in 1986, the Mofete field is situated in the Campi Flegrei caldera which is subject to seismic activity causing both uplift and subsidence, e.g. a maximum of 1·6 m of uplift was recorded at Pozzuoli harbour over a two-year period.[52] Therefore long-term testing was delayed for some three years. After 1985, the movement stopped and testing was resumed.

AGIP installed 19 stations and a centralized recording and processing base to monitor the seismic activity. During 1983 and 1984, numerous micro-earthquakes with magnitudes under 2 were recorded, but there were only two weak events in the Mofete area. Consequently, long-term testing proved possible and it had the double aim of confirming that a production–re-injection cycle could be kept up without any harmful effects such as scaling and corrosion, flow rate decline, diminution in pressure or drop in temperature and inducing a disturbance within the reservoir of which the analysis could be utilized to assess the reservoir potential. Appropriate surface equipment was installed to measure the flow rate and other factors.

During three months of production, scale inhibitors gave satisfactory results in regard to calcium carbonate scaling, but turned out to be unsuccessful in the case of silica. Silica is a grave problem because of the temperature and composition of the reservoir fluid. In fact, the negative results obtained by the inhibitors may actually endanger complete exploitation of the energy potential because, at temperature exceeding 160°C, much silica deposition takes place. Corrosion tests substantiated findings elsewhere, namely that stainless steel offers excellent resistance and copper alloys are deficient in this respect. The reservoir test derived data from production from well number 1 and re-injection into well number 7D with accompanying monitoring in three other wells. Production went on for three months, monitoring being initiated a month prior to this time interval and continuing for two months after its termination. Satisfactory information was obtained and is being analysed. Data interpretation will promote assessment of this high-temperature geothermal field of limited volume characterized by a general deficiency of permeability.

The work could be extended to other geothermal prospects in the Campi Flegrei caldera, e.g. San Vito just outside Pozzuoli, but these do not seem to be connected hydraulically with Mofete. The 1986 data implied an electric power-generating capacity for Mofete of 10–15 MW.

In the region near Naples which is discussed above electric field intensities were often rather high, frequently exceeding 5 mV/m and there was no apparent directional regularity. But there was a correlation between areas of low resistivity and areas with maximum values of electric field intensity. It may be added that a local decrease in field intensity was recorded at places of surface thermal manifestations such as Solfatara.

A seismic noise survey carried out at Solfatara Crater was reported by P. Capello and others in 1974.[53] Weak thermal activity had been recorded here at the bottom of the crater and is localized mainly in the southeastern zone characterized by fumaroles and boiling mud pools. Temperatures range between 90 and 150°C. The survey in question was carried out in 1973 using 43 stations and a closer grid was used in the southeast of the crater. Care was taken to distinguish traffic and meteorological noise from true seismic noise. Frequency analyses of the latter demonstrated that the shape of the spectra changes greatly from station to station. However, some characteristics were recorded from which it was possible to group the data into three categories. Type 1 had dominant frequencies higher than 20 Hz. Type 2 had single or double peaks between 8 and 16 Hz, and a dominant frequency occasionally in excess of 20 Hz. Type 3 had a spectrum intermediate between types 1 and 2. The most interesting observation is the lack of low frequencies in all spectra. Type 1 originates in areas with maximum thermal activity and the single- or double-peaked type 2 usually occur in the central part of the crater, i.e. where loose materials attain their greatest thickness. Type 3 may well be determined by intermediate conditions and the mechanical properties of the surface layer.

From all this it may be inferred that fumaroles and mud pools are high-frequency and high-amplitude noise sources, also that seismic noise is strongly affected by the mechanical properties of the outcropping rocks. Perhaps a deeper source exists from which a lower-frequency seismic noise is emitted and this may connect with a yet deeper and large-scale geothermal system.

Curiously, there is no surface activity at Pianura, but high values were obtained anyway. As well as in the Phlegraean Fields, Lipari–Vulcano and Ischia were involved in the exercise. The first two plus five other islands comprise the Aeolian Group of the southern Tyrrhenian Sea. All are volcanic, rising from the sea-floor to a maximum elevation of about 1500 m;

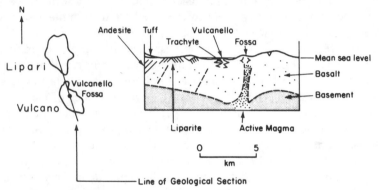

Fig. 4.19. The Italian volcanic islands of Lipari and Vulcano, which are characterized by surface thermal manifestations such as fumaroles and hot springs.

see Fig. 4.19. Outcrops include pumice, tuff and various other pyroclastics. Geothermal activity exists in the form of fumaroles and hot springs, most notably between Fossa and Vulcanello in the northern part of Vulcano. Thirty-eight randomly spaced stations were employed in measuring and the field intensity was high, several millivolts per metre. Again no regularity in the direction of the total electric field vectors was recorded. However, the highest field intensities occurred around areas with maximum geothermal activity at the surface. High-frequency seismic ground noise was observed and is probably connected to the escape of gases at the surface along vents of fumaroles. There is a low-frequency ground noise as well, this occurring over a much greater area and probably reflecting a large-scale geothermal system. Ischia is also a volcano island situated about 30 km west of Naples and possessing outcrops of tuff, loose pyroclastics and lava flows. There are a number of fumaroles and hot springs, e.g. Casamicciola.

Measurements of natural electric field vectors were carried out at 67 stations, mostly near the coast. Similar data were obtained to those recorded at Lipari and Vulcano, i.e. high-intensity and no regularity in field direction. It may be noted, though, that there is apparently no correlation between areas showing surface geothermal activity and the occurrence of maximum field intensity.

In the pre-Appennine belt of western Italy, embracing Tuscany, Latium and Campania, the geothermal reservoirs, under 3000 m depth, are believed to be capable of supplying power plants having a total capacity of approximately 1000 MW. It may be noted that, at Larderello, wells have been drilled as deep as 5000 km. Part of the region lying along the Tyrrhenian coast comprises the above-mentioned Etruscan Swell and is marked by regional uplift of Late Pliocene to Quaternary age with Neogene sediments elevated to as much as 300 m in the central area. Much magmatic activity accompanied this phenomenon, starting in the Messinian and entailing a number of intrusive and volcanic episodes associated with young mineral deposits and, of course, such famous high-enthalpy resources as the dry-steam field at Larderello and Monte Amiata with a combined electric power output of 459 MW in 1984, according to A. Merla in 1986.[54]

Apart from the generation of electric power, there has been rather little usage of geothermal energy. Some greenhouse and domestic heating has been undertaken, these utilizing approximately 6·5 MW over almost 27 000 m^2 and over 100 MW, respectively. In addition, there remains the boric minerals industry, which uses about 27 MW.

(vi) Hungary. In Hungary, thermal springs have been known since ancient times and occur in a number of areas such as the western part of Budapest and around Héviz at the southwestern end of Lake Balaton. Their curative properties have been used for centuries. Since the last century, exploration revealed other thermal manifestations, especially in the Pannonian back-arc Basin which probably evolved in connection with subduction of oceanic lithosphere of the Carpathian flysch trough towards it. Heat-flow measurements in the Mecsek Mountains are anomalously high at 103–139 mW/m^2. There are many thermal water-bearing formations in the Pannonian Basin, e.g. Triassic and Eocene dolomites and limestones, Miocene sandstone and limestone, and Pliocene and Pleistocene sands. Many wells are artesian and lack of hydrostatic equilibrium promotes surplus pressure, causing water migration in the sediments.

Mineralization increases with depth. The dominant geothermal resource is the Upper Pannonian reservoir system which covers Hungary and parts of its neighbours Austria, Czechoslovakia, Roumania and Yugoslavia as well. The waters contained in it are low-enthalpy, but the geothermal energy reserve which it represents is enormous, as P. Ottlik and his associates indicated in 1981.[55] The following data were given by T. Boldizsár in 1978.[56] For the Tertiary, 160 000 km^3 of sediment store 4.4×10^{22} J energy; for the Pannonian down to 1 km, 38 000 km^3 of sediments store 0.8×10^{22} J energy and, for the Pannonian below 1 km, 18 000 km^3 of sediments store 0.3×10^{22} J of energy. The average yield of Pannonian wells is about 1500 litres/min and the total dissolved solid content of the water ranges from 2000 to 5000 mg/litre.

The uses of thermal water are balneological and therapeutic as well as agricultural and, of course, some waters are applied to space and district heating. There are 214 thermal wells for the first purpose, including famous centres such as Budapest, Héviz, Debrecen and Zalakaros. Of these, no less than 28 are in the capital, together with almost 100 thermal springs. Buildings around such places are heated geothermally and the total estimated calorific value of geothermal energy used for district and apartment heating in 1981 was ca. 1.25×10^{14} J annually.

However, uses in agriculture are more important, especially in horticulture. Greenhouses are initially heated at temperatures between 60 and 100°F (15–38°C), after which the emerging lower-temperature water is routed to plastic tents and tunnels, and eventually to soil heating. Altogether some 500×10^3 m^2 of greenhouses and 1200×10^3 m^2 of the plastic facilities are heated by thermal water. In fact, four-fifths of all greenhouses in the southern portion of the Great Hungarian Plain are so heated.

Thermal water is also used domestically. Multi-purpose utilization is also practised, e.g. near Szeged, where two thermal wells supply hot water for heating 1000 apartments, domestic hot water for 400 apartments and hot water for the local swimming pool.

(vii) Greece. Geothermal investigations began here in 1970 and some promising areas have been identified, notably the islands of Milos (where, as noted above in the section on France, the EEC has approved the construction of a testing and comparison laboratory for the study of alternative geophysical investigatory methods) and the well-known Lesbos together with the Sperchios River graben. In 1977 a drilling programme was commenced on Milos and some exploratory wells have been completed. Construction of a 15-MW power plant was mooted also.

However, this project was not included in the development programme for the period 1977 to 1986.[24]

(viii) USSR. Regional exploration during the 1960s showed that thermal areas occur within western Siberia, particularly near the Ural Mountains, as well as in North and Trans-Caucasia, northern Kazakhstan, the Turanian Basin and sectors of the adjacent republics of Tadzhik and Kirgiz together with the Carpathian Mountains, the Kamchatka Peninsula and the Kurile Islands. There is a 5-MW power plant operating in the southern part of the Kamchatka Peninsula at Pauzhetka and a small 680-kW binary cycle plant was tested at Paratunka near Petropavlovsk–Kamchatskiy, the latter using 80°C waters with Freon as a working fluid. In one of the southern Kurile Islands near Yuzhno–Kurilsk, a 12-MW power plant was proposed. Development may continue also at Pauzhetka where the potential is estimated at 30–50 MW. Some volcanic areas of the Soviet Union have 100°C surface waters and, at 100–400 m depth, some parts of Kamchatka and the Kurile Islands have temperatures up to 200°C. It is believed that about half of the entire territory of this vast country is underlain by economically exploitable thermal waters. The geothermal reserve is estimated to be $7\cdot9 \times 10^6$ m^3 daily at a depth of 1000–3000 m and having temperatures ranging from 50 to 130°C. Some of this is utilized in space heating, particularly in the Dagestan Republic and Georgia where, for example, Makhachkala, a town with 250 000 inhabitants, supplies two-thirds of its population with geothermal heating. The town of Izberbash is heated similarly.

(ix) The rest of Europe. In Spain, investigations have been undertaken at Lanzarote in the Canary Islands and in Catalonia on the mainland.

Hot dry rock researches have been carried out in the Molasse Basin in Switzerland and similar work was undertaken in the Bohus granites of Sweden. Investigations have been effected also in the Federal Republic of Germany, Poland, Bulgaria and Yugoslavia. A study has been made of two geothermal areas of the Western Plain of Roumania so as to obtain data regarding the thermal aquifers there. In the first, Oradea–Satu Mare, the Lower Pontain thermal aquifer was found to contain waters deriving from two base components and of different ages, one 35 000 years old and the other fossil. It is unlikely that recharge occurs, therefore, but its waters, ranging in temperatures from 35 to 88°C, could be utilized geothermally. A deep aquifer in the Mesozoic probably undergoes natural recharge. In the second, Arad–Timisoara, Pontian aquifer thermal waters in the north have rather high temperatures even at shallow depth and could be economically exploited.[57]

Large-scale geothermal programmes exist in Turkey, especially in its western part, and these have proved to be so successful that it has been suggested that by the end of this century about 10% of the electricity requirements of the whole country could come from geothermal sources. At present there is a 20·4-MW geothermoelectric plant at Kizildere, although problems of calcium carbonate scaling were encountered. In addition, there are wells near Afyon (this word means 'opium' in English) which met 160°C thermal waters at depths around 900 m. These are intended for use in space heating.

(b) Africa. In the existing Red Sea, a nascent ocean is now dilating at a rate of about 2 m annually and dividing the African plate from the Arabian shield. The former was stationary through most of Phanerozoic time, but is at present moving northwards to ram Europe and simultaneously splitting along the Great Rift Valley system which contains a potentially vast geothermal resource so far exploited only in Kenya. This Tertiary–Quaternary tectonic structure is one of the largest on Earth, extending for almost 10 000 km from Mozambique in southern Africa to Syria in the Near East, a distance covering about one-sixth of the circumference of the planet.

(i) Kenya. In Kenya, geothermal energy makes a valuable contribution to energy needs and the main potential lies within the Great Rift Valley, three main fields having been identified, namely Olkaria (see Chapter 2), Eburru and Lake Bogoria. At the first, located a few kilometres south of Lake Naivasha, geothermal exploration began in the 1950s and by 1958 two exploration wells had been drilled there. They did not produce steam and, because of the intensive development of hydroelectric power, further work was delayed until the early 1970s. Then six additional wells were drilled and showed that power production was feasible. During the past ten years, production drilling has been carried out and a 45-MW power plant constructed. The first unit came on-line in July 1981, the second in 1982 and the third in 1985, as G. S. Bodvarsson and his colleagues indicated in 1987.[58] The total installed capacity here is expected to reach 90 MWe.

(ii) Ethiopia. Exploration commenced in Ethiopia in 1968 and a number of geothermal areas were identified, mostly along the rift system. The first detailed study was made in the Aluto–Lake Langano region, approximately 100 km south of Addis Ababa, where it was believed that reserves should be able to supply geothermoelectric plants to a total capacity of 50 MWe. This had been selected as one of the three priority

areas in the Ethiopian Rift Valley during a reconnaissance investigation which was carried out by Ethiopian–United Nations Development Programme personnel in 1970 and 1971. The first deep geothermal exploratory drilling site for generating electric power is near the Aluto volcano situated between *en echelon* offsets of the Wonji fault belt. The Ethiopian part of the Great Rift Valley consists of several units, of which the most important are the Lake Rudolf Rift, the Lake Stefanie Rift, the Main Ethiopian Rift and the Afar Depression. Direct information came from eight deep exploratory wells.

The geological data from the top downwards include Aluto volcanic pyroclastics of Quaternary age, upper lacustrine sediments, welded and non-welded tuffs with rhyolite breccias and pyroclastics, lower lacustrine sediments, fine-grained non-porphyritic and porphyritic trachytic–basaltic layers interbedded with thin beds of ashes, tuffs and rhyolite (the Bofa basalt, Pliocene in age) and crystalline ignimbrite with minor rhyolite and basaltic layers (Miocene). Convective circulation of fluids was revealed by the presence of hydrothermal minerals in the basalt units presumed to constitute the geothermal reservoir, these including calcite and chlorite. The highest temperature measured in Langano well 6 exceeds 330°C, as Z. Gebregzabher indicated in 1986.[59]

As well as the Aluto–Langano geothermal field, there is the Corbetti geothermal project some 60 km south-southeast of it, and about 80 km southwest of that is the Abaya geothermal prospect.

(iii) The rest of Africa. Elsewhere in Africa only preliminary work has been done in several countries, among them Morocco, Algeria, Egypt (along the Red Sea coast and in the Gulf of Suez), Chad, Djibouti, Zaire and Tanzania. In Uganda, there are numerous areas of thermal discharge such as the Buranga hot springs covering 4 km² of the Sempaya Valley which emit 500 litres/min of water having a maximum temperature of 98°C.[60] The Acholi thermal area is also of potential interest for geothermal development. In addition, there are hot, mineralized springs in Zambia along young, deep major faults cutting Karroo sediments. Others relate to the East African Rift system.[61]

(c) Asia. The Philippines is the largest producer of geothermal energy-derived electrical power in Asia (894 MWe), 17·5% of the needs of that country, but much more is known about the geothermal fields in Japan.

(i) Israel. In the Near East, a comprehensive research programme was initiated in 1976 in the Lake Kenneret and Emeq Israel areas of Israel as well as in the Dead Sea and Negev to the south. In the Negev 40°C waters

are intended for agricultural use. There appear to be no high-temperature reservoirs at depths under 3000 m. Nevertheless, there are aquifers of temperature less than 90°C at reasonable depth in the Dead Sea Rift. It may be possible to employ hot springs in the north in shrimp farms, an export industry.[62] Incidentally, a number of preliminary investigations have been made into geothermal prospects in neighbouring Jordan, but without definite results as yet.

(ii) Iran. In the Middle East, Iran is a promising geothermal region and four areas have been identified to date. The first is Maku–Khoy in the extreme northwest of the country which is bounded to the west by Turkey and to the north by the USSR. Then there are two extinct volcanoes near the carpet centre of Tabriz, namely Mounts Sabalon and Sahand. Finally, there is Mount Damavand lying about 100 km east-northeast of the capital, Tehran; see Fig. 4.20. In 1978 the author worked on these sites, particularly the last, where a 50-MW electrical power-generating plant was planned to supply Tehran. The first stage of this operation developed from an agreement between the Ministry of Energy of the Imperial Iranian Government and the Italian ENEL company, which carried out a reconnaissance survey starting in 1976. Mount Damavand was considered an optimum prospect because it still has occasional fumarolic activity, its local geology is well known and it is relatively easy of access. Many hot springs exist near it and there is an excellent cap rock, the Shemshak Sandstone (basal Jurassic, Liassic in age). Extensive tectonic activity has fractured many of the strata. Mount Damavand lies in the Alborz Mountains where strata up to 10 000 m thick occur and, according to M. Rassekh and his co-workers, comprise Precambrian to Middle Triassic rocks in a platform and a Late Triassic to Neogene parageosynclinal succession.[63]

Two alkaline sub-volcanoes intruded into the middle Nur Valley of the Central Alborz not far from Mount Damavand into sediments of the Shemshak Formation of the Late Triassic to Liassic (Jurassic) were dated by the rubidium–strontium method. For the volcanic plug of the Gel-e-Pir, an age of 3.0 ± 1.3 Ma was obtained and, for the trachytic deposits found near Baladeh, an age of 13 ± 7 Ma. These magmatic rocks are taken therefore as representing equivalents of the Neogene granite intrusion of the Alam Kuh and, in fact, as rather older than effusives in the Mount Damavand volcanic locality.[64] The entire region was affected by the tremendous orogenic events of the Tertiary and, like most of Iran which lies within the site of the extinct Tethys Ocean, it is extremely unstable tectonically; see Fig. 4.21. Indeed, Mount Damavand is a recently formed

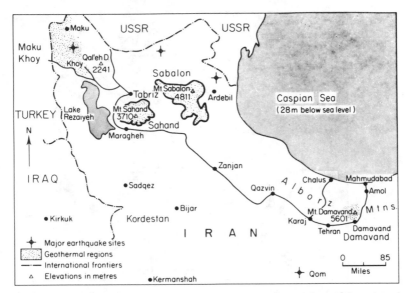

FIG. 4.20.   The geothermal areas of Iran. Sites of former major earthquakes are
shown and lie within extensive areas of recent volcanicity extending into Turkey
(including Mount Ararat), these demonstrating plate movements and the long-term
effects of the Alpine orogeny. On the Maku–Khoy area there is widespread
fracturing and faulting whilst, to the southeast, the Alborz Mountains are of Middle
Tertiary age and represent old Tethyan Ocean sediments. Mount Damavand is not
extinct, as are Sahand and Sabalon-e-Kuh, but fumarolically active. All the areas
are characterized by many mineral and hot springs, and thus constitute potential
geothermal resources. The occurrence of significant amounts of tritium in
groundwater at Mount Damavand shows that a suitable aquifer and circulation
system exist there.

volcano, as is known by the absence of evidence of glaciation upon it. Hence
it attained its present elevation of 5601 m after the last glaciation ended, i.e.
within a few thousands of years of historic times. Observations showed that
there is a first-class circulation system for groundwater and the thermal
water possesses significant quantities of tritium, indicating rapid transit
time for the recharge. The location of the actual geothermal anomaly is to
the north of Mount Damavand, on the Caspian Sea side. It is a great pity
that the development of its geothermal energy was aborted by the advent of
the existing regime in that unhappy country, all geothermal regions of
which are shown in Figs 4.20 and 4.21.

   (iii) Pakistan. East of Iran lies Pakistan: in this country there are
geothermal manifestations in the form of hot springs along the Main

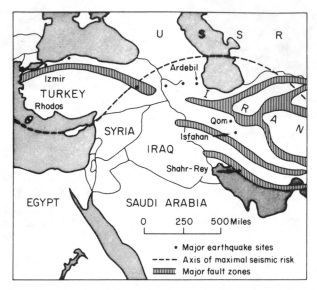

FIG. 4.21.   Maximum seismic risk axis for the Iranian Plateau and the Middle East.
Major fault zones are shown together with earthquake sites. These demonstrate the
seismic instability of Iran which is related to the thrust northeastwards of the
Arabian plate against the Eurasian one. The eastern Iranian area of the Kut Block is
believed to be a stable region, but suffers as much seismic disturbance as any other
part of the country.

Mantle thrust and the Main Karakorum thrust in the Chilas and Hunza
areas, respectively. T. A. Shuja in 1986 alluded to these and mentioned that
the concentration of hot springs in Sind province also indicates geothermal
activity.[65] In addition, there is a string of thermal seepages and springs
along the alignment of the Syntaxial Bend in Punjab province. In southwest
Pakistan, in Baluchistan province, there is a graben structure, Hamun-e-
Mushkel, showing geothermal prospects on the basis of aeromagnetic
studies.

(iv) India. Also in the Middle East is India, where geothermal
investigations commenced in 1973 with the launch of the Puga geothermal
project in Ladakh. The geothermal fields involved are situated across the
Great Himalayan Range in the Indus Valley at altitudes exceeding 4000 m.
The resistivity method was used to locate the geothermal reservoir.
Numerous hot springs occur and have temperatures varying from 30 to
84°C, the latter being the boiling point at the relevant altitude, which is
4400 m at the southern boundary of the Indus Suture Zone, the site of the

collision junction of the Indian and Asian plates. Thus, as C. L. Arora indicated in 1986, the Puga System lies at 4400 m and the thermal manifestations are restricted to the eastern part of the Puga Valley which is broadly aligned along the perhaps faulted crest of an anticline encompassing the Precambrian sequence of paragneisses and schists striking N 70° W and dipping 15° to 45° either way.[66] Resistivity depth probes showed that a conductive zone is present, this having a value of 10–25 ohm m and a thickness varying from 50 to 300 m over an area of 3 km². These data imply a shallow geothermal reservoir. Thermal surveys also indicated a significant anomaly corresponding to this zone which, when drilled, met with a reservoir of wet steam with a temperature of up to 135°C, thus confirming the results of the resistivity survey. Thus wells drilled at Puga are steam-producing and a small-scale thermoelectric power plant is under construction. The geothermal fluids are also earmarked for a small chemical refinery and for space heating. Similar results have been obtained in the adjacent area, in which thicker zones with moderate electrical conductivity have been recorded.

Another application of the method was made in the NNW–SSE-trending West Coast geothermal belt of India, this being covered by Cretaceous–Eocene Traps (basalts). The region is characterized by many hot springs having temperatures up to 70°C, these following a 400-km long alignment associated with steep gravity gradients and an isolated occurrence of native mercury in the zone of a gravity high. Thermal waters close to this western coastal belt in Maharashtra State generally discharge Na–Ca–Cl and Ca–Na–Cl types through the associated basic lava flows. K. Muthuraman in 1986 referred to experimental work made to study reactions between dilute seawater and basalt conducted in static autoclaves at selected elevated temperatures.[67] This demonstrated the possibility of producing chloride waters with relatively high calcium similar to the thermal waters in question. Taking into account the increase in calcium in the resultant solutions during seawater–basalt reactions at higher temperatures, the base temperatures computed by using Na–K–Ca geothermometry would be much lower than the actual temperatures of the system. At lower temperatures of about 100°C, absorption of potassium by basalt is feasible and therefore alkali geothermometry also may not be reliable for such systems. Anhydrite saturation temperature appears to constitute a reliable geothermometer for such coastal thermal water systems which involve a seawater component. Results of computer processing of the chemistry of some of the thermal waters, using the WATEQ program, are interesting. Two are oversaturated with diopside, tremolite, calcite and aragonite, showing a rather low temperature of origin. In two other cases, interactions with ultramafic rocks is demonstrated because the waters are oversaturated with diopside, tremolite, talc, chrysolite, sepiolite and its precipitate. There is no real evidence of west-coast thermal waters having emerged directly either from marine evaporites or oil-field waters. K. Muthuraman proposed that most of them may have originated through interaction of an admixture of seawater and meteoric water with the local basalt flows at some elevated temperatures.[67]

Also in Maharashtra State, actually some 160 km northwest of Nagpur at an elevation of 675 m, there is a group of hot springs (44–47°C) discharging with a total of some 3 litres/s. Water samples from these Tattapani and Salbardi hot springs were classified as sodium bicarbonate in type by V. K. Saxena and M. L. Gupta in 1986.[68] Waters in deep aquifers are associated with quartz, shale and clay terrains. Activity studies made by these researchers on the minerals and waters found in the aquifers implied that the thermal waters are in equilibrium with montmorillonite, kaolinite and quartz at approximately 150°C. The geochemical thermometers estimate $150 \pm 10°C$ as aquifer temperatures for the hot springs in question.

In India as a whole, it has been demonstrated that there are at least 34 geothermal systems capable of generating electric power, all with fluid temperatures exceeding 150°C. The total potential was estimated at 1279 MWe for 30 years. In addition, there are also 42 medium-temperature systems, which have temperatures ranging from 90 to 150°C and are exploitable for non-electrical purposes. Perhaps the most interesting regions are in the north and west of the sub-continent. In the former, as noted above, a very important one is in the Puga Valley in Ladakh District. Others include the Parbati Valley and the Beas Valley in Kulu District, in Himachal Pradesh and Sohna in the Gurgaon District, Haryana. At Manikaran in Himachal Pradesh, the geothermal energy is applied to refrigeration. In the west of India, they are in the Thana District and Maharashtra. A survey of the geothermal resources of India was given in 1977 by M. L. Gupta and R. K. Drolia.[69]

S. K. Guha in 1986 gave an excellent summary of results attained by geothermal investigations in the sub-continent by shallow and deep drilling, the two most promising being Puga and Tattapani.[70] Referring to other likely areas, he mentioned that at the Parbati Valley (Manikaran) project geothermal exploration commenced in 1974 and expanded, with United Nations assistance, in the period from 1976 to 1980. There are six thermally anomalous places in the valley and Manikaran is the optimum one, extending for over 1·5 km and having a maximum temperature of 95°C in the thermal springs (at their altitude the boiling point of water is 88°C). It exists in the jointed and fractured Manikaran quartzites which may be of Precambrian age and underlie carbonaceous phyllites and limestone. Isotopic geothermometry has suggested that temperatures up to 160°C may exist at depth and a maximum of 109°C has been encountered within 100 m depth in shallow drilling work.

The Beas Valley project began in 1976 and demonstrated the existence of a geothermal anomaly extending about 70 km along the valley. The highest surface temperature is recorded as 57°C at Bashisht. Most springs flow through the valley and ten shallow thermal gradient boreholes from 50 to 200 m deep have been drilled. One developing thermo-artesian involves rocks of the Delhi system (late Precambrian) exposed in two longitudinal ridges and separated by an alluvial tract. Studies were effected between 1974 and 1977, and showed a higher-than-normal thermal gradient. Thermo-artesian flows were recorded in four wells with a maximum temperature of 55°C. Cation geothermometry indicated a base temperature of 100°C.

Guha also alluded to geothermal work at Simla and Kinnaur Districts, Himachal Pradesh State.[70] Geothermal activity was initiated in the Sutlej Valley in 1978, where thermal springs extend for some 90 km and emerge from a sequence of interbedded quartzite, gneisses and schists intruded by granites and pegmatites. Temperatures range between 36 and 60°C, but geothermometry implies base temperatures up to 145°C. Appropriate data are given in Table 4.9.

(v) Nepal. Thermal springs have been found in several parts of Nepal, one group north of the Main Central thrust, a second group close to this and a third group near the main boundary fault in the Siwaliks. Those in the second group have the highest temperatures.[71]

(vi) Thailand. The Phrao Basin is well known for its geothermal and seismic activities. It is one of the Tertiary–Quaternary basins in northern Thailand and the main features of the region are north–south-trending ranges separated by it. The Western Mountain Range lies to the west and the Sukhotai Fold Belt to the east. The former comprises Precambrian and Palaeozoic rocks with granitic intrusions and constitutes part of a high range extending from the Tibetan Plateau to the Malacca Peninsula. The Sukhotai Belt is characterized by sedimentary sequences of Palaeozoic to Mesozoic age forming low mountain ranges.

The existing boundary between the two palaeocontinental margins may correspond to a major crustal discontinuity and lies in a narrow belt, the

TABLE 4.9
Some physical parameters of geothermal manifestations and boreholes in various geothermal areas in India (see Fig. 4.22)[70]

Location	Geology	Temperature (°C) Surface	Base[a]	Discharge (litres/min)
Puga	Precambrian–Palaeozoic gneisses and schists with reconsolidated Quaternary breccia	84	225	1 241 (springs, cumulative) 190 tonnes/h (boreholes, cumulative)
Alaknanda	Precambrian gneisses and schists	64	150	800 (borehole) 430 m deep, bottom = 92°C
Sutlej Valley	Precambrian low-grade metamorphics	50	145	200 (borehole) 183 m deep, bottom = 60°C
Beas Valley	Precambrian phyllites and schists	57	180	1 000 (borehole) 450 m deep, bottom = 43°C
Parbati Valley, Manikaran	Precambrian quartzites	95	180	1 800 (springs, cumulative) 1 500 (boreholes, cumulative) 706 m deep, bottom = 109°C
Sohna	Granite and basic intrusives of Precambrian Delhi system faulted in a graben	47	100	25 547 m deep, bottom = 550°C
Unhavre (Khed), Maharashtra	Deccan Trap basalts near dikes and faults	71	120	870 350 m(?) deep, bottom = 85°C
Tattapani–Jhor	Migmatites, gneisses/Archean near boundary of Gondwana rift basin	88	160	400 (springs, cumulative) 600 ≤ 0 (borehole) Bottom = 108°C
Cambay Gujarat	Quaternary and Tertiary sediments in graben	45	176	Not measured Bottom also not measured
Bhimbandh (Monghyr–Rajgir), Bihar	Faulted Precambrian quartzites	65	100	240
Surajkund (Hazaribag), Bihar	Faulted biotite gneisses (Archean)	88	166	240
Bakreswar	Faulted gneisses (Archean) overlain by recent alluvium	72	180	630

[a] Estimated from geothermometry.

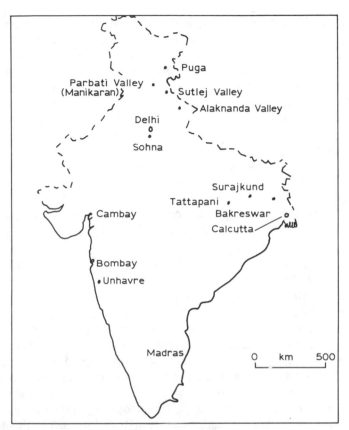

FIG. 4.22. The major hot springs of selected geothermal areas in India.

Chiang Mai-Fang deep Cenozoic Basin and granitic mountains. Recent tectonic movements took place during distension and were accompanied by partial melting of crustal material and also by diapirism of granite magmatism. The thermal springs of the northern part of the country relate to this regional tectonic setting are are scattered throughout the region, rising from a mainly granitic basement along large-scale active regional faults connected with the final phase of distensional tectonics. Northern Thailand actually belongs to a rather old orogenic belt originating after the collision between two continental masses with the deformation of their margins.

There are at least five geothermal areas along the western margin of the Phrao Basin, these probably being controlled by the major fault extending northwards from the San Kampaeng geothermal field. The intermediate depth of the geothermal reservoir here may yield sufficient hot fluids for small-scale usage.[72] Geochemical surveys have been effected in the Phrao Basin since 1984 with the object of identifying the presence of geothermal reservoirs. The temperatures of hot springs were found to range from 110 to 130°C on the basis of quartz concentrations with maximum steam loss, but gas geothermometers gave figures of 235 to 255°C. The thermal waters fall into the alkali–sodium–bicarbonate type. The decline of bicarbonate concentrations, total dissolved solids (TDS) and conductivity from the most northerly hot spring to the southern one may result from the fact that the most northerly spring, CM-13, is nearer to geothermal resource than the others.

(vii) China. There are at least 3000 active hydrothermal systems here, some, for example those in Tibet, related to the plate marginal group and others, such as that at Tianjin, related to intra-plate systems not connected with young igneous activity.[73] Many usually rather shallow wells were drilled after 1970 in various parts of the east and south with the object of evaluating their geothermal potential, if any. The resultant fluids rarely exceeded 100°C in temperature. Up to 1980 some seven small-scale geothermoelectric power plants were erected, all with a capacity under 1 MWe and either operating on a binary cycle or driven by flashed steam under less than atmospheric pressure.[24] The optimum locations appear to be the Xizang Autonomous Region of Tibet, western Yunnan Province (in particular its Tengehong District) and on the island of Taiwan, but, in Taiwan, investigations established that the geothermal fluids are too acidic to be utilized easily. On the mainland, a well drilled in 1976 in Yunnan, an area of recent volcanism, met geothermal fluids having a temperature of 150°C just 26 m below the ground surface. In Tibet, the Yangbajain geothermal field seems to be among the most significant of the Himalayan Belt and it has been estimated to have an enormous potential of approximately 180 MWe (see Fig. 4.23). At least 15 wells have been drilled down to a maximum depth of 600 m and a steam/water mixture was obtained at a temperature of about 175°C.[74] In 1977 a pilot geothermo-electric plant began working with a capacity of 1 MWe and other plants were later under construction.

By 1986 nine small-scale pilot units in China (outside Yangbajain) had been constructed, but cannot be run safely.[75] There is also the Tianjin hydrothermal field, 130 km from Beijing.[75] It covers 3500 km² and its hot waters lie in a four-aquifer/three-aquitard multilayered hydrothermal system. Four production areas have been designated. The field is situated within the North China Basin, *ca.* 150 000 km^2 in extent, and is bordered to the north by the Yanshan Fold zone, to the west by the Taihang Shan swell, to the south by the Shandong Uplift and to the east by the Jiao swell. As well as a Quaternary aquifer, there are two regional ones in the Tertiary and Early Palaeozoic–Sinian, respectively. A lateral interconnection, the Regional Aquifer Branching, exists between these and relates to the limit of the regional aquitard. Isotopic studies have shown that the regional groundwater system is recharged in correspondence to the Yanshan in the north, Taihang Shan in the west and Tai Shan in the south where Lower Palaeozoic–Sinian formations outcrop. Infiltrated water flows first in the

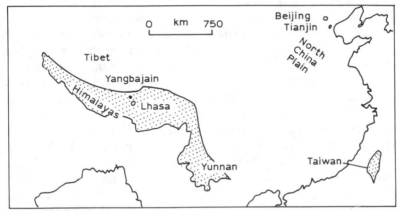

FIG. 4.23. Geothermal resources in China and Taiwan; high-temperature belts (Himalayas and Taiwan) are stippled.

Lower Palaeozoic–Sinian aquifer, then divides at the Regional Aquifer Branching. One part flows into the Tertiary aquifer and the other continues flowing in the same aquifer and, penetrating deep subsidence zones, acquires geothermal temperatures. Because of the hydraulic conductivity of the aquifer, the deep hot water flows to the uplifted zones of the reservoir from which heat is transmitted to the Tertiary reservoir mostly by conduction through the interbedded regional aquitard. The Tertiary upper reservoir is a sandstone and siltstone 150 m thick with low-temperature waters at 35–45°C. The Tertiary lower reservoir is a sandstone and siltstone horizon 300–400 m thick containing medium-temperature waters at 45–65°C. The Ordovician reservoir is a karstic limestone 200–300 m thick and contains rather high-temperature waters at 50–80°C. Finally, the Sinian reservoir is a fractured dolomite and limestone horizon more than 2000 m thick and contains high-temperature waters at 78–96°C. All of the thermal waters are of high alkalinity. During more than ten years of intense exploitation, there do not appear to have been any changes either in the temperatures or in the hydrochemical properties of the thermal waters. To 1986 approximately 200×10^6 m^3 of hot water were extracted from the upper and lower reservoirs of the Tertiary.

(viii) Japan. Japan is richly endowed with geothermal energy; indeed, investigations of it were started 70 years ago, so that by 1919 wells had been drilled in the Beppu geothermal field in a search for steam. By 1925 electricity was being produced from geothermal fluids, but the relevant geothermoelectric plant was very small with a capacity of only 1·12 kW. By 1940 a second small plant had been constructed on the Izu Peninsula on Honshu Island. Systematic researches commenced in the 1960s and by the end of 1977 the country had a total installed capacity of 168 MWe contributed from plants distributed over the geothermal fields at Matsukawa (22 MWe), the first to be exploited in Japan, Onuma (10 MWe), Onikobe (25 MWe), Otake (11 MWe), Hatchobaru at Takinoue (50 MWe) and Kakkonda (50 MWe). Total production in 1977 was 751 128 MW h. Another plant was constructed at the Mori geothermal field in Hokkaido.

The Kakkonda geothermal power plant mentioned above was completed by May 1978 at Takinoue about 50 km northwest of the city of Morioka in an area known as a hot-spring bath centre from antiquity. In fact, it is only 8·5 km southwest of Matsukawa, lies along the Kakkonda River and is situated in the Hachimantai National Park at an elevation of 650 m. Over a least a decade, the Geological Survey of Japan participated in geothermal development there. Also the Japan Metals and Chemical Company made a detailed geological survey together with a rock alteration survey carried out using X-ray diffraction technology. The power plant was preceded by drilling of six exploratory wells from 1972 to 1974 followed by 11 production and 15 injection wells completed by August 1977. Steam collection pipelines (3·5 km) and the plant itself were constructed between May 1974 and November 1977. Test operation commenced in December 1977 and output was gradually increased to 50 MWe by May 1978.

The geology of the Hachimantai Volcanic Region, in the centre of which the Takinoue geothermal area is located, comprises one of the biggest Quaternary volcanic fields in the country and has an area of over 800 km². Five formations are included, namely Palaeozoic sedimentary rocks, Cretaceous granite, Neogene Green Tuff (Early to Middle Miocene in age) and sediments (Middle Miocene to Pliocene in age), Plio–Pleistocene

Tamagawa Welded Tuff and the aforementioned Quaternary volcanoes. The Hachimantai region is considered to be a large-scale tectono-volcanic depression infilled with Tamagawa Welded Tuff and Quaternary volcanics covering 50 km north–south length and 40 km east–west width. There are surface geothermal manifestations everywhere, from which it may be inferred that the area favours hydrothermal systems. A gravity survey indicated possible uplifted and basin structures in the basement. Fracture zones are believed to be mainly northwest–southeast and east–west in orientation.

In the Takinoue geothermal area itself, the Neogene sedimentary sequence outcrops along the Kakkonda River as an inlier in a region covered with Tamagawa Welded Tuff and Quaternary volcanics. The inlier was created by a structural dome with northwest–southeast fold axes and also by northwest–southeast and east–west faults. Geothermal fluid is met with in fractures within the sedimentary layers of the upper Takinoue-onsen and Kunimitoge Formations (Neogene) and also in fractures within the Obonai Formation (Early Neogene). The geology is actually quite different from that observed at Matsukawa, where the geothermal fluid lies near the boundary between the thick Tamagawa Welded Tuff and the Neogene sedimentary sequences, the former acting as a cap rock. On the other hand, in Takinoue, there is no comparable cap rock.

Chemical analyses of the geothermal waters and fumarolic gases as well as river waters was combined with an attempt to utilize the silica geothermometer. However, this was applicable only when a ratio of dilution by fresh water could be determined. The amount of chloride in the hot water being emitted from the GSR-2 site was known to be 610 mg/kg, that in fresh water being zero, and an appropriate equation was presented:

$$SiO_{2hw} = (SiO_{2sp}) \times 610/Cl_{sp}^- - 20 \times (610/Cl_{sp}^- - 1) \qquad (4.4)$$

in which the content of silica in the fresh water is 20 mg/kg, Cl^- is the chloride content and SiO_{2sp} is the silica content in a sample. Calculation by this equation gave a temperature of 180–255°C for the original hot water.

An electrical resistivity survey was carried out over 25–300 m depth and showed a low. As this covered the entire area surveyed, the geothermal reservoir probably exceeds this in extent. Heat discharge was measured as well and gave 4·2 MWt from fumaroles, 3 MWt from hot springs, 31·9 MWt from river waters, 11·6 MWt by conduction and 11·6 MWt by evaporation, totalling 62·3 MWt. A seismic survey produced two possible stratigraphic boundaries, one between the Takinoue-onsen and Kunimitoge Formations and a second between the Obonai Formation and the basement.

Exploratory wells established that the Takinoue region is characterized by a folded structure with axes trending north-northwest–south-southeast. The quantity of steam required to generate 50 MWe is some 500 tonnes/h, according to H. Nakamura and K. Sumi in 1981.[76] To tap this, at least 14 production wells were thought to be necessary. They could not be drilled, however, because the area is a national park and two or three wells at a time had to be localized in small base areas, five of these being outlined in the plan. Directional drilling was employed so that the subterranean gathering volume could be optimized. The geothermal fluid is obtained from the Kunimitoge and Obonai Formations reservoir and the separated hot water is re-injected into the same rocks. To preclude interference between the production and injection wells, casing in the former was installed to between 570 and 850 m depth; thus geothermal fluid could be withdrawn only from the lower part of the reservoir. The disposal water is re-injected into the upper part. This is feasible because there is an impermeable stratum between the upper and lower portions of the Kunimitoge Formation. Consequently, the re-injection wells were drilled to 700–800 m depth.

Among injection problems was the precipitation of silica from separated hot water which contains over 500 mg/kg. It was observed that appreciable precipitation takes place only when water is exposed to the air. Well and reservoir, production and re-injection tests were effected.

The electric power generated at Takinoue by the Kakkonda geothermal plant is consumed, domestically and industrially, in the Iwate Prefecture and, in 1978, accounted for 8·8% of total usage in this district. In the future, another 50-MWe power plant was foreseen, if planning permission from the Environmental Agency was given. The locations of this and other major geothermal regions of Japan are shown in Fig. 4.24.

(ix) The Philippines. The geological history of the Philippines is both complex and interesting. Perhaps the Philippine Archipelagic Complex rotated towards its existing orientation on a counter-clockwise basis at the commencement of the Late Eocene when there was a change in Pacific plate motion. The last stages of the rotation occurred in the Late Miocene which initiated left-lateral movements on the Philippine master fault system traversing the major islands from north to south. Cenozoic crustal movement in the Philippine Island Arc might have generated episodes of intense folding and faulting since the Tertiary. Such intense crustal warping may well have triggered multiple phases of volcanic and plutonic igneous activity, climaxing as intrusions in the final rotational movement of the archipelago in the Plio–Quaternary.

FIG. 4.24.   Some geothermal areas in Japan.

Many of the potential geothermal resources areas in this region, together with related young volcanoes, are sited along the Philippine shear system or its nearby branch faults and some are localized as well along the peripherics of vast intra-Miocene silicic batholiths. Geothermal areas in northern Luzon are mostly confined to the eastern and southern margin of the Agno Batholith, a part of the northern Luzon Cordillera quartz–diorite. This is the largest batholith in the country and its eastern and southern margins are characterized with Plio–Quaternary andesite–dacite volcanic centres possibly connected with the ebbing phases of magmatic evolution in the region. Some important potential geothermal areas such as Batong–Buhay and Daklan are situated near these centres. The main heat source is primarily the residual heat of the batholith which is still restricted to the body of the pluton, but is being gradually released along open channels of young volcanic fissures to produce surface manifestations.

Geothermal exploration in the Philippines commenced in 1962 and, after 1970, intensified to the point at which six geothermal fields were identified, the most advanced being Tiwi, Albay, South Luzon. Here a small-scale geothermoelectric power plant was already in operation by 1967 and, at the end of 1978, a 55-MWe plant was installed and three more had been planned. Most of the landforms in the area are composed of Pleistocene lavas and pyroclastics, the former ranging from pyroxene andesites from Mt Malinao to limited hornblende–biotite–dacite extrusions at Putsan–Bolo. The Milinao volcano covers half of the relevant area and it has lava flows extending at least 6 km from its centre. The Tiwi fracture system controls the permeability of the underlying geothermal system, some components extending down into the basement and acting as major channels for ascending hot fluids. The relevant temperatures range between 200 and 315°C at production depths and, according to R. T. Datuin and A. C. Troncales in 1986, Tiwi was generating 330 MW of geothermal electricity by then.[77]

Near the capital, Manila, there is a second geothermal field, namely Maikiling–Banahaw (Makban) or Los Banos, with a 330 MWe capacity coming from six units. A total of 78 wells have been drilled in the area and 50 of them are steam-producing. A feasibility study was effected at the Tongonan geothermal field in Leyte and a 37·5-MW power plant was scheduled for completion there by 1981. The installed capacity by 1986 was 115·5 MWe. The Tonganon field lies along the eastern tectonic block of the southern extension of the Philippines master fault and comprises a typical fault valley bounded by north-northeast-trending rolling upland and steep ridges following the trend of the fault scarp incision generated by the Philippine major shear zone. Thermal springs and fumaroles occur along the Bao River valley and are localized at the intersections of major fault and subsidiary fractures cutting through the andesitic volcanics and sediments. The thermal springs have source temperatures ranging from 184 to 219°C based on the Na–K–Ca geothermometer.[78]

In addition, exploratory wells have been drilled at several other geothermal localities, e.g. Negros and Davao del Norte. The Southern Negros geothermal field is situated at the southeastern end of Negros Island. Work began there in 1973, the first pilot plant being commissioned in 1980 with an installed generating capacity of 1·5 MWe. Another, with the same capacity, followed in late 1982. M. Fanelli and L. Taffi in 1980 stated that by 1985 this country should have had a total installed geothermo electric capacity of 1320 MWe, this providing almost one-fifth of the electricity requirements for the Philippines.[24] In fact, the installed electric power-generating capacity of Philippine geothermal fields about that time was less, *ca.* 891 MWe (see Table 4.10).

(x) Indonesia. A large proportion of the Indonesian Archipelago is taken up by volcanic manifestations, including young volcanoes comprising active belts. There are at least 500 of these Quaternary features covering about 350 000 km², i.e. 17·5% of the total area of the country, and comprising the largest volcanic region in the world. Usually, the volcanoes can be categorized as andesitic explosive or eruptive centres, and two of them underwent catastrophic explosions with caldera formation fairly

TABLE 4.10
Installed capacity in Philippines geothermal power
plants by 1986[77]

Geothermal power plants	Installed capacity (MWe)
Makban	330
Tiwi	330
Tongonan	115·5
Southern Negros	115·5
Total	891

recently, namely Mt Tambora on Sumbawa Island in 1815 and Mt Krakatoa in the Sunda Strait in 1883. The latter produced approximately 18 km^3 of pyroclastics with a thermal energy loss of 1·8 × 10^{26} erg. Tectonic movements usually precede intense volcanism which is accompanied by the ascent of large quantities of magma to shallow depths. This phenomenon, together with abundant precipitation of as much as 4000 mm annually in the west, and volcanic rocks capable of acting as cap rocks or reservoir rocks, created perfect circumstances for the development of geothermal systems. Thus at least 90 thermal areas are found, most of them being hyperthermal and some marked by fumaroles, geysers, hot springs and hot mud pools with surface temperatures at or above boiling point and almost invariable over 70°C.

In Indonesia, geothermal exploration began in 1926, but it has been restricted largely to the bigger islands of Java and Bali, although it appears that the greatest potential is actually in the peripheral islands. The most advanced geothermal field is probably Kawah Kamojang, a vapour-dominated hydrothermal system in western Java which yields dry steam from its wells, at which a 30-MW geothermoelectric power plant was scheduled for construction. This output has been far exceeded since geothermal contractors financed by the New Zealand Government have added 110 MW to the total generating capacity so that it is now 140 MW. However, there is work proceeding also in the Dieng Plateau geothermal field in central Java at which site water at 173°C exists at a depth just under 100 m. The geothermal potential of the Dieng volcanic complex is estimated as around 2000 MWe. Indeed, 11 production wells in this area have been prepared as an initial step in an exploitation programme scheduled for commencement in 1987 with the construction of two units of

110 MWe. Exploration studies were being made as well at three sites in western Java, namely Prabatki Salak, Palabukan–Rattu and Bantan. Finally, a 13-MWe geothermoelectric power plant was proposed for the island of Sumatra.[24,34,79]

In addition to the above, progress has been made in developing geothermal resources at Lahendong in North Sulawesi Province, the field being located about 40 km southeast of the capital city of Manado in a region containing most Quaternary volcanoes (see Fig. 4.25). The most prominent volcanic area is Minahasa, which includes the field itself. As well as Quaternary pyroclasts and eruptive andesites, rhyolites and basalts,

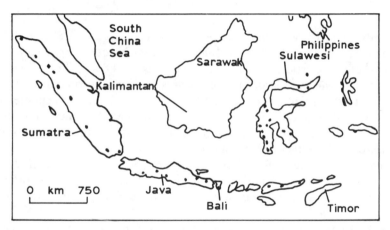

FIG. 4.25.   Geothermal resources in Indonesia on which general surveys show 5000 MWe potential.

there are Tertiary claystones, sandstones, tuffs and breccias. The Tertiary is, in fact, the basement of the Lahendong geothermal field of which the geological structure is the remnants of a caldera, with Lake Linow in it, and two major faults.

Extensive geophysical prospecting has been carried out and low resistivities were recorded in the Lahendong area. Near Lake Linow low resistivity zones under 22 ohm m are distributed to depths of 300–400 m, showing the presence of groundwater. An underlying layer with a resistivity of 65–75 ohm m is believed to constitute a cap rock, the geothermal reservoir underlying that in a layer 800–1000 m deep. Gravity surveys demonstrated high anomalies to exist northeast and southwest of the lake because of thick andesite lavas. The geophysical data led to the conclusion that the Tertiary in the region is in the form of a basin structure with the lake at its centre. Temperature readings were made in 13 shallow-gradient boreholes extending down from 30 to 125 m and bottom-hole values ranged from 45 to 113°C. Two exploratory wells had been drilled to depths between 300 and 400 m by 1986 and, in one, a wellhead temperature of 165°C was recorded.[80] A zone of anomalous temperatures coincided with the area of low resistivity shown by the seismic survey. The deep geothermal reservoir extends over 10 km^2 and its upper part is taken to be water-dominated (temperatures exceeding 200°C), this overlying a hot chloride–water zone at an undetermined depth. The potential of the Lahendong geothermal field is estimated as approximately 90 MW. In fact, under a fourth five-year plan, the State Electricity Public Corporation intends constructing a 30-MW geothermal power plant during the quinquennium 1984–1989.

(d) Australasia. Only one country is of interest within this vast region at the moment, and that is New Zealand.

Here the first geothermal wells were drilled in the Rotorua geothermal field in the 1930s and produced thermal water for domestic heating. Industrial usage began during the next decade and, when an energy shortage arose in 1950, intensive exploration began at Wairakei, which has a maximum reservoir temperature of 270°C (see Fig. 2.8). This work was extended later to the geothermal field at Kawerau (see Fig. 4.26). The hydrothermal system of the former is liquid-dominated and has supplied a 192-MWe geothermoelectric power plant since 1958. Geo-thermal fluids from the latter are utilized to process cellulose and to supply a small-scale 10-MWe geothermoelectric power plant. The total installed geothermoelectric capacity of the two is 202·6 MWe. It must not be forgotten, however, that the Wairakei plant (Figs 4.27–4.29) has been operating over the drawdown limit for a number of years and the resultant progressive fall in output has been partly disguised by improved (dual) separators and other modifications. Nonetheless, output may well continue falling and J. W. Elder in 1980 suggested that it may drop below 50 MW by AD 2000.[81]

The Broadlands geothermal field (Ohaki) was thoroughly explored during the 1960s, but work on it was suspended because of the discovery of large hydrocarbon deposits; see Fig. 4.30. In 1974 it was further investigated and geothermoelectric power plants totalling 150 MWe capacity are on-line. Its geothermal fluids have a temperature of 300°C. The

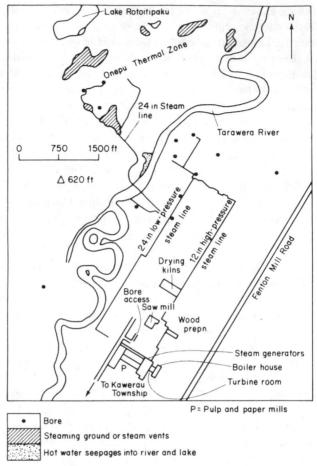

FIG. 4.26. The Kawerau (New Zealand) geothermal (vapour-dominated) steam field.

chemical geothermometer based upon the solid solution of aluminium in quartz crystals gave values ranging from 177 to 645°C, hence confirming its unreliability. By contrast, the sodium–potassium geothermometer has given good results there. Its electrical capacity is $10–11\,MW/km^2$, compared with $13–14\,MW/km^2$ at Wairakei. There exploitation has induced gravity changes, as R. G. Allis and T. M. Hunt indicated in 1986.[82] Also seismic ground noise has been recorded. The pipelines have become corroded with a lining several millimetres thick and, from this, P. Blattner and his associates in 1983 developed a magnetite–water fractionation curve with an isotopic cross-over at about 50°C.[83] They derived oxygen isotope calibration points for the two below the previously available experimental range, namely (175°C) -7.9 ± 0.5 and (112°C) -3.7 ± 0.5.

Finally, reference may be made to another geothermal field in New Zealand, a hydrothermal, vapour-dominated one located at Ngawha Springs, approximately 200 km north of Auckland. It is anticipated that this could contribute as much as 400-MWe geothermoelectric power.[84] There are many other geothermal fields, for instance Orakeikorako, Reporoa, Te Kopia, Waiotapu, Waikiti and Tikitere in the thermal belt of the North Island which is 50 km wide and 250 km in length extending from the White Island Volcano in the Bay of Plenty to the centrally placed volcanic mountains. All fall within the Taupo Volcanic Zone (see Fig. 2.5). The renewable potential of the thermal energy of the Taupo–Rotorua District was estimated as 5000 MW by J. W. Elder in 1980.[81] He pointed out that this corresponds to the energy in a liquid water flow at 100°C, relative to 0°C, of approximately 12 m³/s. Also electricity generation is a highly inefficient process, the gross efficiency of a power plant devoted to this being around 10%. The spare hot water could be re-injected, although relatively little such work has been done in New Zealand. Relevant information regarding some of the above-mentioned geothermal fields is summarized in Table 4.11.

From the table it is apparent that the average area of a field is 12 km². If its exploitable life span is given by $T = y\tau$, where τ is the hydrological time-scale and y a dimensionless factor, then $\tau = RC$, where R is the net resistance of the geothermal reservoir to discharge and C is the reservoir capacitance. C may be taken as equal to εA, ε, the open porosity, being approximately 0.1 and A the area of horizontal section of the reservoir at a depth determined from resistivity lows or deep isotherms. The exploitable power, F, is given by $F = \dot{\iota} C$, where $\dot{\iota}$ is a power flux which is constant. In the case of Wairakei, $\tau = 10$ years $= RC$; therefore $R = 7$ years/km², a value which has been applied to all of the geothermal reservoirs under discussion. An estimate of $\dot{\iota}$ may be made for this same geothermal field, which has an exploited area of 15 km², $C = 1.5$ km², and mean power $= 150$ MW. In fact, $\dot{\iota} = F/C = 100$ MW/km².

As regards the long-term management of these geothermal resources, J. W. Elder in 1980 indicated that the potential benefit arising from geothermal resources exploitation in New Zealand is great enough to justify substantial investment in project-orientated research and development plus relevant geothermal power plant and associated equipment.[81]

FIG. 4.27.   General view of Wairakei. Reproduced by courtesy of The High Commissioner for New Zealand, London.

FIG. 4.28.   Vertical twin-tower silencers forming part of the standard equipment at each wellhead (Wairakei). Reproduced by courtesy of The High Commissioner for New Zealand, London.

FIG. 4.29. View of the power house showing low-pressure pipework above and high-pressure/intermediate-pressure pipework below. Reproduced by courtesy of The High Commissioner for New Zealand, London.

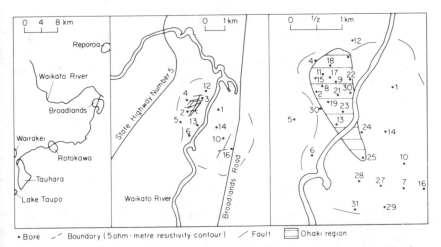

FIG. 4.30. Broadlands geothermal field. Left, its general geographical setting; centre, some faults within it; right, the Dhaki region. Numbers are well numbers.

TABLE 4.11
Geothermal reservoir parameters in New Zealand[81]

Name	Area (km^2)	Capacitance (km^2)	Resistance (s/m^2)	Thermal volume (km^3)	Natural heat output as flow (MW)	Maximum reservoir temperature $(°C)$
Atiamuri	5	—	—	—	(30)	—
Kawerau	8	1	10	10	200	290
Ketetahi	10	—	—	—	120	—
Mokai	5	—	—	—	(30)	—
Ngatamariki	5	—	—	—	120	—
Ngawha	40	—	—	—	(30)	300
Ohaki–Broadlands	11	0·2	500	1	140	300
Orakei–Korako	8	—	—	—	600	260
Rotakawa	10	—	—	—	400	300
Rotorua–Whaka	(15)	—	—	—	500	—
Tauhara	15	—	—	—	300	280
Te Kopia	5	—	—	—	250	240
Tikitere–Taheke	12	—	—	—	500	270
Tokaanu–Waihi	4	—	—	—	80	—
Waikite	(5)	—	—	—	170	—
Waimangu	12	—	—	—	250	270
Waiotapu–Reporoa	10	—	—	—	800	295
Wairakei	15	1	100	5	500	270
Waitangi	(5)	—	—	—	40	—

The major need seen by him then was organizational, in the shape of strong coordination with a single geothermal budget. He suggested that individual projects ought to be constructed on a stage-by-stage basis so as to permit adaptive development as the key parameters of geothermal reservoirs are measured in a pre-production mining phase, in which such reservoirs are collapsed to a condition appropriate for permanent stations. The rationale behind this is that, since hydrothermal systems readjust to exploitation on a rather long time-scale, say around a decade for the hydrology and a century for the thermal state, such an approach is advisable.

REFERENCES

1. BABA, K., 1975. In: *Assessment of Geothermal Resources of Japan*, ed. J. Suyama, K. Sumi, I. Takashima and K. Yuhara, Proc. United States–Japan Geological Surveys Panel Discussion on the Assessment of Geothermal Resources, Tokyo, Japan, 17 October 1975, Geological Survey of Japan, pp. 63–119.
2. CATALDI, R. and SQUARCI, P., 1978. Valutazione del potenziale geotermico in Italia con particolare riguardo alla Toscana centrale e meridionale. *Proc. AEI Conf.*, **32**, 1–8.
3. MUFFLER, L. J. P. and CATALDI, R., 1978. Methods for regional assessment of geothermal resources. *Geothermics*, **7**(2–4), 53–89.

4. NATHENSON, M. and MUFFLER, L. J. P., 1975. Geothermal resources in hydrothermal convection systems and conduction-dominated areas. In: *Assessment of Geothermal Resources of the United States*, ed. D. E. White and D. L. Williams, US Geol. Surv. Circ. No. 726, pp. 104–21.

5. BROOK, C. A., MARINER, R. H., MABEY, D. R., SWANSON, J. R., GUFFANTI, M. and MUFFLER, L. J. P., 1979. Hydrothermal convection systems with reservoir temperatures ≥90°C. In: *Assessment of Geothermal Resources of the United States—1978*, ed. L. J. P. Muffler, US Geol. Surv. Circ. No. 790, pp. 18–85.

6. BODVARSSON, G., 1972. Thermal problems in the siting of reinjection wells. *Geothermics*, **1**, 63–6.

7. BODVARSSON, G., 1974. Geothermal resource energetics. *Geothermics*, **3**, 83–92.

8. NOGUCHI, T., 1970. An attempted evaluation of geothermal energy in Japan. *Geothermics*, Sp. issue **2(2)**, Part 1, 474–7.

9. SMITH, R. L. and SHAW, H. R., 1975. Igneous-related geothermal systems. In: *Assessment of Geothermal Resources of the United States*, ed. D. E. White and D. L. Williams, US Geol. Surv. Circ. No. 726, pp. 58–83.

10. RAMEY, H. J. JR, 1970. *A Reservoir Engineering Study of The Geysers Geothermal Field*. Document submitted as evidence, Reich and Reich, Petitioners, vs Commissioner of Internal Revenue, 1969 Tax Court of the USA, 52, T.C. No. 74, 28 pp.

11. NATHENSON, M., 1975. Physical factors determining the fraction of stored energy recoverable from hydrothermal convection systems and conduction-dominated areas. In: *Assessment of Geothermal Resources of the United States*, ed. D. E. White and D. L. Williams, US Geol. Surv. Circ. No. 726, pp. 104–21.

12. CELATI, R., NOTO, P., PANICHI, C., SQUARCI, P. and TAFFI, L., 1973. Interactions between the steam reservoir and the surrounding aquifers in the Larderello geothermal field. *Geothermics*, **2(3–4)**, 174–85.

13. PETRACCO, C. and SQUARCI, P., 1975. Hydrological balance of Larderello geothermal region. *Proc. 2nd UN Symp.*, Part 1, pp. 521–30. US Govt Printing Office, Washington, DC, USA.

14. LEARDINI, T., 1977. Geothermal resource assessment and cost of geothermal power. *Proc. 10th World Energy Conference, Istanbul, 14–24 September 1977*.

15. SCHURR, S. H. and NETSCHERT, B. C., 1960. *Energy in the American Economy, 1850–1975*. Johns Hopkins University Press, Baltimore, USA, 774 pp.

16. NETSCHERT, B. C., 1958. *The Future Supply of Oil and Gas*. Johns Hopkins University Press, Baltimore, USA, 134 pp.

17. FLAWN, P. T., 1966. *Mineral Resources*. Rand McNally, Chicago, USA, 406 pp.

18. MCKELVEY, V. E., 1972. Mineral resource estimates and public policy. *Am. Sci.*, **60**, 32–40.

19. SCHANZ, J. J. JR, 1975. *Resource Terminology: An Examination of Concepts and Terms and Recommendations for Improvement*. Electric Power Research Institute, Palo Alto, California, USA, Research Project 336, August 1975, 116 pp.

20. CATALDI, R. and CELATI, R., 1983. Review of Italian experience in geothermal resource assessment. *Zbl. Geol. Paläont.*, **1(1/2)**, 168–84.

21. EEC (EUROPEAN ECONOMIC COMMUNITY), 1980. Assessment of EEC geothermal resources and reserves. 2nd Meeting of the Working Group, Brussels, 26 March 1980.

22. DONALDSON, I. G. and GRANT, M. A., 1977. An estimate of the resource potential of New Zealand geothermal fields for power generation. Proc. Larderello Workshop on Geothermal Resource Assessment and Reservoir Engineering, 12–16 September 1977, Larderello, Italy. *ENEL Studi e Ricerche*, pp. 413–28.

23. DONALDSON, I. G. and GRANT, M. A., 1978. An estimate of the resource potential of New Zealand geothermal fields for power generation. *Geothermics*, 7(2–4), 243–52.

24. FANELLI, M. and TAFFI, L., 1980. Status of geothermal research and development in the world. *Revue de l'Inst. Français du Pétrole*, 35(3), 429–48.

25. MEIDAV, T., SANYAL, S. and FACCA, G., 1977. An update of world geothermal energy development. *Geothermal Energy Mag.*, 5(5), 30–4.

26. BARBIER, E. and FANELLI, M., 1983. International Institute for Geothermal Research, Pisa, Italy.

27. DI PIPPO, R., 1985. Geothermal electric power, the state of the world—1985. *International Symp. on Geothermal Energy*, International Volume. Geothermal Res. Council, pp. 3–18.

28. BARBIER, E., 1986 (1987). Geothermal energy in the world's energy scenario. *Geothermics*, 15(5/6), 807–19.

29. ANON., 1983. *UN Statistical Yearbook, 1981*. United Nations, New York, USA.

30. ANON., 1985. *International Symp. on Geothermal Energy, Hawaii, August 1985*, International Volume. Geothermal Res. Council, pp. 3–18.

31. KANEKO, M., 1983. *National Outlook in Japan for Geothermal Energy Development*. Ministry of International Trade and Industry, Tokyo.

32. BERTELSMEIER, W. and KOCH, J. A., 1986. Geothermal energy for development—the World Bank and geothermal development. *Geothermics*, 15(5/6), 559–64.

33. BARTEL, L. C. and JACOBSON, R. D., 1987. Results of a controlled-source audiofrequency magnetotelluric survey at the Puhimau thermal area, Kilauea Volcano, Hawaii. *Geophysics*, 52(5), 655–77.

34. GLASS, I. I., 1977. Prospects for geothermal energy applications and utilization in Canada. *Energy*, 2, 407–28.

35. ANON., 1978. NATO Committee challenges modern society, 1978. *Geothermal Pilot Study Final Report: Creating an International Geothermal Energy Community*. Berkeley, Lawrence Berkeley Laboratory, LBL-6869, 102 pp.

36. ANON., 1979. *Training Needs in Geothermal Energy*. Rept of Workshop, Langavartn, Iceland, July 1978. UN University, NRR-3/UNDP-17, 51 pp.

37. ZUNIGA, A. and LOPEZ, C., 1977. Development of geothermal resources of Nicaragua. *Geothermal Energy Mag.*, 5(7), 8–13.

38. OLADE, undated. *Vision del Estado Actual de la Geotermia en America Latina*. Ser. Do. Olade, 2, 345 pp.

39. GUDMUNDSSON, J. S., 1976. Utilization of geothermal energy in Iceland. *Appl. Energy*, 2, 127–40.

40. STEFÁNSSON, V., 1981. The Krafla geothermal field, northeast Iceland. In: *Geothermal Systems: Principles and Case Histories*, ed. L. Rybach and L. J. P. Muffler, pp. 273–94. John Wiley & Sons, Chichester, 359 pp.

41. ARNÓSSON, S., RAGNARS, K., BENEDIKTSSON, S., GISLASON, G., THÓRHALLSON, S., BJÖRNSSON, S., GRÖNVOLD, K. and LINDAL, B., 1975. Exploitation of saline high-temperature water for space heating. *Proc. 2nd UN Symp.*, Part 2, pp. 2077–82. US Govt Printing Office, Washington, DC, USA.

42. GARNISH, J. D., VAUX, R. and FULLER, R. W. E., 1986. *Geothermal Aquifers*, ed. R. Vaux. Department of Energy R&D Programme 1976–1986, vii. ETSU (Energy Technology Support Unit), AERE Harwell, Oxfordshire OX11 0RA, UK.

43. ANON., 1986. *Principaux Resultats Scientifiques et Techniques*, pp. 39–43. BRGM (Bureau de Recherches Géologiques et Minières), Paris and Orléans, France, 301 pp.

44. TEISSIER, J. L., 1986. Station thermale de Balaruc-les-Bains (Hérault—France). Détection de failles drainantes par prospection des teneurs anomales de radon dans le sol. Agropolis-Verseau. *2es Rencontres Internationales Eaux et Technologies Avancées, October 1986*, Montpellier, France, pp. 325–8.

45. CRIAUD, A. and FOUILLAC, C., 1986. Etude des eaux thermominérales carbogazeuses du Massif Central français I: potentiel d'oxydo-réduction et comportement du fer. *Geochim. Cosmochim. Acta*, **50**, **4**, 525–33.

46. FOUILLAC, C., FOUILLAC, A. M., CRIAUD, A., IUNDT, F. and ROJAS, J., 1986. Isotopic studies of oxygen, hydrogen and sulfur in the Dogger Aquifer from Paris Basin. *Proc. 5th International Symp. on Water–Rock Interaction, August 1986, Reyjkavik, Iceland*, pp. 201–5.

47. DALL'AGLIO, M., DA ROIT, R., ORLANDI, C. and TONANI, F., 1966. Prospezione geochimica del mercurio. Distribuzione del mercurio nelle aluvioni della Toscana. *L'Industria Mineraria*, **17**, 391.

48. KLUSMAN, R. W., COWLING, S., CULVEY, B., ROBERTS, C. and SCHWAB, A. P., 1977. Preliminary evaluation of secondary controls on mercury in soils of geothermal districts. *Geothermics*, **6**, 1–8.

49. CATALDI, R., 1967. Remarks on the geothermal research in the region of Monte Amiata (Tuscany, Italy). *Bull. Volcanol.*, **30**, 243–70.

50. GUGLIELMINETTI, M. and SOMMARUGA, C., 1979. Distribuzione e usi non elettrici delle risorse geotermiche in Italia. *Energia Materie Prime*, **2**(5–6), 47–55.

51. DEL PEZZO, E., GUERRA, I., LUONGO, G. and SCARPA, R., 1975. Seismic noise measurements in the Monte Amiata geothermal area. *Geothermics*, **4**, 40–3.

52. GUGLIELMINETTI, M., 1986. Mofete geothermal field. *Geothermics*, Sp. issue **15**(5/6), 781–90.

53. CAPELLO, P., LO BASCIO, A. and LUONGO, G., 1974. Seismic noise survey at Solfatara Crater. *Geothermics*, **2**, 29–31.

54. MERLA, A., 1986. Post-orogenetic magmatic activity and related geothermal resources in southern Tuscany, Italy. *Geothermics*, Sp. issue **15**(5/6), 791–5.

55. OTTLIK, P., GÁLFI, J., HORVÁTH, F., KORIM, K. and STEGENA, L., 1981. Low enthalpy geothermal resource of the Pannonian Basin, Hungary. In: *Geothermal Systems: Principles and Case Histories*, ed. L. Rybach and L. J. P. Muffler, pp. 221–45. John Wiley & Sons, Chichester, 359 pp.

56. BOLDIZSÁR, T., 1978. Geothermal energy from hot rocks. In: *Proc. Nordic Symp. on Geothermal Energy, Göteborg*, ed. Ch. Svensson and S. A. Larson, pp. 46–51.

57. TENU, A., CONSTANTINESCU, T., DAVIDESCU, F., NUTI, S., NOTO, P. and SQUARCI, P., 1981. Research on the thermal waters of the western plain of Romania. *Geothermics*, **10**(1), 1–28.

58. BODVARSSON, G. S., PRUESS, K., STEFÁNSSON, V. and BJÖRNSSON, S., 1987. East Olkaria geothermal field, Kenya. 1: History match with production and pressure decline data. *J. Geophys. Res.*, **92**(B1), 521–39.

59. GEBREGZABHER, Z., 1986. Hydrothermal alteration minerals in Aluto Langano geothermal wells, Ethiopia. *Geothermics*, Sp. issue **15**(5/6), 735–40.
60. BAZAALE-DOLO, A. S., 1986. Preliminary interpretation of chemical results of thermal discharges of Western and Northern Uganda. *Geothermics*, Sp. issue **15**(5/6), 749–58.
61. DOMINCO, E. and LIGUORI, P. E., 1986. Low enthalpy geothermal project in Zambia. *Geothermics*, Sp. issue **15**(5/6), 759–64.
62. EISENSTAT, S. M., 1978. Geothermal resource development in Israel. *Geothermal Energy Mag.*, **6**(1), 8–17.
63. RASSEKH, M., THIEDIG, F., VOLLMER, T. and WEGGEN, J., 1984. Zur Geologie des Gebietes zwischen Chalus- und Haraz-Tal (Zentral-Elburz, Iran). *Verh. Naturwiss. Ver. Hamburg*, (NF) **27**, 107–96.
64. BAUMANN, A., RASSEKH, M., THIEDIG, F. and WEGGEN, J., 1984. Rb/Sr-Altersdatierungen von Vulkaniten aus dem mittleren Nur-Tal (Zentral-Elburz, Iran). *Verh. Naturwiss. Ver. Hamburg*, (NF) **27**, 91–106.
65. SHUJA, T. A., 1986. Geothermal areas in Pakistan. *Geothermics*, Sp. issue **15**(5/6), 719–23.
66. ARORA, C. L., 1986. Geoelectric study of some Indian geothermal areas. *Geothermics*, Sp. issue **15**(5/6), 677–88.
67. MUTHURAMAN, K., 1986. Sea water–basalt interactions and genesis of the coastal thermal waters of Maharashtra, India. *Geothermics*, Sp. issue **15**(5/6), 689–703.
68. SAXENA, V. K. and GUPTA, M. L., 1986. Geochemistry of the thermal waters of Salbardi and Tatapani, India. *Geothermics*, Sp. issue **15**(5/6), 795–814.
69. GUPTA, M. L. and DROLIA, R. K., 1977. Assessment of geothermal resources of India. Seminar on Development, Utilization of Geothermal Energy Resources, Hyderabad, India, p. I-1.
70. GUHA, S. K., 1986. Status of exploration for geothermal resources in India. *Geothermics*, Sp. issue **15**(5/6), 665–75.
71. BHATTARAI, D. R., 1986. Geothermal manifestations in Nepal. *Geothermics*, Sp. issue **15**(5/6), 715–17.
72. SURINKUM, A., SAELIM, B. and THIENPRASERT, A., 1986. Geothermal energy in Phrao Basin, Chiang Mai, Thailand. *Geothermics*, Sp. issue **15**(5/6), 589–95.
73. WEI, T., ZHIJIE, L., SHIBIN, L. and MINGTAO, Z., 1986. Present status of research and utilization of geothermal energy in China. *Geothermics*, Sp. issue **15**(5/6), 623–6.
74. WU, F., WEI, T., SHIBIN, L. and ZHIFEI, Z., 1986. First decade of geothermal development in Yangbajain field, China. *Geothermics*, Sp. issue **15**(5/6), 633–8.
75. BRANDI, G. P., CECCARELLI, A., DI PAOLA, G. M., FENGZHONG, L. and PIZZI, G., 1986. Model approach to the management of the Tianjin hydrothermal field, China. *Geothermics*, Sp. issue **15**(5/6), 639–55.
76. NAKAMURA, H. and SUMI, K., 1981. Exploration and development at Takinoue, Japan. In: *Geothermal Systems: Principles and Case Histories*, ed. L. Rybach and L. J. P. Muffler, pp. 247–72. John Wiley & Sons, Chichester, 359 pp.
77. DATUIN, R. T. and TRONCALES, A. C., 1986. Philippines geothermal resources: general geological setting and development. *Geothermics*, Sp. issue **15**(5/6), 613–22.

78. GLOVER, R. D., 1972. *Report on Visit to the Philippines, May 1972*, Part 4. Phil.– New Zealand Technical Cooperation Project, PNOC-EDC and DSIR, 27 pp.
79. MUFFLER, L. J. P., 1975. Present status of resources development. *Proc. 2nd UN Symp.*, Part 1, pp. 33–44. US Government Printing Office, Washington, DC, USA.
80. SULASDI, D., 1986. Exploration of Lahendong geothermal field in North Sulawesi, Indonesia. *Geothermics*, Sp. issue **15**(5/6), 609–11.
81. ELDER, J. W., 1980. *Geothermal Energy Exploitation in New Zealand*. New Zealand Energy Research and Development Committee, ISSN 0110-1692, Univ. of Auckland, Private Bag, Auckland, New Zealand, 26 pp.
82. ALLIS, R. G. and HUNT, T. M., 1986. Analysis of exploitation-induced gravity changes at Wairakei geothermal field. *Geophysics*, **51**(8), 1647–60.
83. BLATTNER, P., BRAITHWAITE, W. R. and GLOVER, R. B., 1983. New evidence on magnetite oxygen isotopes geothermometers at 175°C and 112°C in Wairakei steam pipelines. *Isotope Geosci.*, **1**, 195–204.
84. BIRSIC, R. J., 1978. Geothermal steam paths of New Zealand. *Geothermal Energy Mag.*, **6**(11), 29–32.

CHAPTER 5

Exploitation of Geothermal Fields

'Thou preparest the waters under the Earth and Thou bringest them forth
at Thy pleasure to sustain the people of Egypt even as Thou hast made them
live for Thee'.

Hymn to the Aten, PHARAOH AKHNATON, 14th Century BC

5.1 BACKGROUND

Exploitation is a critical matter, one relating to the quantity of heat
available, its quality and for how long it can be obtained from a geothermal
field. There are several relevant aspects: one concerns the purely physical
processes involved, a second relates to the engineering work and a third
covers the economic factors. Extraction of thermal energy may be achieved
by mining fluids out of a geothermal reservoir with accompanying natural
replenishment, or fluids may be circulated artificially through a reservoir. In
the latter case, the object is simply to mine only the heat energy itself. The
great majority of geothermal fields under exploitation are vapour- or, more
frequently, liquid-dominated; that is to say, the dominant mobile phase of
the undisturbed reservoir is one or the other category or a mixture of both.
In the vapour-dominated fields, there is often a liquid-dominated
'condensate layer' occurring near the surface.

From the above considerations, it follows that the subterranean pressure
gradient approaches that of a static column of the dominant mobile phase,
whichever it is. The vapour-dominated hydrothermal fields may include
either completely immobile or only slightly mobile water and the liquid-
dominated ones may contain liquid water alone or perhaps a mixture of

steam and water. The latter type may be split into warm water, hot water and two-phase reservoirs on the basis of the various possible reactions to exploitation.[1] There is a fundamental difference between the first two, single-phase, systems and the third water–steam system regarding the extraction of heat and geothermal fluids from their respective geothermal reservoirs. Thus the warm water systems do not boil under anticipated pressure drawdowns, the hot-water ones undergo boiling only at levels above those of expected exploitation and the two-phase ones are exploited within the steam–water zone.

As noted earlier, the total thermal energy of a geothermal reservoir can be calculated when the necessary information regarding its volume, fluid and rock properties, and general attributes is available: clearly, the higher the temperature (effectively, the enthalpy) is, the more heat is present. However, only some of this heat may be derived by mining and the amount depends upon the efficiency both of the sub-surface operations and of the engineering activities on the surface. Both relate to the final use proposed for the thermal energy, but heat extraction from the sub-surface can be investigated almost independently of this final use. As each reservoir type behaves differently when exploited, removal of thermal energy from most of these types may be correlated with a wet to dry spectrum of feasible situations, e.g. warm-water systems obviously fall into the wet class and vapour-dominated ones into the dry. Naturally, there is a gradation from one type to another in the field, so that, where a two-phase system develops, it may become progressively drier. Of course, some special geothermal phenomena such as geopressurized, magmatic and hot dry rock must be excluded, but they have less economic significance. Nevertheless, with all reservoir types, the extraction of heat is essentially a mining job and entails depletion of a source to which replenishment may be slow.

An important caveat on flow in geothermal fields, both horizontal and vertical, is advisable and stems from the extrapolation of aquifer groundwater data to these fields. In the undisturbed state, geothermal reservoirs demonstrate a marked vertical circulation pattern and this permits the inference of vertical permeability to be made. It seems that horizontal effects are of less impact generally, although perhaps dominant in regard to single well behaviour, but the entire situation is one of three-dimensional flow. Also fracturing can influence this situation, particularly in regard to transmissivity (the product of permeability and aquifer thickness). Although small fractures can be treated as two-dimensional in nature, their interconnections in a geothermal reservoir combine into a three-dimensional network. There is no doubt that, as free flow channels, they act as preferential geothermal supply sources when transgressed by a well.

Before exploitation, it is feasible that, in the two-phase region of a geothermal reservoir, the main physical mechanism could consist of a vertical movement of both phases under gravity. Fracture permeability appears to be almost universal in geothermal reservoirs so that the supply of geothermal fluids to a well is not a simple and uniform influx over a section of a particular aquifer, but rather a complex feeding from several fractures. It is to be expected that one fracture may predominate and then comprise a feed zone or point, a feature controlling pressures throughout the length of the well. Above a saturated zone in the rocks, there may exist a two-phase layer. If fluids derive from below the water surface, a water-fed well results and the liquid may be either water or fluid which was liquid and boiled en route to the well. Then a localized two-phase zone may surround the well and flow through it will be isenthalpic (the enthalpy possessed by the fluid mixture entering the well matching that of the original water at a distance). On the other hand, two-phase wells supplied from an existing two-phase zone produce a fluid with an enthalpy higher than that of liquid water at the temperature of the feed point. At Wairakei, both types of well have been recorded and have enthalpies which are sometimes only slightly more than that of the appropriate liquid water and sometimes match that of wet or dry or even superheated steam.[1]

Polluting gases or dissolved solids may be found as chemical contaminants in geothermal reservoirs; this matter is explored in more detail in Chapter 6.

Generally, the solids in question do not exert much influence on heat extraction, although, at Kizildere in Turkey, calcite clogged some wells fast enough to reduce production appreciably. Kizildere constitutes a liquid-dominated geothermal reservoir now producing through at least six wells to supply a 240-MW power plant; it is located at the junction of Büyük Menderes and Gediz graben in western Anatolia between Aydin and Denizli. A similar adverse material, salt, was recorded in Imperial Valley, California, USA. In some cases, as much as 300 g/litre have been recorded from wells and have been responsible for corrosion and encrustation in geothermal power plants.

Gaseous pollution is far more dangerous to exploitation because it can change the characteristics of a geothermal reservoir and thus affect the performance. As a general rule of thumb, for higher-temperature systems, the greater the gas content is, the more widely distributed are the two-phase conditions.

5.2 GENERAL MODEL

In 1975 A. McNabb presented a model for the Wairakei geothermal system in New Zealand.[2] The reservoir is envisaged as a vertical cylinder surrounded by cold water and consisting of an underlying hot-water-saturated zone and an overlying two-phase zone; see Fig. 5.1. Although McNabb did not include it, on top of this system may be imagined a near-surface zone in which mixing with cold groundwater can take place.[1,2] The units employed are practical, i.e. if the two-phase zone be taken as lying within the hot zone, then comparison with actual geothermal fields, such as Broadlands, can be obtained (cf. Section 5.2.2). In Broadlands, it is 1·5 km down to the liquid surface. The boundaries of the zones are idealized as simple geometric forms, but, in nature, they are diffused with accompanying peripheral distortions, e.g. of the pattern of temperatures. The separation between the underlying single phase and the overlying two-phase zones must also be regarded as diffuse. However, at Wairakei, it seems that the water level is unusually smooth, but this is not the case at either Broadlands or Kawerau.

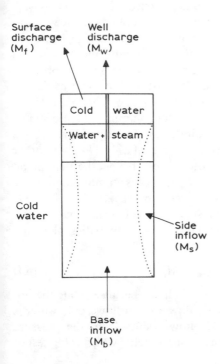

FIG. 5.1. Basic model for Wairakei, adapted from A. McNabb (1975).[2]

The reason for the irregularities at these boundaries is the great variability in the permeability of rocks. This particular factor is neglected in simple models, i.e. the rocks in each unit are regarded as uniform with respect to this parameter. The boundaries must also be envisaged as moving. Decompression (boiling) occurs within the two-phase zone. In the case of a hot-water field in which the main production is from the single-phase (water) zone, then the two-phase zone can be disregarded except in relation to the fall in water level.

5.2.1 Single-Phase Model

This is a reduction of the above-mentioned model to one lacking both the two-phase and uppermost zones and with wells tapping the hot water below the free surface. It is assumed that a continuous supply of hot water maintains this situation from below, its infiltration mass rate being M_b and its enthalpy H_b. If steady-state conditions obtain, as they usually do before exploitation of a geothermal field, and there is no conductive heat loss and no mass flux from the sides, then the mass and heat influxes at depth are equal to those discharged at the surface. In such an event, the reservoir is isenthalpic and, apart from small-scale boundary effects, isothermal as well. The base flow is taken as arising from percolation of cold meteoric water to great depth where it is heated and, because of its lowered density, later rises upwards towards the surface. The process entails movement along an extremely long flow path in which flow is practically unaltered by pressure changes induced through exploitation. For instance, at Wairakei, if a circulation depth of 10 km is assumed, the natural flux of 400 kg/s is driven by the 30-MPa pressure difference between water columns at 20 and 300°C. In this situation, it is clear that the exploitation-induced pressure drop is then merely 2·6 MPa and this could produce a rise of only 37 kg/s, in other words a minute contribution to the output of the geothermal plant. Withdrawal of geothermal fluid at a rate M_w produces an internal pressure drop promoting a flow M_s from the side which can affect the base flow M_b plus the surface outflow M_f. In making reservoir-scale calculations, water is regarded as an incompressible liquid and it is possible to express the mass balance as follows:

$$M_b + M_s = M_w + M_f \tag{5.1}$$

It is more difficult to formulate the heat balance because, while the hot water still enters and leaves through base flow and outflows, cold water is introduced by the side flow. This side flow results from the pressure difference inside and outside the hot reservoir, a factor which is highest at the well feed level. Such subterranean recharging with cold water draws inwards the cold boundary of the geothermal reservoir and causes some heating of the water involved. Initially, this does not affect the enthalpy since the incoming cold water does not immediately penetrate the central reservoir. However, it removes heat from hot rocks at the side boundaries and so causes a cold front to develop and one that, although slow-moving, reduces the size of the reservoir. As the internal enthalpy characteristics remain unchanged, the interior stays hot, isenthalpic and isothermal.

The well discharge depends upon several factors, e.g. the presence or absence of adjacent wells, field pressure, etc. Whilst drilling more wells might produce increased discharge initially, later drawdown may lower it.

Observations on the fluid discharge will indicate whether cold water is entering the reservoir or not since, if it does, then the enthalpy will decline. If the well discharge be increased beyond a certain point, the water level in the geothermal system will drop, but this is not regarded as serious unless the wells are too shallow.

Re-injection of fluid into a single-phase system increases pressure within it and this aids in maintenance of the mass discharge from the wells. As relatively cooler water is injected, the discharging wells will also cool. If re-injection takes place within the reservoir, the flow path for cooler water to the wells will be reduced. On the other hand, if it is effected at the boundary, the field pressure is maintained and the already-developed cold front is strengthened locally. Re-injection is a useful method of disposing of unwanted effluent, both liquid and gaseous, and can be regarded as a key management strategy. Most knowledge of its long-term consequences has been obtained from Japan and New Zealand. Therefore, although there is little difficulty in carrying out the operation, valuable results for the prediction of reservoir behaviour will be available only following a massive re-injection exercise over a long time, say 10 million tonnes annually for five years. The subject is examined in greater detail in Section 5.3 below.

5.2.2 Two-Phase Model

This has to include all systems consisting of two phases, even if one of these is immobile as in the vapour-dominated type. Unlike the single-phase system in which some fluid arises by decompression, this is an insignificant event in the two-phase model. A diffusion front traverses a compressed liquid rapidly; for example, at Wairakei it can occur within a few days,[1] so the period of production by decompression, when a bore is switched on or off, endures for only a few days. Thereafter an almost steady state is attained, together with a declining water level and continuing cold recharge from the side boundaries of the system. This recharge supplies mass, and energy comes from the cooling of the peripheries of the geothermal field. Such an exploitation is associated with long-lived wells because they can exploit the entire resource. However, a two-phase steam–water mixture is much more compressible than either of these components separately. Thus it takes far longer for a pulse of pressure to spread out to the boundaries of the geothermal field to induce recharge. Greater compressibility is associated with more fluid yield from storage from any given drop in pressure. In addition, two-phase conditions entail a drop in temperature in tandem with the pressure drop, so that energy is mined from the rock which so cools. M. A. Grant and M. L. Sorey in 1979[3] assessed the compressibility of a two-phase steam–water mixture by postulating that the pressure in a fluid in a unit volume of porous rock is diminished by Δp which causes a corresponding drop in temperature (ΔT) along the saturation curve $p = p_{sat}(T)$:

$$\Delta T = -\Delta p/(\mathrm{d}p_{sat}/\mathrm{d}T) \qquad (5.2)$$

The effect is that heat is freed from the rock and boils some of the water to form steam. The latter occupies more volume than its parental water and so there is an accompanying increase in volume, ΔV, which can be related to the drop in pressure, Δp, so as to obtain the effective compressibility, c, in a manner expressed in 1981 by I. G. Donaldson and M. A. Grant:[1]

$$c_2 = \frac{1}{V}\frac{\Delta V}{\Delta p} = \frac{1}{\phi}\frac{1}{(H_s - H_w)}\frac{(\rho_w - \rho_s)}{\rho_s\rho_w}\frac{\overline{\rho c}}{(\mathrm{d}p_{sat}/\mathrm{d}T)} \qquad (5.3)$$

where the subscript 2 was employed in order to emphasize that this is only valid in two-phase situations. In the complete expression for the compressibility (eqn (5.3)):

ϕ = porosity of the medium;
H_s = enthalpy of the steam phase;
H_w = enthalpy of the water phase;
ρ_s = steam density;
ρ_w = water density;
$\overline{\rho c}$ = heat capacity of the wet rock.

Complete derivations of the compressibility of the two-phase system are available, e.g. in a paper published by A. P. Moench and P. Atkinson in 1978.[4] As might be expected, it exceeds that of water or steam by orders of magnitude, and it is apparent from the form of derivation of c_2 that it is not compression of each or either of the phases that counts, but rather the transfer of matter between them. Pore fluids respond to drops in pressure by sacrificing fluid to flow and replacing lost volume by boiling water to steam. If $1 m^3$ of fluid is removed from pore space in rock at 250°C, an additional $0·025 m^3$ boils in order to fill the enlarged requirement of $1·025 m^3$ with steam.[1] The quantity of boiling and therefore of cooling or decompression is exactly the same whether the fluid removed is steam or water. Because the two-phase steam–water mixture is so much more compressible, it is clear that, for any given drop in pressure, a proportionately greater fluid volume is derived from each rock volume. Boiling entails heat removal from the rock and, if a two-phase system exists in it, there has to be some steam present initially. Consequently, when decompression forces boiling to take place, all the resultant extra steam comes from the water component and the thermal energy for the actual boiling comes from the rock.

In the case of the single-phase system, energy is taken from the perimeter of the geothermal reservoir after only a few days because of the rapid transmission of any change in pressure. In the case of the two-phase system, years may elapse before this can be accomplished and it is important to determine where energy comes from in the meanwhile. If one well penetrating and deriving fluid from a two-phase zone be considered, and its surrounding area is taken as being initially at uniform pressure p and uniform water saturation S, then a diffusion equation for p is calculable and was given by M. A. Grant and M. L. Sorey in 1979.[3] It is expressed by

$$\frac{1}{\kappa_h} \frac{\partial p}{\partial t} = \nabla^2 p \qquad (5.4)$$

in which

κ_h = hydraulic diffusivity $(k/\phi\mu_t c_2)$;
k = permeability of the medium;
μ_t = total dynamic viscosity of the fluid $(= F_w(S)\mu_w + F_s(S)/\mu_s)$.

In the last (bracketed) expression, μ_w and μ_s are the dynamic viscosities of the independent phases water and steam, while $F_w(S)$ and $F_s(S)$ are the water saturation-dependent relative permeabilities of the water and steam phases, respectively, in the mixed flow system. The relative permeability to either one of these phases is that fraction of the actual permeability which has to be stipulated if Darcy's Law is to be applied to each phase independently.

The possibly greater impact of horizontal structure in a single well was mentioned earlier. Hence, in respect of transient flow around such a well in a uniform horizontal aquifer in which a two-phase system is present, solutions to the equation exist; for instance, taking one well drilled at a time $t = 0$ and discharging at a constant mass rate W:

$$p = (Wv_t/4\pi\kappa h)[-\exp(-r^2/\kappa t)]$$ (5.5)

In the above, the total kinematic viscosity, v_t, is related to the kinematic viscosities of both water and steam, v_w, v_s, by the following expression:

$$\frac{1}{v_t} = \frac{F_w(S)}{v_w} + \frac{F_s(S)}{v_s}$$ (5.6)

and the mass of fluid stored per unit of rock volume is

$$m = \phi[\rho_w S + \rho_s(1 - S)]$$ (5.7)

The rate of fluid loss from storage or surrendered to the flow in the reservoir is the rate of change with time and this is proportional to the following expression:

$$c_2 \frac{dp}{dt} \propto c_2 \frac{1}{t} \exp(-r^2/\kappa t)$$ (5.8)

After the lapse of a time t, the pressure front will have diffused as far as a distance of the order of $(k_h t)^{1/2}$, i.e. the diffusion radius, and the situation then is that fluid is removed from storage within this radius at a rate diminishing with increase in t, only a negligible quantity being obtained from beyond the diffusion radius. The rate of yield per unit volume is thus available: the rate of yield per unit area may be derived from it by multiplying it by $2\pi r$. The product attains a maximum at $r = (\kappa_h t)^{1/2}$, this being small both for small and large values of r. Most of the fluid reaching the well comes, therefore, from a broad zone around the diffusion radius.

Broadlands geothermal field in New Zealand represents a real instance of the liquid-dominated two-phase model under discussion and, in this type of resource, years may go by before the diffusion front can reach its periphery. Consequently, thermal energy and mass are extracted from a region around an active well and one which continuously increases until a semi-steady state is attained. This is in effect the same situation as would exist if a liquid field were continuously being influenced by a cold inflow. Such an inflow balances the mass flow from the field, the ultimate mass source being located outside it. Heat flow from a well is sustained by heat which the incoming fluid derived from rock cooling on the periphery of the system, i.e. the well actually mines heat from this rock. If pressure declines with time, both mass and heat are continuously given from the two-phase zone. As well as the horizontal propagation of the pressure wave and of the withdrawal front for mass and heat, there will be vertical effects as well. Eventually, the pressure wave reaches the base of the steam–water zone and propagates rapidly through the water-saturated part. As a result, water boils at the interface, the level of which drops, energy being extracted from nearby rock.

Another variant of the model is applicable to vapour-dominated systems in which the reservoir comprises steam in association with immobile water. Should an all-liquid zone exist, it is taken to lie at such great depth that it can be neglected. Pressure transmission is a case of two-phase propagation, steam alone moving. With the drop in pressure, some water boils and adds more steam to the flow. The process entails heat removal from the rock; thus mass and heat are mined. Movement of the diffusion front is so influenced that years may pass before a kilometre has been traversed. It is therefore justifiable to dispense with the sides of the model and the well may be regarded as existing in an infinite medium. If there is a dry-steam zone present, the matter becomes more complex. The possible origin of this zone is connected with the fact that the water saturation in the two-phase zone is linearly related to temperature.

Therefore, depending upon the initial conditions, both pressure and temperature may decline sufficiently to lower the water saturation as far as zero. This can produce the above-mentioned dry-steam zone. In it there is not much increase in mass yield through steam expansion, but as the steam pressure falls a small additional energy yield produces a slight superheating of the steam. The drawdown profile is hardly affected and so the steam well has an expanding drawdown zone with the diffusion radius $(\kappa_h t)^{1/2}$. One potential arrangement is an outer two-phase portion and an inner dry-steam region. The well will derive both mass and energy from the former, the latter constituting an inert nucleus in the exploitation area of the well.

5.3 RE-INJECTION

The re-injection of used geothermal fluids is a means of waste disposal; for instance, geothermal brines have been injected into large fractures. Millions of tons of condensate have been so re-injected since 1969 at The Geysers and at Valles Caldera. From 1970 to 1982 several thousand tons of such fluids were re-injected at Ahuachapán in El Salvador. Similar work was effected at Otake in Japan, in Hilo, Hawaii, at Larderello in Italy and at the Salton Sea in California.

Where geological formations without large fractures are concerned, brine disposal has not been so successful. Thus, at East Mesa, many problems were encountered, particularly at the beginning, as S. M. Benson and others indicated in 1978.[5] However, in many other porous aquifer localities round the world, vast quantities of brine have been injected without trouble—for example in the Gulf States, where the water derived from gas and oil wells.

Geothermal experience is more restricted and entails consideration of temperature effects regarded as unimportant in oil and industrial applications except in the case of steam flooding. Most of the available data come from Larderello and will be discussed later in this section. Re-injection pumps are described in Section 5.8.5.

The physics of the flow in the course of re-injection of fluids can be considered either with reference to a uniform porous medium, in which Darcy's Law may be taken as a good approximation, or in terms of a reservoir dominated by at least one and perhaps several large fractures. The differences between these models have significant thermal and hydro-dynamic implications. In the former, fluid temperatures in the pores stay close to that of the porous matrix itself. This is not invariably so in fractured media. Also, in homogeneous porous media, hydrodynamic flow is relatively slow and approximately isotropic, whereas in fractured rock it can become turbulent in the fractures and is usually anisotropic.

R. C. Schroeder and his colleagues in 1982 investigated the two possibilities.[6] They pointed out that, where a fluid having a temperature of T_1 is injected into a porous rock formation with a temperature of T_2, two fronts start moving away from the point of injection. One is the hydrodynamic front which is situated at the furthest distance travelled by injected fluid, and the other is the thermal front at which temperatures jump from T_1 to T_2.

This problem was solved for linear one-phase flow by G. Bodvarsson in 1969.[7] Others have achieved a solution for radial flow. If cool liquid is injected into a porous rock which is completely saturated with a two-phase fluid having a particular steam saturation, then the hydrodynamic front can be quite broad, the injected fluid promoting steam compression, condensation and increasing pressure in the two-phase zone. All this produces an outward flow of mobile water outside the swept volume. Now the viscosity of all fluids is strongly temperature-dependent and this fact manifests itself in cases where the injected fluid temperature T_1 is significantly different from the temperature of the country rock, T_2. Studies have shown that, early during the injection of cold water, the pressure response at the injection well is determined by the viscosity of the reservoir fluid at temperature T_2. This response later rapidly changes to a steeper growth rate which is determined by the far higher viscosity of the cold fluid being injected.

This discussion assumes that radial flow occurs away from the injection well. However, different fluid densities will occur at the two temperatures T_1, T_2, causing heavier fluids to gravitate to the base of the reservoir. Clearly, the result will be that the front becomes ever more inclined as time passes. Such gravitation segregation effects have been considered in connection with numerical studies of injection by M. J. Lippmann and D. Mangold with various associates.[8,9] These are significant when cold liquid is injected into a two-phase area of a reservoir because of the larger density differences between water and steam.

The influence of the thermal conductivity of the host rock must be examined too. Usually, this is so low, even when the rock is saturated with liquid, that the broadening of the thermal front caused by heat flow in front of ingressing cold water can be neglected. However, there can still be quite large heat loss over a large boundary, especially from a thin reservoir (aquifer).

Two other physical phenomena are important. One is a fingering which develops where the invading fluid and the in-situ fluid are miscible. Where cool brine encounters hot brine, the effect is related only to the fluid chemistry of the liquids because mixing takes place at the hydrodynamic front and possesses strictly chemical, not thermal, importance. But in the case where liquid injection penetrates a two-phase fluid, both the miscibility and the phase interaction of water and steam may result in an appreciable broadening of the boiling zone in front of the hydrodynamic front.

The other physical phenomenon to be described in regard to the advance of fronts is hydrodynamic dispersion, which applies to both the thermal and the hydrodynamic fronts. It causes a spreading of the thermal front and has not yet been fully explored.

In the special case of injection into fractures, some special considerations must be taken into account relating to the anisotropy of the flow due to orientation of the features. Numerical data for plane parallel fractures show that, for typical geothermal rocks, fractures can be very widely spaced (over 50 m) and still exhibit the thermal behaviour of an equivalent porous medium. Fractures may be modelled as discrete channels, but fluid and heat flow from rock to fluid must be accurately treated. The fluid flow in parallel smooth planar channels apparently follows the following relationship:

$$Q = 2\pi \left(\frac{w^2}{12} \frac{w}{\mu} \right) \left[\frac{p}{\ln (r_e/r_w)} \right] \qquad (5.9)$$

in which w is the fracture aperture, μ is the viscosity, p is the pressure, r is the radius and Q is the flow rate. The subscripts e and w refer to a reference radius and to the well-bore radius, respectively. The equation implies that a 'fracture permeability' ought to be defined as $w^2/12$. The relationship was shown to be valid for rough irregular fractures, even for different effective applied stresses across the fracture faces, by P. A. Witherspoon and his co-workers in 1979.[10] It holds true only for relatively impermeable rock. As soon as some matrix permeability occurs and fluid therefore 'leaks' into the fracture, the equation is no longer applicable.

Obviously, from the numerical modelling standpoint, cold-water injection into steam or two-phase zones is dominated by movements of fronts. Re-injection of cold water into a hot-water reservoir should lead to the creation of a cool-water zone around the well, itself possessing an outer warm-water zone resulting from the extraction of heat from rock by the cooling re-injected water. Thus, in the case of one-dimensional radial injection into a two-phase reservoir, there will be at all times a cold zone around the relevant well and this will be surrounded by the swept zone close to the original reservoir temperature. Outside that will be the natural geothermal waters, as G. Bodvarsson pointed out in 1972.[11]

The most basic problem will be that of one-dimensional radial flow in a thin reservoir, for which semi-analytical solutions have been derived using the similarity solution method (see M. J. O'Sullivan and K. Pruess in 1980).[12] R. C. Schroeder and his associates in 1982 also simulated the problem and obtained good agreement with their solutions for vapour saturations and pressures, employing various grid spacings.[6] However, their numerical simulation did not do so well for the temperature front which, while predicted in the proper location, nevertheless had been smeared out to a large extent. Despite this, the simulation was regarded as seeming to 'work'. A further consideration of the approach led to the conclusion that comparisons with the similarity solution method and application of various grid spacings showed that numerical simulation of injection can give accurate results.

Better understanding is achieved from a lumped parameter approximation and also from the invariance properties of the governing equations in finite difference form. R. C. Schroeder and others in 1982 analysed injection well tests using simple, constant-rate injection and obtained a semi-log plot of the pressure build-up curve, this showing two straight-line sections.[6] The first corresponded with the movement of hot water and the second with the movement of cold water. In both cases, the mobility of the water could be calculated: this is not normally possible for production tests in two-phase reservoirs because the mobility changes during the course of testing and depends in an unclear manner upon the relative permeabilities. From the straight-line parts of the pressure plot, kinematic mobilities may be derived from equations given by S. K. Garg in 1978:[13] the values of 8.23×10^{-7} s for the 100°C water and 1.79×10^{-6} s for the 234°C water values compare favourably with the exact ones of 8.24×10^{-7} s and 1.73×10^{-6} s, respectively. The location of the thermal front facilitated estimation of porosity.

Optimum use of re-injection in a vapour-dominated geothermal reservoir was investigated[13] through a five-spot configuration of production and injection wells, a production well spacing of 1 km being assumed and reservoir parameters typical of Italian reservoirs being employed. A production rate of 0.025 kg/s was selected to provide a fluid supply in the reservoir sufficient to sustain about 30 years of production. Since the configuration is symmetrical, only one-eighth of a typical five-spot was examined. Three cases were considered, namely no injection, an injection rate equalling the production rate and an injection exceeding the production rate by a factor of two. Throughout the following discussion, an idealized homogeneous, isotropic and thin reservoir is considered; that is to say, one different from real reservoirs in which factors such as degree of fracturing and gravity will cause preferential movements as regards both direction and depth of injected fluid.

In the first case, vapour saturation is high enough for the water to be immobile and hence vigorous boiling is necessary in order to keep up the rate of production from steam alone. Consequently, the pressure must fall because of boiling. The reservoir permeability is great enough to permit quite rapid spread of boiling and the associated pressure decline. As time passes, the pressure drop is practically continuous and uniform across the reservoir, and this is accompanied by an almost uniform temperature drop and increase in vapour saturation. After the reservoir has totally superheated, i.e. dried out, the residual mass is small and the pressure decline accelerates. Then the reservoir temperature is still high (~ 220°C), so that there is much heat *in situ*. When unexploited, the reservoir contains 57.4×10^{13} J of energy, of which about 3×10^{13} J are within the fluid. Obviously, not all of this energy is available for exploitation. After its utilizable life terminates, the reservoir still comprises 50×10^{13} J, approximately 4.4×10^{13} J having been transferred from the rock in order to maintain the boiling.

In the second case, where all of the fluid produced is re-injected, the injected fluid will not affect the production in a meaningful manner until after 30 years or so, that stage at which the reservoir would reach exhaustion if no injection occurred. Since the injected fluid is denser than the original fluid, with 75% vapour saturation, it takes up less space. The pressure gradient necessary in order to push more viscous hot water through the reservoir is not kept up beyond the condensation point where the boiling fluid encounters the hot water. Large effective compressibility in the two-phase region does not allow pressure changes at the injection well to influence the production well. After 30 years, approximately half of the reservoir has dried out, a tiny portion is totally liquid and the remainder is boiling. Subsequently, production derives from boiling in and enlargement of the two-phase region. In the boiling region, some of the water is mobile and trespasses into the previously superheated region. While the pressure within the reservoir drops, some of the condensed hot water begins to boil. In addition, pressure declines so as to produce sufficient steam and the gradient in the superheated region around the well steepens once the mass supply near the well is exhausted. Both effects produce unacceptably low down-hole pressures at about 40 years. However, there is then still plenty of heat left in the reservoir (48.9×10^{13} J).

In the third case the results qualitatively resemble those of the second, with the reservoir at 30 years possessing a smaller superheated zone, a boiling zone and a large liquid zone. Pressure throughout the reservoir declines steadily, but the temperature drops more rapidly in the two-phase boiling zone. At the end of useful production, the residual energy is still 46.9×10^{13} J. Therefore it is clear that higher re-injection rates would further increase the length of life of the field.

The cases described above make it apparent that re-injection can appreciably lengthen the life of a two-phase system, but does not increase the power output. Actually, the re-injected fluid reduces the volume of the boiling zone available for the production of steam and this promotes a slightly increased drop in production pressure as more fluid is introduced.

Referring again to real wells, if re-injection takes place into a two-phase well, there will be a similar effect—cold water by it, hot water and natural two-phase conditions. It is to be expected that, in circumstances of normal permeability and rate of re-injection, the re-injected liquid will disperse radially from the well, although buoyancy effects will produce some distortion.

In a vapour-dominated system in which vertical pressures are more intimately connected with steam conditions, water downflow will be more significant. In performance, re-injection wells have shown that permeability variations play an important role and radial dispersal of re-injected liquid is non-uniform. This was demonstrated at well BR7 at Broadlands, from 1979 to 1981, by re-injection testing. There are three feed zones. At one, re-injection fluid was accepted at 90–150°C. Another, 50 m deeper, still contained fluid at 280°C, as P. F. Bixley has stated (in Ref. 1).

Returning to the subject of Italian reservoirs, R. C. Schroeder and his fellow-workers in 1982 discussed a one-dimensional vertical system approximation of Larderello.[6] The reservoir properties and thermo-dynamic conditions resemble those of the depleted zones there. Figure 5.2 shows the idealized model as compared with the real reservoir. The field has been exploited for a very long period of time without re-injection, the wells

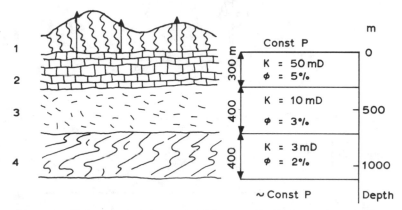

FIG. 5.2. Linear flow model for re-injection studies.[6] The rock and thermo-dynamic characteristics resemble those existing in some zones at Larderello, Italy. 1, Cap rock; 2, carbonate formations; 3, fractured quartzites and phyllites; 4, phyllites.

producing from a fracture system at the top of the reservoir. The assumption is that this makes the pressure uniform at the top and equal to bottom-hole values in the productive wells. Steam produced by boiling water in deeper layers ascends to collect in the said fractures. Pressure is maintained almost constant throughout by connecting the system with fictitious elements with large volumes. In these conditions, the system stays almost steady, having a steam production rate of $17 \, kg/s \, km^2$. An injected flow rate of $20 \, kg/s \, km^2$ of water was simulated at various depths. Oscillating trends may occur where $P \ll P_{sat}$ and derive from finite discretization, and also from a possible thermal equilibrium of rock and fluid at each point within the reservoir. A finer space discretization reduces both the amplitude and frequency of these oscillations. Relative permeability of the two phases is a significant parameter for re-injection because it affects both the pressure gradient and liquid propagation throughout the rock volume. At the moment, there appear to be no sure criteria for attribution of a given relative permeability curve to the various reservoir rocks. An instance from Wairakei is shown in Fig. 5.3. From this it may be seen that the 3-m space discretization used in the injection zone gave greatly reduced oscillations.

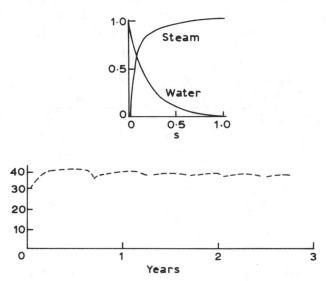

FIG. 5.3. Simulation with appropriate relative permeability from Wairakei permeability data: the relative permeability curves for water and steam there are shown above.

A stable isotope study of re-injection processes at Larderello was carried out by F. D'Amore and his co-workers in 1987.[14] They examined the results of a number of re-injection tests effected over several years in two areas of the Larderello geothermal field. One of these is situated in the central portion of the intensively exploited zone of Larderello which is characterized by a reservoir temperature exceeding 250°C and a pressure of *ca.* 5 bar in the fractured media together with a high fracture-based permeability in the dolomitic limestone rocks. The re-injection well was sited centrally. The process of re-injection commenced in January 1979, continuing, with two short intervals, until April 1982. The rates of injection ranged from 30 to 180 tons/h. The second area was located in the

Serrazzano zone some 4 km southwest of the central part of Larderello. This peripheral area is surrounded by wells which are non-productive because of a lack of permeability. The re-injection well is sited off-centre with regard to the monitored wells and the injected water flow rate ranged from 409 to 90 tons/h. The reservoir temperature here is lower, approximately 220°C or less, with the same pressure in the fractured media. As regards the gas/steam ratio, carbon dioxide constitutes over 90 mol% of the uncondensable gas content and the species was taken as non-reactive. Consequently, the small variations in dry gas composition, mostly for hydrogen and hydrogen sulphide, noted during the re-injection tests and believed to be due to chemical reactions are disregarded here.

There is an associated caveat, namely that the situation in other geothermal fields with a different gas composition is not analogous to this one. For instance, at The Geysers, the percentage of species in the dry gas other than carbon dioxide can exceed 40 mol%. The wells in both areas vary in depth between 400 and 600 m.

During the three years of shallow re-injection in the intensely exploited area of Larderello, almost 85% of the water was estimated to have been recovered as superheated steam and, in spite of the close spacing of the wells, the wellhead temperatures stayed constant, suggesting a high vertically orientated permeability permitting deep penetration of water and efficient fluid–rock heat exchange. High flow rates were found to cause appreciable enrichments in the oxygen-18 and deuterium contents, and also a strong decrease in the gas/steam ratio. The closer wells showed stronger enrichments. Table 5.1 summarizes important data.

During the three periods of re-injection, namely January to November 1979, December 1979 to September 1980 and October 1980 to April 1982, the composition of the re-injected water altered. Fractionation effects were examined. Observed deviations on $\delta^{18}O/\delta D$ diagrams imply straight mixing lines between deep steam and completely vaporized re-injected water giving steam produced by its boiling. It is feasible to calculate, at any given temperature, the fraction of re-injected water so vaporized by using single-stage or continuous steam separation. In the first case, the average value of $\delta^{18}O$ for re-injected water can be expressed as a balance between generated steam, s, and residual liquid water, l, in a plume under the re-injection well:

$$\delta^{18}O = f_s\delta^{18}O_s + (1 - f_s)\delta^{18}O_l \qquad (5.10)$$

and the following approximation is applicable:[15]

$$\delta^{18}O_l - \delta^{18}O_s = 10^3 \ln \alpha \qquad (5.11)$$

TABLE 5.1

The maximum oxygen-18 enrichments, in $\delta^{18}O$ (‰), for various re-injection periods

Well	Period			Well	Period		
	1	2	3		1	2	3
Central				Peripheral			
W1	4·7	3·3	4·6	W11	2·5	2·1	—
W2	4·2	3·1	6·0	W12	2·3	2·5	—
W3	5·1	5·0	6·9	W13	2·6	4·9	—
W4	≤2·8	3·1	5·2	W14	3·0	3·3	—
W5	2·0	2·1	3·8	W15	2·2	3·2	—
W6	1·9	2·7	3·4	W16	1·6	2·5	—
W7	0·7	0·8	3·0	W17	1·5	3·2	—
W8	≤0·7	1·9	1·5	W18	0·6	1·2	—
W9	0·0	2·4	3·3	W19	0·1	0·6	—

in which α is the fractionation factor, a function of temperature. Thereafter

$$f_s = [(\delta^{18}O_s - \delta^{18}O) + 10^3 \ln \alpha]/10^3 \ln \alpha \qquad (5.12)$$

and, if a continuous steam separation is assumed at depth, the following equation becomes applicable:

$$(\delta^{18}O_s - \delta^{18}O)10^3[(1 - f_w)^{\alpha - 1} - 1] \qquad (5.13)$$

The reality of the proposed plume under the well is established by the fact that, in spite of a boric acid content exceeding 1000 ppm in the re-injected water, the concentration of boron in the condensate of the steam produced by the wells around the central re-injection well did not alter during the test (the average value is *ca.* 100 ppm). When the re-injection tests ended in April 1982, the boron content increased constantly in the surrounding wells. At 200°C the distribution coefficient between vapour and liquid for boric acid is $B = C_v/C_l = 0.019$, where C is the molar concentration.[16,17] Therefore the boric acid of the re-injected water remains concentrated in the fraction of the liquid water under the central test well. Calculations of mass balance of this system demonstrate that some 2500 tonnes of boric acid are stored in the said plume. As the volume of liquid water declines after the end of re-injection, boron is volatilized in the vapour phase and hence its concentration rises in most of the monitored wells.

Analysis of samples taken during the tests in the various periods of re-injection gave some interesting results. The theoretical isotope patterns of steam produced in the course of continuous boiling of re-injected water as obtained at three selected temperatures were examined, together with values of $f_w = 1 - f_s$. The first period of re-injection is characterized by progressive enrichment of the oxygen-18 and deuterium contents due to the mixing of deep steam with steam produced by boiling of some 60% of the re-injected water, the plume boiling temperature being taken as about 200°C. At the end of the initial period of re-injection, isotopic ratios were progressively lowered as they shift back in the direction of the deep steam along a straight line intersecting the composition of the injection well water. This implies an absence of fractionation or complete boiling of the residual water so that $f_w = 0$. During the second period, a mixture of reservoir steam and steam arising from boiling of 60% of the re-injected water occurs as before, but, after re-injection, steam is produced by boiling of some 80% of the original water and $f_w = 0.2$. During the third period, there may well be mixing of deep steam with approximately 45% of the steam phase of the injected water, with $f_w = 0.65$.

For re-injection in the peripheral zone, δD–$\delta^{18}O$ diagrams for three selected wells of the area clearly showed that fractionation of the re-injected water occurred in all of them. As in the central region of the geothermal field, the fractionation effects result from incomplete boiling and increase with increasing distance from the re-injection well itself. The ensuing chemical and isotopic compositional variations at Larderello may result from pre-exploitation lateral flow of steam from centrally located boiling zones towards the margin of the field.[16,17] In the course of this, conductive heat losses to the surface causes partial condensation. Gases such as carbon dioxide become concentrated in the residual steam which is depleted in oxygen-18. At a reservoir temperature approaching 220°C, there will be almost no deuterium fractionation between vapour and liquid, as A. H. Truesdell and others indicated in 1977.[18] Any slightly volatile salts such as H_3BO_3 are removed in the condensate.

The process can be modelled as a Rayleigh condensation in which condensate is drained downwards and thereafter continuously removed so that the residual steam becomes ever more chemically and isotopically fractionated as its mass lessens during lateral flow. Obviously, the original fluid in the reservoir is not homogeneous in composition and can be affected in a number of ways by re-injection. Centrally around the re-injection well in the central area of the geothermal field, there are no high $\delta^{18}O$ areal variations (values between -2 and -3‰) in the pre-injection state (except for two wells, W5 and W8, which are furthest from the re-injection well). However, there is a detectable areal variation in gas content. In the central area, there is a linear correlation between $\delta^{18}O$ versus gas/

steam and δD versus gas/steam for well W1, but not for the other wells. The δD data tend to approach the average δD value of re-injected water for high re-injection rates (lowest gas/steam ratios). There is then no apparent fractionation for deuterium, the observed shift to lower values of gas/steam being attributable to a redistribution of flow connected with an injection-induced pressure increase. The reason why this redistribution is not obvious in well W1, sited nearest to the re-injection well, is that, unlike W2 and the other wells, its initial gas content was similar to that of the re-injection well before the re-injection process commenced. A simple mixing between deep original fluid of well W2 (with gas/steam $= 12\cdot2\%$) with steam arising from boiling of the injected water with a gas/steam ratio of 0 would give a straight-line relationship on a graph of δD versus gas/steam (% by weight). In fact, there is some departure from this simple linear relationship because the situation is more complex and may involve not two but three component mixtures. The hypothetical third component could be a fluid with a gas/steam ratio appreciably lower in well W2.

D'Amore and his colleagues suggested that it may be the fluid feeding the well W1 (situated between the injection well and W2) which has a gas/steam ratio of $4\cdot4\%$, but the same deuterium content as W2.[17] These workers computed the fraction of steam coming from well W1 and feeding well W2 and, by simple mass balance arguments, showed that part of the deep steam from each well undergoes displacement. However, as they themselves admit, this type of calculation is a vast oversimplification of reality because of the complexity of flow paths.

In summary, it is justifiable to regard both deuterium and oxygen-18 as natural tracers in the fluids of the vapour-dominated geothermal field at Larderello and they can be utilized elsewhere in the same way. In the course of re-injection and subsequently, strong short-term variations of the isotopic composition of the steam were recorded close to the re-injection area: these arose mostly because of mixing between the injected and deep fluid components. Large isotopic fractionations took place at depth during the evaporation of the re-injected water and can affect evaluations of the recovered fluid, using a simple mixing model. Stable isotope and gas/steam ratios were closely correlated in the fluids collected in the monitored wells. This accords with the hypothesis that an effective process of mixing takes place at depth and is accompanied by the formation of a liquid plume which undergoes various degrees of evaporation. The non-linear correlation between isotope and gas/steam ratios may be interpreted by a model in which the deep areal chemical differences in the original fluids are modified by re-injection.

5.4 ARTIFICIAL STIMULATION

There are formations covering huge areas and containing enormous volumes of hot dry rock characterized by abnormally high thermal gradients, some at temperatures as high as 600°C, at shallow depths of less than 3 km, and these could be made productive by introducing surface water under pressure, i.e. by artificial stimulation. Many of them are in the USA, where most geothermal energy regions have average formation temperatures exceeding 150°C within a depth of 10 km. In addition, artificial stimulation is sometimes required in situations where, with the passage of time, a decline in steam pressure from producing wells occurs, e.g. at The Geysers and Larderello. Attempts to halt this decline involve injection of surface waters or the re-injection of condensates from the turbine exhausts, but they cannot arrest it. On the other hand, increased permeability might do so, and this can be achieved by employing small-charge, optimally-spaced, high-energy chemical explosives. Their efficiency will relate to the very variable conditions on the site, e.g. anhydrite, dolomite and limestone at Larderello, graywackes at The Geysers and volcanic debris at Wairakei.

In 1971 C. F. Austin and others listed five fundamental models of geothermal deposits, namely granite stock, basaltic magma, wet geo-thermal gradient, metamorphic zone and dry geothermal gradient.[19] An eastern Californian granite stock area was selected for an experiment into how explosive stimulation might be used. One potential application is borehole enlargement, springing a hole in order to make one of larger diameter at a chosen horizon. Such work can create a cavity much larger than the original diameter of the borehole. Explosives need only a light service rig and there is relatively little chance of trapping debris or equipment in the hole. Also the vibration and shock effects of the explosion can assist in improving production rates. Structural factors are significant; for instance, in a granitic or metamorphic host, an effort to enlarge a borehole by under-reaming may result in loss of tools or a fine-grained dike of considerable strength might impose serious eccentric loads upon any tool which rotates during operation. Explosives would solve these problems. Thus springing is best carried out using brissant explosives. Clearly, the technique has to be applied cautiously in thin-bedded and steeply dipping strata. Blasting can destroy grain-to-grain bonding as well as physical blockages of fractures and pores, and can create new fractures as well.

Permeability may be increased by stress waves and also from

perforations resulting from the use of a conical-shaped charge, which is substituted for bulk explosive charges so as to increase the effective diameter of a well by making perforations at right angles to the axis of the borehole. Such conical-shaped charges are able to perforate casing and cement in order to provide access from the reservoir to the production tubing. Also, under the right conditions, they can increase the effective diameter of a well by a factor of as much as three. Bullets have been employed, too, in order to perforate rock in place of the jet from a conical-shaped charge.

5.4.1 Conventional Explosives

The explosive stimulation of a geothermal well entails understanding of a number of relevant factors, the most important being the means of delivery, the type of host rock and the fluid phase in the formation, i.e. whether it is steam, gas, liquid or a mixture. Detonation of an explosive charge in a well produces two recognizable zones in the host rock, namely a close-in one in which the fabric of the rock is destroyed by crushing and compaction, and a more distant one in which grain-bond failures and fracturing occur. In the latter, stress-wave passage is the main phenomenon; it has been studied by a number of investigators such as W. Goldsmith and his associates in 1968.[20] They found that basalts subjected to stress waves show slight attenuation and virtually no dispersion, the coefficient of attenuation and Young's modulus tending to decrease as the number of shocks received by a sample at a given initial shock level increases. Strength seems not to be affected and basalt appears to be a good example of the capacity of a brittle material to attain damage saturation with a given severity of repeated impacts. This implies that efforts to produce fracturing in a basalt reservoir would be unsuccessful beyond a single detonation, as C. F. Austin and G. W. Leonard indicated in 1973.[21]

In porous media, stress-wave-induced damage is localized, tests on scoria having shown that it and other similar materials are excellent energy-absorbing media.

Explosive charges can be made for temperatures up to 400°C, but, if the temperatures are to be higher, the costs become unacceptable. The explosives chosen have to be appropriate. The geothermal well conditions may include side effects of high temperature such as melting, and this can affect the calculated explosive geometry. Another point is that well gases or fluids may interfere by interacting with the explosive, maybe either desensitizing or sensitizing it. Therefore, if feasible, explosives should be isolated from such gases and fluids. An allied danger is that materials used in well boring, e.g. lubricants, may also react with the explosives. Contamination may take place after the shot is effected. This means that charges with high outputs of toxic chemicals, for instance mercury, have to be strictly controlled, especially in regard to the quantities employed.

Then also the disposal of failed or unused charges has to be considered: two approaches are available. One is to use uncased explosives; when in liquid form, these establish a close wall contact and can be injected into rock pores. However, their behaviour is difficult to predict. For instance, in the metal mining industry, it was found that using some types of slurried explosive in drill holes having pyrite in their walls produces instability. Another drawback is that they are expensive to produce in a form capable of surviving temperatures up to 400°C. One DuPont heat-resistant explosive is TACOT. The density and solubility of uncased explosives is critical because they may induce them to float or sink within the borehole. Cased explosives are much more predictable and hence attractive in artificial stimulation work. They can be pre-designed and supplied with complete operating instructions. This simplifies matters greatly and ensures that no particular expertise is necessary on site. TNT (trinitrotoluene) is an example; it is easily available as well as being fairly cheap and rather stable. Although it melts above 80°C, it can often be used because most geothermal wells fall into the temperature range 50–200°C, many being at the lower end of this range.

TNT can be detonated through a lead azide detonator and an HMX booster. To prevent intermixing of the well fluids and TNT, it is sometimes utilized in a light casing, but, where a well tapping a higher temperature fluid is concerned, it is employed in an insulated casing. Unfortunately, this may modify the pattern of energy transfer into the walls of the borehole and could contribute more debris as well. Tests on TNT at the required temperatures were carried out at the US Naval Weapons Center in China Lake, California, in 1972 and Teflon casing was used.[21] This is compatible with TNT, has a suitably low thermal diffusivity and resists rapid decomposition at the relevant temperatures. In addition, its destruction avoids the introduction of metallic debris into the well. The results demonstrated that a suitable thickness for the Teflon wall guaranteed reasonable handling times and facilitated the emplacement and detonation of TNT charges in unquenched wells at up to 400°C. It was suggested that, where longer times are necessary than can be attained using encased TNT, in the temperature range 200–300°C it might be feasible to employ a main charge of TACOT.[21] This would act as a heat sink and insulator for a lead azide detonator with a NONA booster and, in the temperature range 300–

500°C, main charges of NONA could be utilized up to 400°C. To protect the detonator, a vented ablative system is required.

5.4.2 Nuclear Explosives

If nuclear explosives remain acceptable, their use could ensure great subterranean fracturing efficiency capable of stimulating geothermal resource areas in order to increase the generation of electrical power significantly. As early as 1957 the US Atomic Energy Commission (USAEC) established the Plowshare Program, which later acquired extensive theoretical and experimental information regarding underground nuclear tests which are still going on (see Fig. 5.4). Most of these underground tests took place at the Nevada Test Site, although a few were conducted at other locations, e.g. 'Rulison' was held near Grand Valley in Colorado in 1969.

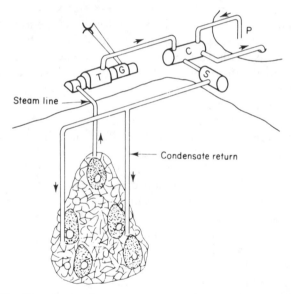

Steam line

Condensate return

FIG. 5.4. The Plowshare concept, the fracturing of subterranean hot dry rock underlying ten western states in the USA by means of hydrogen bombs. After this has been done, water would have been piped in and converted to steam below 3 km depth at a temperature of approximately 180°C. The steam would have flowed back to the surface through fractures and would have been superheated. Finally, it would have been piped to a turbine generator, the condensate being recirculated back to the underground cavities. T, turbine; G, generator; C, condenser; S, surge tank; P, cooling pond.

The Plowshare geothermal concept was described in 1971 by the USAEC and the American Oil Shale Corporation.[22] It entailed generating power from energy contained in hot dry rock. This would be fractured by a nuclear explosion, an array being fired sequentially, after which water would be injected and steam withdrawn to a surface facility. The figure shows the type of plant which was envisaged as recovering energy from a plutonic mass at, say, 3 km depth using a hydrogen bomb. Masses of hot dry rock occur in subterranean localities in at least ten of the western states of the USA and all could have been fractured using such an approach.

The relationship between the precise number of nuclear explosives necessary and the energy recovered by their use from energy of the host rock is given by

$$N = (E - NJW)/\varepsilon \tau V_{\mathrm{f}} \qquad (5.14)$$

in which N is the number of nuclear explosives required, E is the energy necessary to run a 200-MW power plant for 30 years ($= 1 \cdot 91 \times 10^{11}$ kW h for an 80% plant factor and a 22% conversion efficiency), J is the energy recovered per kilotonne yield ($= 1 \cdot 05 \times 1\,000\,000$ kW h/kt), W is the nuclear explosive yield in kilotonnes, ε is the fraction of energy recovered from the rock (assumed to be 0·9), τ is the sensible heat ($= 180$ kW h/m^3) and V_{f} is the fractured rock volume in cubic metres.

The amount of energy requisite for the operation of such a power plant through its expected lifespan must be supplied by the total of the energy obtained from the nuclear explosives and that derived from the fractured rock. The quantity V_{f} will be the product of the cavity volume and a fracturing-efficiency coefficient, M. M itself is a relationship expressed by

$$M = [r_{\mathrm{f}}/r_{\mathrm{c}}]e_{\mathrm{h}} \qquad (5.15)$$

in which r_{f} is the fracture radius, r_{c} is the cavity radius and e_{h} is the enhancement factor. This last has never been accurately determined, but there is evidence from experiments carried out in Lithonia granite in which fractured volumes increased by a factor of 36 when blasting was effected to a free surface. However, a pre-existing cavity will not quite act as a free surface, as J. B. Burnham and his colleagues indicated in 1973.[23] This is because the geometry intercepted is actually limited and the change in density between the cavity in question and the host rock will be much less than in free-surface blasting.

Experiments have shown that stresses necessary to widen existing cracks are far lower than those required to initiate new ones. Nevertheless, difficulties remain in such work, including that of finding the correct array in order to solve the matter of hydraulic flow. Water and steam must be transmitted through low-permeability regions without excessive channelling and the dissolution of silica must be taken into account as well. A fair amount of this silica may be transported with steam into the turbine system at the prevailing high temperatures and pressures, and, if the amount becomes too great, then some intermediate heat exchanger may have to be installed at the surface. L. A. Charlot in 1971 suggested that, under such conditions, radionuclides usually found in the molten rock may be soluble enough in the steam to rise to the surface.[24]

Obviously, there would be environmental effects from underground nuclear explosions, including ground movement. When such a nuclear test takes place, energy release vaporizes rock to form an expanding spherical cavity, a resultant shock wave progressing outwards. When this reaches the surface of the ground, it induces vibrations which can cause damage over a considerable area. In the event that the technique was to be used, a regional survey extending as far as 100 km in radius from the detonation centre would have to be effected. J. A. Blume in 1969 presented a very interesting study of such ground motion effects.[25] It seems that even a 100-kt device could produce significant damage at relatively great distances and this would obviously preclude employing megatonne nuclear devices. For instance, in the Rulison experiment mentioned earlier, a mere 40-kt explosion was involved and this produced $90 000-worth of damage to buildings in a rather isolated area in Colorado. Nuclear explosions at the Nevada Test Site have created small-scale seismic after-shocks resulting from minor movements along pre-existing fault planes due to the release of natural strain energy. In regions of natural seismicity, such nuclear detonations might trigger seismic events before they would otherwise occur and G. M. Sandquist and G. A. Whan in 1973 argued that this may constitute a method for reducing the environmental impact of a natural earthquake since it would promote the liberation of natural strain energy prior to its accumulating sufficiently to cause a natural, larger-scale earthquake.[26]

Of course, the above discussion might be considered to be rather academic after the famous accident at 01 23 (local time) on Saturday, 26 April 1986, at Chernobyl in the Ukraine, USSR, when for four to five days, an outmoded, badly built nuclear power plant with only a rudimentary emergency system was mishandled by the technical staff. Loss of coolant in the core a day earlier had caused a temperature rise which turned water into steam, this then penetrating the walls of pressure tubes conveying water through the core. The steam reacted with graphite blocks surrounding these pressure tubes and gas explosions shattered the building, ignited the graphite and the core of the reactor was blown open. In contact with air, the graphite fire intensified and the uranium fuel heated and melted. An enormous cloud of smoke and gas containing radioactive particles rose into the atmosphere and dispersed, subsequently spreading through Scandinavia to reach the United Kingdom and northern Italy.

In connection with this event, and against the possibility that nuclear explosions may be applied to the artificial stimulation of geothermal systems (after all, they are still in use for weapons testing), it is important to list the potentially volatile radionuclides remaining after 180 days have elapsed from the detonation of an appropriately sized device, i.e. 1000 kt, in igneous rock (see Table 5.2).

Of the above radioactive contaminants, all are γ-emitters except for rubidium-106, tritium, phosphorus-32, sulphur-35 and argon-37. Thermonuclear explosives have a relatively low level of γ-radiation for volatile radionuclides, but a very high level of β-activity due to tritium. The induced activities listed in Table 5.2 refer to such an explosive without neutron shielding. For every 15 cm of boric acid neutron-absorbing shielding, it has been calculated that there is a factor of about ten reduction in activity of the subterranean formations.[27]

Induced activities depend upon the nature of the geology as well as upon the detailed design of the explosive. Non-condensable gases in steam are separated in the condenser and, in natural geothermal fields, exhausted from it by venting to the atmosphere. In a nuclear-stimulated geothermal power plant, such gases could be collected and their radionuclides separated or concentrated and later stored underground. If thermonuclear explosives were used, more argon and tritium would be produced, and once the latter exchanged with hydrogen in the steam or water, it would be extremely difficult to remove.

Nuclear explosives need a fission trigger which is composed of 3 kt or more of fission energy from uranium-235 or plutonium-239, this providing the high temperature necessary to initiate fusion. Fissioning of the trigger produces no less than 200 different isotopes of some 36 elements of atomic masses ranging from *ca.* 75 to 150. Most of these are radioactive, but short-lived. Induced radioactive materials form as a result of neutron capture in stable elements inside the nuclear device and also in elements in environmental soil materials such as sodium, silicon, aluminium, iron, manganese and so on. Surrounding the nuclear device with neutron-absorbers such as boron reduces the effects and would be desirable in situations where geothermal stimulation is to be undertaken.

TABLE 5.2

Potentially volatile radionuclides remaining after 180 days have elapsed from the detonation of a 1000-kt nuclear explosion in igneous rock (Hardhat granite)

Nuclide	Half-life[a]	Fission (kCi)[b]	Fusion (kCi)[b]	Radiation concentration guide[c] (pCi/ml)		Volatility in steam at 80 atm, 623 K
				Air	Water	
^{85}Kr	10·8y	20	0·06	0·3	—	Permanent gas
^{90}Sr	28·8y	150	0·45	3×10^{-5}	0·3	Gaseous precursor
^{103}Ru	40d	1150	3·45	0·003	80	Oxidizing high, reducing low
^{106}Ru	1y	1000	3·00	2×10^{-4}	10	Oxidizing high, reducing low
^{125}Sb	2·7y	60	0·18	9×10^{-4}	100	Apparently high
^{127}Te	109d	90	0·27	0·001	50	High
^{137}Cs	30y	180	0·54	5×10^{-4}	20	With CO_2, low; without CO_2, high (30%)

Radioisotopes induced in soil

^{3}H	12·3y	220	20 290[d]	0·2	3 000	Permanent gas and tritiated vapours
^{22}Na	2·6y	—	0·6	3×10^{-4}	30	—
^{32}P	14·3d	2	2·5	0·002	20	—
^{35}S	88d	29	40	0·009	60	—
^{37}Ar	35d	70	200	100	—	Permanent gas
^{134}Cs	2y	14	18·3	4×10^{-4}	9	With CO_2, low; without CO_2, high (30%)

[a] Abbreviations: d, days; y, years.

[b] These quantities assume linear scaling of 50-kt fission and fusion devices. Assume a nominal 3-kt fission trigger for the fusion device.

[c] Taken from CFR Part 20 for unrestricted areas; they are maximum values which do not take into account the solubility of the compounds.

[d] Assumes 2 g of residual tritium per kilotonne of fusion yield.

Probably the most important radiological problem is caused by tritium; this enters the water molecule and hence travels at the same velocity as groundwater (the basis of its use as a dating tool). Since a time-lapse of ten half-lives for a specific radioisotope produces a reduction in activity by a factor of 1000 and a lapse of 20 half-lives reduces it a million times, then, even in groundwater flowing at 300 m/year, tritium activity would be reduced by a factor of 1000 within 40 km or so from the site of the detonation. Tritium is not usually a great biological hazard because of its relatively short biological half-life of about 12 days, but it can become so if released in large amounts.

Leakage of radiation into the atmosphere from underground nuclear explosions is another hazard, but, where the work has been carried out at depths exceeding 1200 m, this has not been recorded.

The other possibility of radioactive contamination of groundwater aquifers by seepage is considered improbable. Radionuclides most likely to be involved if such seepage actually took place are gaseous ones such as xenon, argon, krypton and tritium. Even these would have to traverse complex paths through fissures in order to enter groundwater or reach the terrestrial surface. It has been calculated that not more than 5% of total radioactivity would escape, even under the most adverse circumstances.[26]

There is no doubt that nuclear stimulation for geothermal steam production is pre-empted in populous regions for the indefinite future, but, in view of the predictability of effects from nuclear explosives of yield less than 100 kt, their application in unpopulated and remote areas is not so unimaginable. The environmental impact of resulting geothermal steam production is usually limited to the immediate vicinity of detonation and nuclear explosions have been used to produce large volume fractures in subterranean formations. An interested company was Batelle-Northwest, acting as subcontractors to the American Oil Shale Corporation; they found the technique to be economically attractive. A related programme was carried out by the University of California in an effort to identify problems arising from utilizing geothermal steam in desalination of seawater. High costs arise because of coping with corrosion, scaling and the waste disposal of brines. There was another project in the USA to stimulate natural gas production from deep, low-permeability gas fields using the 29-kt Gasbuggy experiment, as well as the Rulison one mentioned above. In the USSR, nuclear fracturing was applied to stimulation of oil production. To obtain geothermal energy the optimum (low yield) range of nuclear explosives would be within a range of 5–100 kt.

Nuclear explosive stimulation of geothermal aquifers was suggested in

1971 by R. Raghavan and his colleagues, who proposed that the same approach might be applicable to stimulation of hydrocarbon production.[28] Low-yield nuclear explosives would have been utilized in order to create a large-diameter well bore or rubble chimney in a geothermal aquifer considered capable of replenishing its own fluids. The differences between a natural geothermal field and one which has been artificially stimulated in this way are illustrated in Fig. 5.5.

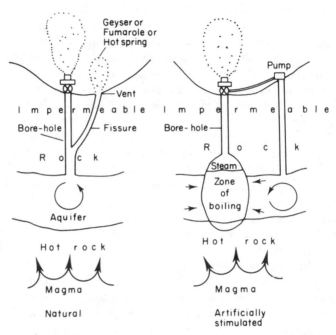

FIG. 5.5.   Comparison between a natural geothermal field and one which has been stimulated artificially.

As regards corrosion control, steam reacts with several minerals to form volatile gaseous compounds which can be transported and deposited on piping in steam turbines as well as in the condenser of such a nuclear-stimulated geothermal power plant. F. G. Straub in 1964 mentioned that salts and silica plus associated radioactivity may accumulate on turbine blades to an extent related to the quantities of these impurities in the steam and also to its pressure.[29] Soluble salts can be washed off with water, but silica, iron oxides and calcium carbonate are insoluble and constitute a serious problem. This is because their build-up may distort the configuration of turbine blades, thus reducing efficiency, promoting turbine imbalance and inducing vibrations capable of damaging the turbine. Some water-insoluble materials can be removed by washing with sodium hydroxide, if their rate of deposition is not too high, but the operation is both time-consuming and sensitive. Radionuclides such as antimony-125 and caesium-134 and -137, together with silica, should be taken out of the steam by scrubbing it with pure water before it enters the turbine. The disadvantage is that this lowers the temperature, and hence the enthalpy, of the steam and, by degrading its energy, reduces the efficiency of the power-generating process. Also, after scrubbing is completed, the steam is at its saturation point and its employment in the turbine might cause erosion of the blades by water droplet impingement. Scrubbing is feasible for steam purification up to temperatures of around 300°C (corresponding to saturation pressures of 80–100 atm). A number of salts and hydroxides can be removed by this treatment. However, there are alternative methods: for instance, a solid-phase scrubber, e.g. limestone, could be an effective remover of silica. A heat exchanger might be added and water boiled in it, the steam from the nuclear outlet being condensed. Ultimately, it would be necessary to reheat the steam, as O. H. Krikorian indicated in 1973.[30] A more detailed examination of corrosion and scaling is given in Section 5.8.6.

5.5 HOT DRY ROCK

Hot dry rock geothermal energy projects commenced at Los Alamos and at the Camborne School of Mines in Cornwall, UK, in the 1970s with the aim of developing a technology adequate to produce an alternative source of energy which could be exploited over a significant proportion of the upper crust of the Earth. The importance of the matter lies in the fact that almost all of the so-far-unused planetary geothermal heat resources occur in such hot dry rock.

5.5.1 The Los Alamos Project (USA)
The Los Alamos experiment began in 1975 and comes under the aegis of the US Hot Dry Rock Geothermal Energy Program and its predecessor agencies. This programme is sponsored by the US Department of Energy and, from 1980 to 1986, included participation by the Kernforschungs-anlage Jülich GmbH representing the Federal Republic of Germany and the New Energy Development Organization representing the Japanese

Government. The involvement of both of these bodies was through an International Energy Agency implementing agreement. Informal cooperation entailing information exchange continues, but the partial financial support offered by the West German and Japanese governments has ceased.

The area selected for development was outside the Valles Caldera in north–central New Mexico, where the most recent volcanic activity took place probably 50 000 years ago and left a large heat resource not far below the ground surface.[31] This permitted instant access to rocks in the appropriate temperature range (180–350°C) for economic exploitation by means of the generation of electrical power.

As early as 1970, an informal group of Los Alamos Scientific Laboratory staff started investigating the feasibility of developing a rock-melting drill, the Subterrene, invented five years before. In 1971 the Los Alamos Scientific Laboratory Geothermal Energy Group was formed and made heat flow measurements around the caldera. Maximum values were recorded west of this in an area where the geology was least disturbed and the Precambrian basement is at shallow depth. Subsequently, in 1972, a 775 m-deep borehole, GT-1, was drilled in Barley Canyon, 5 km west of the caldera. In 1973 the geothermal energy group became redesignated as Group Q-22 and obtained USAEC funding which facilitated drilling a second borehole, GT-2. The site selected was chosen because it had an above-average geothermal gradient, relatively simple geology and structure, and near-surface competent rocks suitable for a heat reservoir. While drilling was proceeding, at 588 m, drilling fluid was lost and air had to be used thereafter. A third borehole to reach 2000 m was slated for 1974 and reached over 3000 m in granitic rock, a temperature of 185°C being recorded. The geothermal gradient was calculated as approximately 50°C/km.

A dual-project plan was suggested for 1975 involving either drilling an initial-energy experimental borehole to be called EE-1 or carrying out more experiments in the existing GT-2. The first was effected and Fig. 5.6 illustrates the GT-2 experimental project.

A hydraulic fracture network was created with a possible diameter of about 3 km and centred at some 2600 m depth. This was done by using water under pressure to produce flow passages in the crystalline rock with low initial permeability. Further directional drilling intersected the fractured reservoir and hydraulic fracturing enlarged the reservoir in 1979. The results of testing the reservoir over a one-year period of circulation of water were given by Z. V. Dash and his colleagues in 1983.[32]

In 1979 this first phase was succeeded by Phase II of the project, which

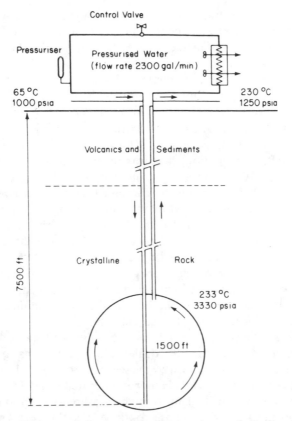

Fɪɢ. 5.6.   The Los Alamos experiment involving a dry geothermal source (hot dry rock system).

widened the scope of the work by constructing a larger, deeper and hotter system. Two new boreholes, some 50 m apart at the surface, were directionally drilled, the deeper one, EE-2, reaching a vertical depth of 4460 m to a rock temperature of 327°C. The upper well, EE-3, lies 300 m above EE-2 and both are inclined 35° to the vertical in their lower 1000 m. The intention was to produce planar vertical hydraulic fractures intersecting both holes. A micro-seismic network comprising geophone sondes located in other boreholes was set up so as to detect and pinpoint micro-earthquakes triggered in the course of hydraulic fracturing, as L. House and his associates indicated in 1985.[33] In 1982 hydraulic fracturing experiments were effected at the bottoms of the two new wells and, unexpectedly, the resulting stimulated zones were shown by the micro-seismic network to be three-dimensional rather than planar and inclined rather than vertical. They neither met each other nor did they connect the two wells hydraulically.

The inclination is assumed to relate to changes in earth stresses caused by the presence of a cooling magma body under a volcanic caldera some kilometres east of Fenton Hill. In December 1979 a large-scale fracturing operation was carried out using 21 000 m^3 of water injected at 3·5 km in EE-2 at a down-hole pressure of 83 MPa and an average flow rate of 100 litre/s.[33,34] After the injection, there was an inducement of seismicity over a rock volume of 0·05 km^3, i.e. approximately 3000 times greater than the actual volume of water involved. The conclusion was reached that this showed shear–slip motion, most likely along existing rock joints, and also that a zone of joint stimulation had been created rather than a single fracture.[33,35] These shear–slip events took place on planes some 10 m apart: this is a span comparable with that of the major joints involved in the well survey. Consequently, the theory of a single planar tensile fracture resultant from hydraulic fracturing had to be abandoned.

Actually, the massive hydraulic injection did not produce a hydraulic connection between the two wells and a later large fracturing operation also failed to do so. Thus, in March 1985, the upper well was sidetracked at 2900 m depth and directionally drilled along a new path, EE-3A, to intersect the fracture zone shown by the seismicity connected with the lower well, EE-2, as R. Parker and R. Hendron indicated in 1987.[36] A hydraulic connection was obtained using a high-pressure stimulation at almost 3600 m depth in EE-3A, drilling was continued and three extra stimulations conducted with open hole packers. Two out of the total of four were unsuccessful and the flow rate produced by the initial stimulation was enhanced by repeatedly shutting in the well EE-2 for a time and then quickly venting it.[37]

The two successful stimulations were sited in zones which were located centrally in the seismicity zone induced during the massive stimulation of EE-2 in 1983. In fact, the unsuccessful ones too were within, but closer to the periphery of, the 1983 seismic zone. In May and June of 1986 a 30-day initial closed loop flow test was effected on the new reservoir, cold water being injected into EE-3A and hot water being produced from EE-2. The latter was cooled down to 20°C in a heat exchanger prior to re-injection. The major flow parameters are given in Table 5.3.

The final impedance almost equalled that measured in the Phase I reservoir after 30 days' circulation. Two tracer experiments during this time

TABLE 5.3

Circulation data from Fenton Hill (compared after 30 days of operation in Phase I and Phase II)[36]

	Enlarged Phase I reservoir	Current Phase II reservoir
Depth of reservoir, m	2 800	3 550
Temperature of reservoir, °C	195	240
Modal volume, m³	160	350
Surface temperature of production water, °C	135	191
Thermal power, MW	3	9
Production flow rate, litre/s	7	13
Injection well pressure, MPa	10	30
Impedance, MPa s/litre	1·7	2·2
Water loss, %	16	33

interval showed that the modal volume (thought to be a characteristic measure of the void volume of major flow paths) rose to 350 m³. This might well indicate that flow channelling (movement of water through fewer paths of larger aperture) did not take place. Such flow channelling would cause inadequate extraction of heat. The water loss rate exceeded that of Phase I, but numerous shut-in tests were conducted during the 30 days. Also there were flow rate changes, from 700 to 1100 litre/s injection rate, and production well pressure changes as well. Hence it was not possible to obtain steady-state water losses, but a declining trend manifested itself. A circulation test lasting a whole year is scheduled to commence in 1988, using the Phase II reservoir. The current (1988) project manning at Los Alamos is equivalent to 40 full-time staff.

In 1988 a special section of the *Journal of Geophysical Research* (in Volume 93, Number B6, pp. 5597 to 6167) was devoted to various aspects of this caldera. Among the topics discussed were the region in general and the Emerging Continental Scientific Drilling, the latter referring to the discussion of the feature as a high priority site for the investigation of fundamental processes in magmatism, hydrothermal systems and ore deposit mechanisms. In their coverage of this, F. Goff and J. N. Gardner (1988)[37a] note that the relevant high-temperature geothermal system is associated with a resurgent caldera and constitutes a dynamic, integrated magma–hydrothermal system having a complex evolution lasting over 13 million years. Conductive heat flux in the VC-1 core hole and the thermal regime of the caldera were discussed by J. H. Sass and P. Morgan (1988),[37b] who gave a heat flow value above 335 m of 247 ± 16 mW m^{-2}. This same core hole verifies the existence of the Valles hydrothermal outflow plume, according to F. Goff, L. Shevenell, J. N. Gardner, F.-D. Vuataz and C. O. Grigsby (1988).[37c] This particular core hole was drilled in 1984 and represents the first attempt of the Continental Scientific Drilling Program (CSDP) in the area. It penetrated an unusual breccia sequence in its lower 30 m (total depth 856 m) and this records a complicated hydrothermal history culminating in hydraulic rock rupture and connected alteration at the edge of the Quaternary caldera, as J. B. Hulen and D. L. Nielson (1988)[37d] indicated. M. Sasada (1988)[37e] studied the microthermometry of fluid inclusions from this same core hole, these revealing the cooling history of the fluids with temperatures above 90°C, the maximum palaeo-temperature having been some 275°C, *ca.* 115°C more than now. Among other matters studied in the borehole were uranium-series age determination of calcite veins by N. C. Sturchio and C. M. Binz (1988),[37f] light lithophile element (boron, lithium) abundances by M. D. Higgins (1988),[37g] and the state of stress and relationship of mechanical properties to hydrothermal alteration by T. N. Dey and R. L. Kranz (1988).[37h]

5.5.2 The Camborne School of Mines Project (UK)

This UK Hot Dry Rock Geothermal Energy Project has been supported primarily by the Department of Energy with a contribution by the European Community which terminated in 1986. A good deal of informal collaboration has occurred between the Camborne School of Mines and Los Alamos groups, Camborne having at present (1988) 65 full-time staff.

The project differs from the Los Alamos one in that a depth was selected which would permit easily available drilling technology to attain rock temperatures such that measurable thermal drawdown resulting from circulation over a reasonable experimental interval of time would be produced. In the American case, fieldwork was done in a reservoir deep enough to afford access to rocks hot enough to generate electrical power. In Cornwall, the accent is on comprehending the parameters which affect development of a full-scale reservoir and hence the decision to avoid the extra difficulties arising from high temperature. It is germane to add that the geothermal gradient at Rosemanowes is much lower than that existing at Fenton Hill.

The first phase of the work entailed the drilling of four boreholes at a depth of 300 m at Rosemanowes between 1977 and 1980: this was described by A. S. Batchelor in 1982.[38] The relevant rock is the Carnmenellis granite, which is actually part of the batholith which underlies the southwest peninsula of England.[39] A horizontal reservoir was created and connection between the boreholes was established with results satisfactory enough to justify beginning a second phase. Phase 2A lasted from 1980 to 1983 and two boreholes were drilled right through the granite at Rosemanowes at an angle of 30° to the vertical at a depth of 2100 m. The surface separation of the holes was 10 m and they were separated by 300 m in a vertical plane at the full depth. The lower well, RH12, was intended for use as the injection one and the upper well, RH11, was slated to be the producing one. A report on the work was made by A. S. Batchelor in 1983.[40]

While R12 was being drilled, several hydraulic fracturing stress measurements were effected using saddle packers and high-pressure pumps run on the drillpipes. The data acquired proved to be important in the analysis of project results. The temperature measurements for the production well gave a bottom-hole value of 79°C which showed an increase in geothermal gradient from 31°C/km at the surface to 35°C/km at 2000 m.

The Camborne project employed explosive stimulation in order to pre-treat the well prior to hydraulic fracturing. This was intended to provide improved access from the borehole to the orthogonally jointed reservoir system for the water used to stimulate the joints in the granite.[41] Later hydrofracturing work was aimed at jacking vertical fractures apart so as to increase the permeability of the reservoir. However, this did not take place: 200 000 m^3 of water were injected at flow rates up to 100 litre/s and large water losses took place. Data analyses from a comprehensive micro-seismic network installed at the site were employed to investigate why this occurred. Micro-seismic emissions from the rock mass were detected by sensors in shallow holes surrounding the site and also by other sensors in the deep system. Most of the growth was downwards, below the bottom of RH12.

The system possessed a high impedance and tracer testing showed a diffusive type of connection. Further prolonged circulation in it induced more seismic activity in a downwards direction. The seismic events demonstrated shear-type fracturing and the downwards growth is related to the variation with depth in the in-situ stress field in Cornwall. Out of this phase came the recognition that the created reservoir was not appropriate for demonstrating the feasibility of reservoir development for hot dry rock (HDR) exploitation.

Phase 2B was effected from 1983 to 1986.[42] A third well, RH15, was drilled on a spiral track in order to intersect the seismically active region below the original wells. It had a true vertical depth of 2600 m and the rock temperature at its bottom is 100°C. The connection with the new well turned out to be unsatisfactory and so the system had to be stimulated. In place of water alone, $5500\,m^3$ of water with viscosity raised to an intermediate value of 50 cP by addition of a substance causing a viscous gel to form were injected into RH15 at an average flow rate of 200 litre/s. The intention was to facilitate tensile stimulation of the natural joints, rather than shear slippage. The results proved successful, micro-seismic activity being much less and confined to a tubular envelope extending vertically mostly between RH15 and RH12. A more compact, more permeable reservoir having less water loss was produced. Circulation of this new reservoir has been continuous since August 1985 and various operating conditions have been used (Table 5.4). The flow rate has been increased

TABLE 5.4
Circulation results at Rosemanowes[36]

	Phase 2A reservoir	Phase 2B reservoir
Depth of reservoir, m	2 100	2 600
Temperature of reservoir, °C	80	100
Average modal volume, m^3	1 800–3 000	136–614
Range of surface temperature of production water, °C	≤56 (RH11)	58–72 (RH15)
Range of thermal power, MW	≤2·3	≤4·5
Range of injection well pressure, MPa	≤14·0	4·0–11·8
Range of impedance, MPa s/litre	1·8	0·53–1·00 (RH12–RH15)
Average water loss, %	66	20
Range of injection flow rate, litre/s, circulation	≤33	5–38

slowly in steps long enough for steady-state conditions to be reached within each one. The hydraulic performance is gradually improving. Large-scale flow rate oscillations or changes in injection flow rate are capable of causing permanent physical changes in the reservoir and these result in a drop in impedance for a given injection flow rate.

Phase 2C of the Camborne project commenced in October 1986 and involved a two-year contract awarded by the UK Department of Energy. The injection flow rate is being raised continually from 20 litre/s and was scheduled to reach approximately 50 litre/s by the middle of 1987. In other parts of the granite, temperatures are higher, around 200°C in fact, and could yield greater amounts of heat energy.

5.5.3 Ancillary Aspects

While in operation both projects aim at carrying out a number of supporting jobs, among these logging and seismic measurement which are both significant in understanding and controlling reservoir development and circulation. The Los Alamos National Laboratory has developed a high-temperature borehole acoustic televiewer in collaboration with the Westfälische Berggewerkschaftskasse of Bochum. It is designed for operation at 275°C to enable viewing of locations of fractures and apertures as well as for observing break-out depths and orientations, and determining of stress directions and ratios.[36] It can be used also to inspect casting interiors.

In this connection, it is interesting to mention that hydraulic fracturing stress measurements and a similar borehole televiewer survey were effected in a 1·6 km-deep well at Auburn, New York, and were reported by S. H. Hickman and his colleagues in 1985.[43] This particular well was drilled at the outer margin of the Appalachian Fold and Thrust Belt in the Appalachian Plateau and penetrated *ca.* 1540 m of Early Palaeozoic sedimentary rocks, terminating 60 m into the Precambrian basement. Drilling was done by the New York State Energy Research and Development Authority (NYSERDA) to evaluate the geothermal potential of central New York State.

Some details regarding the televiewing technique may be appropriate. The borehole televiewer is a wirelogging tool which yields a continuous orientated ultrasonic image of the wall of the well. It comprises a transducer mounted on a motor-driven shaft and aimed at the borehole wall. The transducer rotates three times per second while generating an approximately 1·2 MHz pulse 1800 times a second. The tool is pulled up the hole at a speed of 1·5 m/min on a standard wirelogging cable. The reflected energy returning to the transducer modulates the intensity of a trace on a cathode-ray tube at the surface so that a bright trace corresponds to a good reflection and a dark one indicates a scattered or absorbed signal.

One revolution of the transducer corresponds to one trace on the cathode-ray tube and the initiation of each trace is controlled by a flux gate magnetometer. Successive traces ascend the cathode-ray tube as the tool is pulled up the borehole. The display is photographed and the unprocessed sonic signal from the tool together with the flux gate magnetometer signal are recorded simultaneously on the video-tape for later inspection.

A comparison of the borehole televiewer with a four-arm dipmeter in the Auburn geothermal well was made by R. A. Plumb and S. H. Hickman in 1985.[44]

For its part the School of Mines at Camborne is developing a three-axis sonde for employment down-hole in deep systems (200°C) to permit accurate location of micro-seismic events in deep reservoirs without any necessity of shallow sub-surface instrumentation. In addition, there is a programme there for the characterization and testing of electronic components appropriate for constructing high-temperature down-hole instruments.

Both groups, Los Alamos and Camborne, are investigating techniques for the detection of lineaments in the seismic zones of their respective reservoirs and these could be useful in showing the presence of natural fractures most susceptible to stimulation. Several diagnostic techniques have been used in order to characterize their respective reservoirs, including hydraulic, chemical and thermal analyses together with tracer tests. Attempts are being made to use the accumulated data to develop mathematical modelling methods. Especially at Los Alamos, with its higher water temperatures and more active waters, the corrosion of both borehole and surface plant materials is being examined. Both there and at Camborne, environmental monitoring is in progress and results to date are benign. Clearly, there is much still to be done, but continued experimentation at both places should facilitate their attainment of the ultimate goal, namely commercial electrical power generation.

5.5.4 Future Prospects for Geothermal HDR in the UK
As seen, the basis for the hot dry rock concept is drilling of a minimum of two boreholes into the Earth to a depth sufficient to reach rock at a useful temperature and then to fracture a heat-exchange surface between them. Water is pumped down through the rock fracture zone to acquire heat energy, and afterwards recovered. The on-going hydrofracturing experiments are mainly concerned with establishing the principles of rock mechanics and engineering which are involved at 2–2·5 km depth. Unfortunately, the UK Department of Energy believes that, in order to obtain valuable heat for generating electricity, it is necessary to penetrate to at least 6 km depth.[45] Since the third well at Camborne, RH15, proved successful, the next step should be to drill one to this depth. This would be a major step, taking the HDR technology from the research-and-development to the deployment stage. Of course, prior to attempting to do this, it would be wise to gain a better understanding of the hydrofracturing process at shallower depth.

Whilst the status of geothermal HDR did not change appreciably from an economic standpoint after the strategic review of 1982, the earlier emphasis on generation of electricity has altered in favour of heat-only experiments. This would be cheaper because drilling can be restricted to shallower depths and, in some cases, it might be feasible to link the thermal product into existing district heating schemes. The 1982 strategic review demonstrated that, on technical grounds, the exploitable HDR resource in the UK could generate approximately 2500 TW h of electricity, i.e. about 10% of 1985 consumption. A later assessment increased this estimate by a factor of ten.

However, the resource is far from widely distributed, the existing main area being in southwest England. There the highest temperatures are in the granite and these suffice for CHP (commercial heating purposes) applications. Since there are few appropriate heat-loads in Cornwall, electricity-only generation may have to be enforced. The British Geological Survey is effecting an HDR characterization of the resource elsewhere, e.g. in areas of basement rock under population centres.

As mentioned above, some of the Cornish granites have temperatures of about 200°C and, at this level, the organic Rankine cycle could be used. However, there has been a suggestion that the flashed steam cycle might be substituted since, whilst this would give less electricity, it would provide more useful heat. In this cycle, some of the water leaving the borehole is flashed to steam and passed into a turbine, 90% of the heat being retained in the water.

The likely cost of generating electricity from geothermal HDR stations was envisaged as in the range 3–6 p/kW h (1984 value).[45] No estimates have

TABLE 5.5
Assessment of future contribution of geothermal HDR in the UK[45]

Characteristics	Provide electricity and/or heat from fractured rock by injection of water under pressure
Technical potential	$\geq 25\,000$ TW h ($\geq 10\,000$ Mtce[a]) might be exploited over a century
Economics	In one scenario, uneconomic
Development state	Successful test drilling at a depth of *ca.* 2 km demonstrated feasibility, but further drilling to 6 km is necessary in order to pass from R&D to deployment
Constraints on deployment	Availability of drilling services, technical risks, large capital investment

[a] Primary energy equivalent (million tonnes of coal equivalent).

been made of the costs of CHP or heat-only distribution systems. Pertinent data are listed in Table 5.5.

In an assessment of September 1987, geothermal hot dry rock in the UK was categorized as a promising, but uncertain, resource, the relevant technology being described as not yet proven either scientifically or economically.[46] If it is, then the HDR technology may well become economically attractive if predicted costs can be achieved or fuel prices rise in real terms. The object is to obtain cost reductions and improvement in performance. Clearly, the existing constraints on widespread deployment are geographical bias, insufficiently advanced technology and the risk of failure to drill productive wells.

5.6 FUTURE PROSPECTS FOR GEOTHERMAL AQUIFERS IN THE UK

These future prospects were mentioned in Chapter 2 and, as the subject relates to the end of the preceding section, it is appropriate to consider it here.

The aquifer resources in the UK may be divided into three categories, namely those with waters exceeding 60°C which could be directly used to provide low-grade heat, those in the range 40–60°C which could be exploited by means of heat pumps, and those with waters at temperatures under 40°C.

As regards the first, the deep aquifer resource under urban areas where boreholes were drilled turned out to be poor. In fact, most of the potentially economically viable resource is located around the Bournemouth area of the Wessex Basin and may add up to as little as 0·25 Mtce/year. This is the region in which the Marchwood borehole was drilled near Southampton. Water was drawn from a limited-size reservoir there and so the pumping power requirements would rise to unacceptable levels after about three or four years. The sole danger signal was that the water levels did not recover as rapidly as anticipated after pumping tests. The risk to commercial exploitation is evident. A developer who started pumping at 30 litre/s would have managed very well for a couple of years and then rising power pump needs would have become prohibitive. In fact, the Southampton City Council presented an alternative scheme based on a flow rate of approximately 10–11 litre/s in winter and 5 litre/s in summer. This would supply heat to fewer buildings and satisfy requirements for sustainable heat flow over a quinquennial period of time. Whilst it covers running costs, it would not defray the sunk capital costs of the borehole itself.

In regard to the East Yorkshire and Lincolnshire Basin, the exploratory borehole at Cleethorpes was mentioned in Chapter 2: the anticipated resource in the Permian sands turned out to be disappointing. There may be a good Permian resource in rural Lincolnshire, but it would not coincide with the major heat loads around Humberside. However, the extensive Sherwood sandstone has a high permeability and a maximum fluid temperature of 49°C. This could supply heat loads in Grimsby and one heat load in the form of flats and local authority housing is under investigation as a possible recipient.

5.7 EUROPEAN EXPERIENCE

In France, 40 schemes were developed for using geothermal aquifers between 1970 and 1985, and the rate of installation has reached between 20 and 30 annually, while in Italy three schemes are almost completed in the Po Valley.

The aim in France is also to reduce dependence upon imported oil and the French subsidize geothermal exploitation to keep such energy prices 10–30% below oil prices. Incentives offered include a grant of 30% of the cost of the initial borehole, guarantees regarding the geological exploration risk of up to 50% of the cost of the initial borehole and insurance cover on the long-term risk of aquifer performance. The insurance is initiated by a five-million franc premium payment by the French Government plus an annual levy on successful geothermal schemes.[45] Other differences between France and the UK include the fact that established district heating developments already exist, e.g. Paris is centred over the optimum part of the Dogger aquifer, a fractured limestone at reasonable depth providing good transmissivity and temperature. Also the competing fuel in France is mostly light oil whereas in the UK it is the cheaper natural gas. Then, too, drilling costs are lower in France as a consequence of the large amount of geothermal drilling going on, while in the UK prices are governed by hydrocarbon drilling rates. Nevertheless, R. Harrison in 1984 concluded that the economics of the Paris Basin schemes are very fragile and vulnerable both to technical and market uncertainties.[47] This is especially true for schemes employing heat pumps to improve the matching between aquifer and demand temperatures. In such schemes, the net present values are negative if energy prices remain constant in real terms. The developments in France are supported by cheap loans and outright grants for political and energy policy reasons (see Table 5.6).

TABLE 5.6
Assessment of future contributions from geothermal aquifers[45]

Characteristics	Yield low-grade heat (assessment based on constant heat output at 60–90°C with a heat output of 3–4 MW per well)
Markets	Large heat loads in commercial and institutional buildings, district heating and low-grade process heat:

Year	Allocation of low-grade heat between categories of buildings (%)			
	Commercial		Housing	
	Old	New	Old	New
1980	57	0	43	0
1990	48	9	38	5
2000	40	17	34	9
2010	32	25	29	13

Technical potential	0·6 Mtce/year until AD 2025
Conventional systems replaced	Coal-fired district heating as well as coal, oil or gas local heating
Economics	Not yet cost-effective
State of development	Technically feasible as of now, but specific usable resources have to be found
Constraints on deployment	Large capital investment and perceived risk of dry holes, institutional factors, retrofitting low-temperature heating systems
Estimated contribution in AD 2025	0·25 Mtce (60 million therms, 6 PJ) annually

5.8 EQUIPMENT NECESSARY FOR EXPLOITATION

Any geothermal plant using hot water for generation of electricity with binary cycles or for other uses must include production pumps, surface pipelines, heat exchangers, heat pumps and re-injection pumps.

5.8.1 Production Pumps
These will not be necessary, at least initially, where the geothermal reservoir is artesian in nature. However, in most cases they are necessary, either

because such artesian flow is inadequate or because the geothermal reservoir is unconfined. Then the geothermal waters have to be pumped in order to attain an adequate flow rate which may rise to as much as 250 m³/h (70 litre/s). An advantage accrues if a self-flowing well is pumped, in that the pressure of the liquid is constant, this obviating flashing in the borehole and ultimate scaling. Also prevention of flashing ensures that the pump discharge temperature is elevated above the surface temperature of a self-flowing well, which is significant where high-temperature applications are involved.

There are two types of production pump, namely that having a motor at the wellhead and a vertical shaft (the line shaft type) and the down-hole (submersible) type.

The line shaft or vertical turbine type is normally fitted with an electric motor at the wellhead, this reducing the difficulties and expenses of maintenance. In any event, the maintenance is closely linked with pumping of the fluid. In low-temperature and low-solids content situations, a line shaft pump can function for years without any maintenance at all. In corrosive situations, however, operation may be possible only for a year or so, after which time cleaning and repair work are mandatory. The flow rate can be varied as well, either by means of a mechanical regulator (with little variation in power consumption) or through an electric regulator which eliminates part of the stator coils from the motor circuit. The pump is therefore operable at two flow rates, corresponding to two different powers.

Clearly, the borehole has to be almost or completely vertical for the pump shaft to function properly. Such line shaft pumps have been utilized successfully in geothermal wells and increase fluid pressure by centrifugal force, which is imparted to the liquid by means of a shaft-driven impeller. Sometimes several stages are used in series, this being required where high pressure has to be attained by pumping against very high heads; although each stage possesses the same flow, every stage cumulatively increases the pressure of the geothermal liquid. Pump suction occurs at the bottom of a bowl assembly, the fluid passing upwards and emerging in an annulus between the column pipe and the shaft-enclosing tube.

Tube bearings support the shaft and are regularly spaced, say at 2-m intervals. Rubber bearings are constrained by temperature maxima around 120°C and, in Iceland, optimum results have been obtained employing Teflon bearings every 1·5 m and steel shafts with diameters of 30 mm down to 250 m depth; below this, down to 300 m, a diameter of 43 mm was used and flow rates between 100 and 180 m³/h (30–50 litre/s) were attained. Filtered geothermal water pumped down the shaft-enclosing pipe acted as lubricant for the Teflon bearings in question. These lasted for as long as four years in Iceland, compared with the three- or four-month lifespan of alternative types such as rubber, bronze or babbitt (see T. Karlsson[48] in 1982).

When hot fluids are being pumped, the metal components comprising the pumping system expand and, because the line shaft is normally made of different materials from the column and tubing, this will entail differential rates of such expansion. The axial expansion of the shaft increases with its length, i.e. with the depth at which the impeller is inserted within the borehole. During assembly it is desirable to take into account the probability that the shaft will have to be raised or lowered after the system has equilibrated at the temperature of the pumping operation.

The actual power of the pump motor at the wellhead can be derived from empirical equations using manufacturers' tables. One used in hydrodynamics is that of V. L. Streeter (1961):[49]

$$P(\text{kW}) = \frac{\text{pump capacity (litre/s)} \times \text{head (m)} \times \text{specific gravity}}{76} \times 0.735$$

(5.16)

E. Barbier in 1986 quoted a personal communication in 1984 from the Icelandic National Energy Authority, who informed him that, for example, flow rates of $100 \, \text{m}^3/\text{h}$ (30 litre/s), with a wellhead pressure of $2 \, \text{kg/cm}^2$ and water level at 220 m below ground level, need a pumping power of approximately 120 kW, this figure exceeding that obtained from the above equation by a factor of two.[50] From this it may be inferred that it is probably better to install a pump with a greater capacity than foreseen to be needed and operate it at reduced power than to attempt to employ one of theoretically correct, but actually insufficient, capacity.

The major disadvantage of the rather noisy line shaft pumps is their long drive shaft, and they also require a vertical well which means extra care and expense in drilling. Their maximum operational depth is 300 m or so.

The other type of pump mentioned earlier is the down-hole variety: this is in common use in geothermal service. Its electric motor pump section is under water within the well, suspended from a production casing and connected to the surface by electric cables. The diameter of the pump sections range from just over 10 cm to just under 36 cm and therefore they have to be emplaced in boreholes having wider diameters than are necessary with the first type. Usually, wells of 34 cm diameter suffice and facilitate the entry of the 22·9 cm-diameter motor pump section together with the production casing (18 cm). A sealed induction three-phase engine is cooled and lubricated by oil. Motor cooling is obtained by transferring heat to the circulating well fluid and motors of this kind have been used in situations where fluids with temperatures exceeding 150°C are involved. The pump section comprises a multi-stage centrifugal system rather like that of the line shaft pump and capacities range up to 530 m³/h (147 litre/s) and lifts up to 4500 m. Because of limitations in diameter and hydraulic factors, the lift per stage is relatively low so that sometimes as many as 500 stages have had to be employed in order to meet high head needs. Pump wear and corrosion can be minimized by means of suitable corrosion-resistant materials.

A French company, Pompes Guinard, has developed a down-hole pump called the Turbopress, claimed to be capable of overcoming problems encountered in geothermal wells. It was first intended for employment in 34-cm casings (a popular casing diameter in geothermal wells), but other diameters are also available. In the down-hole set, there is a hydraulic turbine and a centrifugal pump, both being driven by the same shaft. The operating fluid comes from the well fluid which the pump lifts. It is supplied to the turbine by means of a surface charge pump through a standard tubing from which the down-hole pump set hangs. One advantage is that electricity does not have to be sent down the well, this obviating the possibility of damage to power cables in the corrosive conditions found there. Another is that the pump works up to temperatures around 270°C. Also no plastics materials are utilized. In addition, the design of the pump set requires no mechanical seals between turbine and pump (a frequent cause of failure in normal down-hole pumps).

Down-hole pumps are excellent from the standpoint that they can better well production rates attained by line shaft pumps since they are high-speed and, in Iceland, it has been found that they are more economical for wells in which the water level is more than 250 m below the ground surface. Manufacturers of production pumps include KSB (FRG), Pompes Guinard (France) and, in the USA, Byron Jackson, Reda, Floway, Peirless and Centrilift, according to E. Barbier in 1986.[50]

5.8.2 Piping

Production and transmission pipes above ground are vital ingredients of a geothermal plant because they transport hot water as cheaply as is practicable. To conserve thermal energy, they must be well insulated. The type and thickness of such insulation is a function of location, i.e. whether pipes are above ground or buried. In both cases, the temperature of the geothermal fluid being transported is significant. Where the pipes are above ground, care must be taken to guard against damage to thermal efficiency and by people. Where the pipes are buried, factors to be considered include resistance to moisture absorption and superincumbent load. Also they have to be protected against corrosion. This is particularly important if acid soils or very wet soils are involved. Either metallic or non-metallic piping is used. The commonest metallic pipes are made of carbon steel and have high strength as well as the ability to resist high pressures. Steel pipes are easy to emplace and joints may be threaded, welded or joined by gland-type couplers with elastomeric gaskets, e.g. tapered joints with O-rings. The gaskets have to be selected with care in order to incorporate sufficient pipe expansion and resistance to operational temperatures and pressures.

The disadvantage of steel piping is its cost and rather high heat loss if inadequate or no insulation is utilized. An uninsulated steel pipe loses ten times more heat than a similarly sized and pre-insulated asbestos-cement pipe. Suitable insulating materials include spun fibreglass and rock wool. However, as E. Barbier indicated, these are not suitable for buried pipes because of adverse effects of moisture.[50] Alternatives include field-wrapped carbon steel and pre-insulated carbon steel.

Of course, copper may be used instead and pre-insulated copper piping is available in small diameters of less than 8 cm, these being below the sizes available in cement or synthetic plastic. Costs are favourable, too, and so are the installation characteristics. Proper insulation ensures low heat loss and the main drawback is the high cost of pre-insulated copper piping, as S. W. Goering and others pointed out in 1980.[51] Another disadvantage is the possibility of electrochemical interaction occurring with the rest of the plant, particularly with steel. The lower the oxygen content, the lower the damage, but in this situation the expensive copper piping would be unnecessary in any case.

Non-metallic piping includes asbestos-cement, fibreglass-reinforced polyester (FRP) and some patented plastic varieties such as polyvinyl chloride (PVC), chlorinated polyvinyl chloride (CPVC), high-density polyethylene (PE) and polypropylene (MOPLEN).

Asbestos-cement (concrete) piping is widely employed for transmission lines and is cheap compared with metallic pipes. However, it is porous and there is the possibility that oxygen will enter the water during transit. Also it is subject to much greater heat loss than insulated steel pipes because, apart from a soil cover if buried, it is normally uninsulated. Corrosion of asbestos-cement pipes can take place when geothermal fluids pass through them, calcium from their walls slowly dissolving. However, the effect lessens with time and it has been suggested that the pipes will last at least 20 years.

or so before their deterioration becomes significant.[48] In addition, asbestos-cement piping is structurally weaker than steel piping and needs especial care when it is bedded, this entailing extra expense. Consequently, such piping is recommended for employment when geothermal fluid production costs are minimum and the geothermal water must be carried for fairly long distances of several kilometres.

Where carbon steel is necessary in order to overcome pressure restrictions, pre-insulated asbestos concrete has proved useful.[50] It comprises two concentric asbestos-concrete pipes having a polyurethane foam insulation between them, is easy to emplace and has small heat loss because the foam is a first-class thermal insulator. Unlike a single such pipe, it is relatively strong and does not need to fulfil any special bedding requirements. Any interior corrosion can be minimized by coating with epoxy.

Fibreglass-reinforced polyester pipes can survive temperatures up to 150°C and, at least with some types, pressures up to 8 kg/cm^2 without appreciable maintenance costs for over a decade and without any apparent damage. However, they cannot transport steam, which breaks them down. On the other hand, they are resistant to both corrosion and scaling, and, being so light, they are cheap to install. Their major disadvantage is that their emplacement must be done very carefully in order to avoid damaging their outermost epoxy coating because, if this is broken, soil water can infiltrate and reduce the resistance of the pipe. E. Barbier regarded using different types of this piping in the same circuit as undesirable.[50]

Patented plastic pipes have not been widely used in geothermal work to date, but may be in the future because of their attractive properties of ease of joining and resistance to chemical attack, although they do have limits imposed by temperature. Thus, although polyvinyl chloride is enticing because it is so cheap, it sustains a high heat loss and weakens at temperatures exceeding 27°C. At 60°C its strength is reduced by 22% below the initial level. Unsurprisingly, therefore, it tends to bend at geothermal water temperatures. Chlorinated polyvinyl chloride is better, but also dearer and again it is subject to heat loss. It weakens above 38°C, but not so badly as PVC does. As regards strength, it loses 20–40% at 65°C and so requires careful installation. High-density polyethylene piping possesses high resistance against both corrosion and scaling, but is adversely affected by high pressures. Finally, MOPLEN pipes also have a high resistance to chemical attack, but are rather unstable in the presence of oxygen and particularly when the temperature is elevated above the normal atmospheric level.

TABLE 5.7
Heat loss in °C per 305 m of pipeline buried in soil[52]
(conditions: temperature of geothermal fluid, 93°C; temperature of the ground, 1·6°C; depth of burial, 0·9 m; fluid velocity, 1·8 m/s)

Piping material	Nominal pipe diameter (inches)			
	4	8	12	16
Bare carbon steel	1·9	0·6	0·3	0·3
Insulated carbon steel	0·3	0·1	0·1	0·0
Bare fibreglass	1·7	0·4	0·3	0·3
Insulated fibreglass	0·2	0·0	0·0	0·0
Bare asbestos-concrete	0·8	0·2	0·1	0·0
Insulated asbestos-concrete	0·1	0·0	0·0	0·0

Table 5.7 shows heat losses for piping of various materials: from the results it appears to make little difference in terms of heat loss whatever material is utilized for pipelines. For instance, carbon steel and fibreglass demonstrate the same behaviour when thermally insulated, which implies that it is the insulation which is significant. Consequently, it is highly desirable always to insulate supply pipes for geothermal water and recirculating water.

5.8.3 Heat Exchangers
A heat exchanger must be installed between a production well and the users: its function is to transfer heat from the primary geothermal water to clean secondary water, which is afterwards transported to the plant utilizing the heat.

The two fluids in question do not contact each other and the normally corrosive geothermal water is subsequently either re-injected into the geothermal system through special wells or discharged on the surface. It is thus apparent that a heat exchanger is critical in maintenance of geothermal fluid circulation through the equipment and protects piping and plant from its corrosive power. Of course, the pressure exerted by the geothermal fluid is independent of that of the clean water involved. The inevitable disadvantage is that there is a temperature drop between the primary and secondary fluids, and the greater the drop, the more efficient is the exchanger. It is usual for such a drop to be kept down to 2 or 3°C.

There are two types of heat exchanger, surface and down-hole. Among the former are the shell-and-tube and the plate exchanger.

The shell-and-tube heat exchanger comprises a set of parallel tubes within a cylinder, the geothermal fluid circulating outside the tubes and inside the cylinder. The tubes may be U-shaped or straight with removable heads at both ends to make cleaning easier. Baffles are usually placed in the shell side to produce turbulence and a tortuous pathway for the fluid contained there. Materials employed depend upon the nature of the geothermal fluid; if mild steel or copper or silicon bronze are utilized, the exchanger will be cheaper than the plate one.

This plate exchanger is composed of a number of plates clamped into a frame, primary and secondary fluids being passed through alternate passages between them. The arrangement is such that fluids being heated and cooled flow in alternate channels so that each plate has hot fluid on one side and cold fluid on the other. Appropriate stamping of the plates permits a number of flow patterns and sizes. Annual disassembly is feasible and then any scale can be removed.

Plate exchangers allow three-fold higher heat transfer coefficients than shell-and-tube ones because of the high fluid velocity reached in the rather narrow flow channels. This results in high shear forces which improve heat transfer and also reduce the rate of plate contamination. The high heat transfer coefficients facilitate construction of these exchanges with smaller heat transfer surface areas. In turn, this reduces both costs and weight, although there is a corresponding pressure drop. It appears that there is a linear response of heat transfer to decreasing flow rate from the rated maximum, a significant advantage in matching geothermal heat supply to demand of the system. It has been found that titanium is preferable to stainless steel for such exchangers because of its greater resistance to corrosion and also because it improves the reliability of the devices. The disadvantage is that it is some three times more expensive than steel.

Operational limits for plate heat exchangers are 150°C and 20 kg/cm^2 pressure. They need less space than the shell-and-tube, measuring 1·5 m × 1 m × 1 m for a geothermal flow rate of 150 m^3/h.

Down-hole heat exchangers have the advantage that disposing of geothermal fluids is not a problem since they stay down the production well. They are cheaper than surface heat exchangers and usually they are installed no deeper than 150 m. An annular space is left between casing and borehole wall so as to maximize efficiency in transferring heat from the primary fluid to the secondary fluid circulating in the piping in the well. The water circulates down naturally within the casing and through lower perforations, thereafter up through the annulus and back within the casing through upper perforations. When necessary parameters, e.g. diameters of bore and casing, length of the down-hole heat exchanger, tube diameter, number of loops, flow rate and inlet temperatures, are taken into account, velocity and flow rate of the natural convection cell within the well approaches that of a conventional surface shell-and-tube heat exchanger. Corrosion normally occurs only at the air–water interface at the level of static water. E. Barbier and M. Fanelli in 1977 demonstrated that this type of heat exchanger is economic, entails little corrosion and eliminates environmental pollution as well as aiding in preservation of the geothermal aquifer.[53]

Manufacturers of heat exchangers include APV (British), GEA (West German), Vicarb (French) and Alfa-Laval (Swedish), according to E. Barbier in 1986.[50]

5.8.4 Heat Pumps

Sometimes it is necessary to increase a low temperature, say of 15 to 50°C, in a geothermal water, and this may be accomplished by using heat pumps. The principle is to utilize the characteristics of a volatile working fluid such as Freon or isobutane circulating within the pump in an evaporator and in a condenser so as to increase the temperature of the secondary water to the users. Clearly, this process entails extraction of heat from a low-temperature source and its usage at a higher temperature, the heat pump requiring electrical energy to accomplish this transformation. Since the electrical energy is not converted into thermal energy as in an electrical resistance, more heat per unit of electricity used is obtained than would be yielded by direct conversion using such an electrical resistance.

The principal components of a heat pump are the compressor, condenser, evaporator and expansion device. Consumption of electricity enables the compressor to increase the temperature of the working fluid, simultaneously compressing and liquefying it. Then the working fluid delivers its heat to clean secondary water in the condenser, thus increasing the water temperature as required. The working fluid then quits the condenser, passing through the expansion device in which it expands and undergoes flashing and cooling. Finally, it enters the evaporator where it extracts heat from the circulating geothermal fluid, passes on to the compressor and completes the cycle and starts another one.

Thermal energy obtained at the exit of the heat pump is related to the maximum and minimum temperatures of the working fluid. The coefficient of performance of the pump is its maximum thermodynamic efficiency and corresponds to the ratio between the total heating output of the pump and the total electrical input. Values for the coefficient of performance vary with the type and working temperature range of the pump used, as J. Olivet indicated in 1981.[54]

Water-source heat pumps are those using geothermal water and these could be applied to space heating with the source fluids in a temperature range from 0 to 50°C. However, the appropriate units commercially available cover the range 15–35°C. The coefficients of performance of heat pumps at the lower end of the range are reasonable, but are orientated towards a rather small temperature drop in the source fluid. This produces a large mass flow rate and hence large amounts of the fluid, a restriction when this fluid is geothermal and derived from an exhaustible well. At the higher end of the temperature range, i.e. around 35°C, the coefficient of performance is poor. Clearly, therefore, some amendment to design is necessary to correct this, as G. M. Reistad and P. Means indicated in 1980.[55]

Geothermal heat pumps can be either non-isolated or isolated systems (see C. S. Smith and P. F. Ellis II[56] in 1983). In non-isolated pumps the geothermal water circulates directly through the water side of an evaporator, whereas in isolated pumps it is circulated through an isolation heat exchanger in which heat is transferred to treated water flowing in a closed loop between the isolation heat exchanger and the evaporator. The isolated system loses efficiency because of the inevitable drop in temperature across the isolation heat exchanger; it requires an extra heat exchanger and piping, pump and controls as well. Isolation systems can minimize corrosion and scaling problems threatened by most geothermal waters and the above-mentioned titanium plate heat exchangers are applicable to reducing costs for the isolation heat exchanger. They also make cleaning easier. Nevertheless, economics normally dictate the employment of heat pumps which are common in the market and evaporators are usually made of steel, stainless steel, cupronickel or copper. Where these materials must be avoided, an isolation system is recommended.[50]

Manufacturers of heat pumps include Man and Neunert (FRG), Zanussi (Italy), Sulzer (Switzerland), and York, Carrier and Westinghouse (USA), according to E. Barbier in 1986.[50]

5.8.5 Re-injection Pumps

These re-inject withdrawn and used geothermal water into the reservoir or other permeable formation appropriate to receive it and the relevant pumps are usually horizontal centrifugal ones sited at the wellhead. Geothermal water can be re-injected after circulation in the usage system or, if adverse chemical factors exist, immediately after traversing the heat exchanger then located near the production well itself. Of course, the re-injection pump increases the pressure of such geothermal water, sometimes by ten-fold, and this will constitute the re-injection pressure.

The materials out of which the re-injection pump is made depend upon the nature of the geothermal fluid involved. If this is highly corrosive, then stainless steel casing and impellers are required.

Manufacturers of re-injection pumps are listed above under production pumps, which they also make.

5.8.6 Corrosion and Scaling

The solids and gases occurring in geothermal fluids are thermodynamically capable of causing corrosion by attacking metal surfaces or scaling during the utilization cycle. During electric utilization in a double-flash geothermal power plant, temperatures may range from 350°C at the bottom of the well to 60°C or so in the re-injection lines, and associated pressures can vary from 300 to 0·08 bar. Of course, geothermal fluids may contact air in the condenser and when re-injected. However, the chemical composition of these fluids will range from almost pure water to hot brine with a total dissolved solids content of 360 000 ppm or more. Consequently, it is almost impossible to find a solution to corrosion and scaling problems except on a specific basis for particular sites.

5.8.6.1 Corrosion

Corrosion may be general, a sort of 'rusting' which affects all metal surfaces in geothermal equipment, or localized. Into the latter category may be placed pitting (often due to the breaking down of scale) and crevicing (sometimes occurring underneath scale or in cracks in various devices). In addition, a number of other phenomena may be included, among them stress corrosion and sulphide stress cracking; the first is due to stress coupled with the existence of chloride ions in the environment and the second is caused by the presence of hydrogen sulphide in an aqueous phase. Contrary to stress corrosion cracking, the severity of sulphide stress cracking in high-strength steels decreases as the temperature increases and oxygen, which promotes stress corrosion cracking, plays little part in it. Sulphide stress cracking is more common in the petroleum industry than in geothermal activities. In low-strength steels, hydrogen blistering may result from their exposure to aqueous solutions containing hydrogen sulphide; rupturing takes place when trapped hydrogen in voids accumulates at a sufficiently high pressure. It is very interesting to note that the relevant material does not have to be under stress for hydrogen blistering to occur.

As well as the above phenomenon, R. Corsi in 1986 referred to several other types of corrosion and these should be mentioned.[57] Intergranular corrosion is produced by incorrect heat treatment and may occur near boundaries of grains. It is not always accompanied by deterioration of the grains. When an alloy disintegrates, grains drop out and so strength is lost. Galvanic coupling results from electrically connecting two dissimilar metals, the less mobile one being attacked. Of course, careful selection of materials, bearing in mind their increasing nobility in galvanic terms, can aid in avoiding this phenomenon, although the chemical nature of the working environment has to be considered as well. Corrosion fatigue is a consequence of imposition of cyclic stresses on a material in a corrosive environment. Its limit is the highest unit stress applicable under given stress conditions, rate of application of stress, temperature and corrosive environment situation without promoting failure in a given number of stress cycles. Erosion corrosion is an accelerated loss of metal resulting from attack by high-velocity fluid, droplets or particles and it can be a serious hazard in the wet ends of turbines or in sites of two-phase flow. Finally, cavitation is a similar event caused by formation and collapse of bubbles of vapour near a metal surface.

Because of the many types of geothermal fluid which may be encountered, it is practically impossible to predict corrosion in advance and the only way of selecting material for geothermal equipment appears to be on the basis of experience. However, there is a method of determining uniform corrosion rate by weight loss measurement, in which a clean sample of test metal is measured, weighed, exposed to a corroding attack for a known time interval, removed, cleaned and re-weighed. The rate of metal removal by corrosion is calculated from the relationship

$$R = KW/ATD \qquad (5.17)$$

in which R is the rate of corrosion involved, T (h) the time of exposure, W (mg) the weight loss (corrected for loss during cleaning), A (to the nearest cm^2) the area and D (g/ml) the density. If R is expressed in mm/year, the constant K becomes $8·76 \times 10\,000$.

Using this approach, together with visual inspection, it is possible to estimate pitting corrosion. Stress corrosion cracking can be measured by using test coupons of different types. The specimen (observing the ASTM G38 standard) is derived from either bars or sheets of which the thickness always exceeds 6 mm. Materials are charged to a particular value of R and to 70% of this as the $p0.2$ ensuing valuation will refer to normal working conditions and ultimate material stress. The factor $p0.2$ is that stress at which the permanent strain is 0.2%. This test is both long and costly.

There is an alternative electrochemical test available which depends upon analysis of the curve of potential versus current of polarization. A potentiostat is employed to assess anodic polarization behaviour; the metal under test, in a natural state, is cleaned and inserted into a polarization cell and the potential is altered by use of the external circuit in the noble direction. It is maintained at this potential sufficiently long for equilibrium to be attained. Then the current flowing in the external circuit is measured and the operation repeated. The resultant curve of potential versus current density demonstrates where stress corrosion cracking takes place. The technique is widely utilized in evaluation and may be effected in the field by employing special electrodes and equipment able to resist temperatures of 200–250°C and pressures of 20–25 bar.

The corrosive agents include oxygen, hydrogen, chloride ions, hydrogen sulphide, carbon dioxide, ammonia and sulphate ions. Less common are fluoride ions, heavy metals and boron. Their separate and combined quantitative effects are impossible to foresee because of imponderables such as the varying responses of different materials making up geothermal equipment, interaction of several species and the type of attack (uniform, pitting, etc.). In 1981 P. F. Ellis and M. F. Conover discussed the corrosive effects of these chemical species.[58] It is noteworthy that dissolved oxygen is believed to be the most serious contaminant of geothermal fluid steam and condensate, aerated geofluid producing a ten-fold or greater increase in the uniform corrosive rate of carbon steel. Indeed, oxygen at a certain temperature can initiate stress corrosion cracking in the presence of chloride. However, some materials such as aluminium alloys actually need dissolved oxygen in order to maintain their passivation films and they corrode in oxygen-free geothermal fluid. Of course, oxygen contamination must be avoided in re-injection lines where a temperature increase may be anticipated when fluid flows into the related wells. As regards hydrogen, the pH controls corrosion of both carbon and low-alloy steels, of which the rate of corrosion falls as the pH increases. The passivity is frequently pH-dependent and hence the corrosion-resistant alloy may undergo pitting and stress corrosion cracking in low-pH solution.

The highly mobile chloride ion can accumulate in pits or crevices and greatly increases the uniform corrosion rate. At low temperatures and with chloride concentrations of up to 100 000 ppm, this rate is proportional to the square root of the chloride concentration. Chloride ions also cause a breakdown of the passivity of passivated alloys and stress corrosion cracking in austenite stainless steel. In fact, R. Corsi in 1986 mentioned that 5–10 ppm may be enough at temperatures over 50°C to promote stress corrosion cracking in AISI 316 stainless steel.[57]

Interestingly, sulphate ions are usually found in geothermal waters, but hardly ever produce such severe localized corrosive effects as chloride.

The most devastating effects of hydrogen sulphide manifest themselves on some copper and nickel alloys, particularly cupronickels and copper alloys including nickel, such as Monels. Although these are efficient in seawater, they cannot really be used in geothermal fluids which have hydrogen sulphide in them. The actual attack commences with concentrations under 30 ppb of the compound, copper or silver electrical contacts failing in this situation. The effect of hydrogen sulphide on steels is difficult to assess, although high-strength steels are susceptible to stress corrosion cracking at low temperatures. The oxidation of hydrogen sulphide in aerated waters can diminish the pH and increase the corrosiveness.

Allusion may be made also to dissolved carbon dioxide because this frequently occurs in geothermal fluids, lowering their pH values, and has a deleterious effect on carbon and low-alloy steels. Its main effect takes place up to about 80°C, above which stable iron carbonate films are created and inhibit rates of corrosion.

It may be added that ammonia can affect the pH of cooling water as well as inducing stress corrosion cracking in some copper alloys. Heavy-metal ions can corrode, if present in the oxidized state, but usually this is the lowest state in geothermal fluids so that the destructive influence is thus reduced.

Referring to experience gained in Italy, R. Corsi in 1986 asserted that, in that country, the geothermal fluids possess the following characteristics.[57] Most of the water-dominated resources contain between 500 and 16 000 ppm of dissolved solids, of which chloride makes up at least 45%. Steam-dominated resources contain between 50 and 300 ppm of boric acid, 3–30 ppm of chloride, 50–150 ppm of ammonia, 50–500 ppm of hydrogen sulphide and 10 000–20 000 ppm of carbon dioxide. Such non-condensable gases account for a mere 5% of the reservoir fluids and carbon dioxide is the most common of these, hydrogen sulphide and ammonia comprising very minor components. The pH value of the unflashed fluids ranges from 5 to 7.

As regards equipment, between 150 and 350°C no metal appears capable of remaining entirely free of corrosion. The most resistant are the costly austeno-ferritic or super austenitic alloys, but it must be remembered that austenitic stainless steels such as AISI 304 or AISI 316 are liable to stress corrosion at temperatures above 60°C and chloride concentrations exceeding 5–10 ppm. As might be expected, copper and aluminium alloys respond poorly because of dissolved hydrogen sulphide. Naturally, as plastics are unable to resist high temperatures, these materials are not feasible alternatives for carbon steel in piping and separators.

For re-injection piping, where the temperatures are to be around 120°C, the above argument applies. However, if the work is to be carried out at low temperature after flashing to the atmosphere, then fibreglass pipelines may constitute a possibility.

The thermal and hydraulic machines require materials which can withstand stress as well as corrosion erosion resulting from high-velocity geothermal fluid passing through them. Scaling can take place anywhere in the turbines as well, and martensitic stainless steel, e.g. AISI 403, appears appropriate for the blades, alloy steel sufficing for the rotors and casing. Geothermal steam with more than 30 ppm of chloride ions in it is particularly deleterious and may even cause blade failure to take place. Because of this (since 1956 in Italy) steam has been washed by inserting in the pipeline alkaline solutions capable of retaining the acid-soluble component of steam, separating the liquid downstream in separators. Although this procedure has prevented failures and precluded rapid uniform corrosion rates, it is an expensive treatment entailing a loss of approximately 3–5% of electric power production due to de-superheating of steam and pressure drop. In order to obviate this loss, attempts were made to find other materials for the blades and rotors of the turbines, but the results were not as satisfactory as had been hoped. Only titanium alloy and some super-austenitic alloys, e.g. AVESTA 254 SMO, were suitable and indeed some titanium blades were substituted in existing turbines in the dew-point region so as to test the material further. In cases where the chloride concentration is less than 65 ppm, materials such as UNI X5 Cr Ni Mo for rotors and AISI 403 for blades are quite adequate. Stems of turbine-regulating valves may be coated with ceramic material in order to avoid corrosion.

In the condenser and cooling water circuit, carbon steel becomes corroded rather fast and AISI 316 may be substituted. Fibreglass may be utilized for pipelines. Normal cast iron being unreliable, nickel cast iron, for instance ASTM A439D-2B, was recommended by R. Corsi in 1986 for the bodies of large-diameter valves in the cooling water circuits.[57] The cooling water circulation pump bodies should be constructed from ASTM A439D-2B and their inner parts, e.g. the impellers, can be made out of AISI 316. Diffusers are made of AISI 316, too, or sometimes of nickel cast iron.

Finally, regarding the buildings in the plant hydraulic circuit, R. Corsi in 1986 mentioned that in some concrete parts, such as the cooling towers, a catalytic oxidation of sulphides to sulphates occurs and the sulphuric acid attacks the calcium in the cement.[57] It seems that the process is promoted by sulpho-reducing bacteria and means of preventing it are being studied. In the meantime, cooling towers constructed of wood and protective plaster are usually employed.

5.8.6.2 Scaling

The formation of scale is a major difficulty encountered in exploitation of geothermal resources. The chemical compounds involved are silica and silicates in amorphous form, carbonates as low-magnesium calcites, sulphates (mostly calcium sulphate and barite) and sulphides (in many, usually well-crystallized phases; those of lead, zinc, iron and copper often occur, and antimony sulphide has been recorded from the Monte Amiata geothermal field).

How scale forms is not really understood, but there are three potential situations favouring its accumulation. Firstly, deposition may take place from a single-phase fluid saturated with respect to the relevant solid (re-injection pipelines). Secondly, it may occur from flashing fluids (wells, separators, two-phase pipelines). This is probably the most common mechanism, but the process is far from understood. Flashing is caused by drops in pressure or cavitation in turbulent flow and probably produces calcite scale. It enhances supersaturation by steam loss from the liquid phase, increasing concentration of the residual solutes, by temperature diminution during expansion and by loss of stable gases such as carbon dioxide and hydrogen sulphide promoting increase in pH. Thirdly, scaling can result from steam carry-over (separators, turbines, steam pipelines). This can affect turbines badly where they are exposed only to steam. Nucleation and depositional kinetics are a function of the degree of supersaturation, pressure, temperature and catalytic or inhibitory effects due to minor elements.

Calcium carbonate scaling occurs often because almost all geothermal systems contain dissolved carbon dioxide: the quantity in a water solution at equilibrium is proportional to the partial pressure of the gas in contact with this solution, following Henry's Law. Dissolved carbon dioxide involves carbonic acid also, a trigonal planar structure present at H_2CO_3, representing about 3% of the total. At the start, geothermal exploitation has a static carbon dioxide-charged liquid without a vapour phase, but, when production commences, the pressure decreases and the equilibrium shifts. The carbonate concentration rises and may trigger calcium carbonate precipitation, according to the following solubility relationship:

$$K = [Ca^{2+}][CO_3^{2-}] \tag{5.18}$$

The precipitation of calcium carbonate commences with flashing and, where this occurs in the productive well, down-hole scaling may take place; where it involves the formation itself, this may become plugged and, on the other hand, if it is located in surface equipment, then this may become blocked. R. Corsi indicated that the various relationships between the differing ionic species are given by the following equilibria relationships:[57]

$$\frac{[H^+][HCO_3^-]}{[H_2CO_3]_{apparent}} = k_{H_2CO_3} \tag{5.19}$$

$$\frac{[H^+][CO_3^{2-}]}{[HCO_3^-]} = k_{HCO_3^-} \tag{5.20}$$

$$\frac{H_2CO_3]_{apparent}}{P_{CO_2}\alpha_C} = k_{CO_2} \tag{5.21}$$

$$[Ca^{2+}][CO_3^{2-}] = k_{CaCO_3} \tag{5.22}$$

in which the brackets indicate activities, k the equilibrium constants, P_{CO_2} the partial pressure of carbon dioxide and α_C the fugacity coefficient of carbon dioxide.

If these equations are rearranged and the partial pressure is taken as the mole fraction of carbon dioxide in the gas phase, $P_{CO_2} = \gamma_{CO_2}$, the following equation results:

$$[Ca^{2+}] = \frac{k_{H_2CO_3}k_{CO_2}\gamma_{CO_2}P_r\alpha_C}{k_{HCO_3^-}[HCO_3^-]^2\gamma_{Ca^{2+}}} \tag{5.23}$$

in which P_r is the pressure and $\gamma_{Ca^{2+}}$ is the calcium activity coefficient.

Alkalinity is usually expressed by bicarbonate concentration in geothermal waters and the foregoing equations demonstrate that the calcium ion concentration is dependent upon temperature, the partial pressure of carbon dioxide and the ionic strength, I. Consequently, a saturation index, I_s, may be defined, as J. E. Oddo stated in 1982.[59] This is the relationship between the measured calcium ion concentration and that existing under conditions of equilibrium, i.e.

$$I_s = \log F_s \tag{5.24}$$

in which

$$F_s = \frac{[Ca^{2+}]alk^2 k_{HCO_3^-} \gamma_{Ca^{2+}} (\gamma_{alk})^2}{k_{H_2CO_3} k_{CO_2} P_{CO_2} \alpha_C} \qquad (5.25)$$

and, the equilibrium constant being expressible as a function of temperature and ionic strength, I, the following equation may be derived:

$$I_s = \log \frac{[Ca^{2+}]alk^2}{P_r \gamma_{CO_2}} + 10 \cdot 22 + 2 \cdot 739 \times 10^{-2} T - 1 \cdot 38 \times 10^{-5} T^2$$
$$- 1 \cdot 079 \times 10^{-8} P_r - 2 \cdot 52 I^{1/2} + 0 \cdot 919 I \qquad (5.26)$$

in which I, $[Ca^{2+}]$ and alk (alkaline) are expressed as mol/litre, P_r as Pa and T as °C. $I \geq 0$ refers to a supersaturated scaling solution and $I \leq 0$ to an undersaturated one. Calcium carbonate scaling can be predicted in any part of the utilization cycle if the alk, $[Ca^{2+}]$ concentration and CO_2 can be measured, but this is often difficult to do because of the unreliability of down-hole sampling.

Prevention of calcium carbonate scaling can be achieved by alteration of either the partial pressure of carbon dioxide or the pH, and also by adding chemical scale inhibitors. Control is automatically exerted in a well merely by pumping it, and the fluids obtained may be retained in a single-phase system by employing a down-hole mechanical pump, thus avoiding pressure-sensitive scaling of calcium carbonate. The problem is that the temperature of the fluid involved may restrict the use of such a pump and there can be no certainty of one working down-hole at temperatures higher than around 190°C. Another approach is to maintain a high carbon dioxide partial pressure artificially by re-injection into the producing well of some of the carbon dioxide produced: this has been done in the USA by J. T. Kuwada in 1982.[60] However, it appears that the technique is useful only for geothermal fluid with low carbon dioxide partial pressure. On the other hand, temperature-sensitive scaling, e.g. due to barium sulphate, may be either reduced or transferred to more accessible surface occurrences in equipment on the ground.

The pH is easy to change, for instance by adding hydrochloric acid to a geothermal fluid so as to decrease the value of the parameter below the level at which calcium carbonate scaling is feasible. The snag is the cost, the acid in question being more expensive than the electric power produced from the system because of the high buffering capacity of many of the fluids involved. R. Corsi in 1986 mentioned that in an experiment at Torre Alfina 200 ml of 0·1M hydrochloric acid were necessary for every litre of solution to preclude precipitation of calcium carbonate.[57]

Finally, scale inhibitors may be used; R. Corsi and others in 1985 gave an appropriate list which is shown in Table 5.8.[61] All were laboratory tested by R. Corsi and his colleagues in 1985 prior to application in the Cesano and Latera geothermal fields. They obtained the best results with the organic phosphonates, e.g. Dequest 2066 and Sequion 40 Na 32.[61] They found that trace elements did not affect the products' effectiveness, which was retained even when the inhibitors were injected into a two-phase flow with on-going nucleation. Also phosphonates appear to remain efficient until around 200°C, some degradation occurring above 210°C and entailing usage of

TABLE 5.8
Calcium carbonate scale inhibitors

Material	Manufacturer	Chemical type
Dequest 2060	Monsanto	Organophosphonic acid
Dequest 2066	Monsanto	Organophosphonic acid
Ecostabil 4001	Montedison–Arca	Organophosphonic acid
Ecostabil 4004	Montedison–Arca	Organophosphonic acid
Ecostabil EP/85	Montedison–Arca	Organophosphonic acid
Ecopol 4014	Montedison–Arca	Polyacrylamide
Ecopol 4022	Montedison–Arca	Polyacrylamide
Nadar 4053	Nadar	Organophosphonic acid
Nadar 4054	Nadar	Organophosphonic acid
Flocon 247	Pfizer	Polycarboxylic acid
Chelone DPNA	Eurosyn	Organophosphonic acid
Sequion 40 Na 32	Bozzetto	Organophosphonic acid
Procedor ST 90	Procedo	Blend
Sodium fumate	Mitsubishi	—

doubled concentration in order to obtain the same inhibiting action. Finally, the inhibitors in question retain effectiveness for several hours and so trouble-free re-injection can proceed; a small Incaloy pipe (4·76 mm) was employed in the well actually to carry out the process. In the case of well Latera 3D, at a flow rate of 600 tonnes/h, 4 ppm of inhibitor obviated scaling and caused an economic impact of only 0·16 mills/kW h in a double-flash condensing power plant capable of producing about 15 MWe. R. Corsi and others mentioned that, on an annual basis, the cost of the inhibitor represents approximately 25% of the expense involved in reaming a single well.

Silica scaling is commonly found in plants which use steam flashed from high-temperature resources, occurring in re-injection lines as well as in separators and, more rarely, in wells. The behaviour of silica in aqueous solutions was examined by R. O. Fournier and W. L. Marshall[62] in 1983 as well as by A. J. Ellis and W. A. J. Mahon[16] in 1977. Two forms are significant, namely quartz and amorphous silica, and usually in geothermal reservoirs the fluid is in equilibrium with the former at the prevailing temperature. Within the interesting range of pH ≤ 8, the quantity of quartz entering solution in the fluid rises with temperature and falls with salinity. In practice, therefore, the solubility of quartz may be regarded as independent of the pH. As geothermal fluid cools, it becomes super-saturated with respect to quartz, but, owing to the slow kinetics of quartz, silica deposition at lower temperatures is governed by the equilibrium of the amorphous variety which is more soluble than quartz at any given temperature. The potential for the precipitation of silica is present only where temperatures exist which are under the appropriate saturation value for the equilibrium solubility of amorphous silica. Amorphous silica solubility rises with temperature, decreases with salinity and rapidly increases with pH. The actual rate of deposition of amorphous silica is probably controlled by the rate of polymerization of monomeric silica. This is a function of both the temperature and the extent of supersaturation, and the relevant reactions may consist of chains catalysed by chloride and hydroxide; sodium, potassium and sulphate ions apparently play little part. The kinetics of deposition of scale are slow at the ambient temperatures and consequently the location of the deposit depends upon the residence time of the supersaturated fluid.

Acidification is known to lower the rate of deposition; for instance, acidifying Salton Sea fluid (with an initial pH of 5–6) to a pH of 4·5 causes a delay in precipitation of three months at least. On the other hand, alkalinization of pH to 8 or higher ionizes silica and much improves its solubility. This argument implies that changing the pH value of a fluid should prevent silica scaling. Although both hydrochloric acid and sodium hydroxide were tried with geothermal brines in the USA and Mexico with some success, the costs involved were too high. An alternative means using inhibitors was substituted and, in Italy, a technique for inducing silica precipitation was developed.

A completely different philosophy entails attempting to arrange for scale deposition to occur, but to occur in specially designed equipment through selection of incremental stages in lowering pressures and temperatures between the wellheads of the production wells and those of the re-injection ones. The types of equipment employable have been discussed in 1982 by O. J. Vetter and V. Kandarpa.[63] They alluded to a special flash crystallizer of which the basic principle is that brine is suddenly flashed from non-scaling to scaling conditions with simultaneous provision of sufficient seed for precipitation reactions so that the deposits do not accrete on the internal walls of the equipment, but rather on the seeds. Of course, for this to happen, a large internal surface area is provided within the fluid through the addition of seed (suspension) which obviates solids accreting on the walls. The growth of scale on the walls results from a different type of seed, either arising from the walls or from some of the added seeds adhering to them. The flash crystallization through seed-addition with simultaneous flashing entails a complicated control of various precipitation reactions governed by the thermodynamics and kinetics of each reaction, and the general hydrodynamics of the relevant system. The drawback is that the technique cannot solve the difficulty of re-injection pipelines in which silica particles may be found because of temperature drop. Here water clarification was suggested by R. Corsi in 1986, this involving a set of solid/liquid separations, e.g. sedimentation flotation, filtration or cyclone separation.[57] The same author gave the data in Table 5.9 for silica removal.

The snag with all the techniques listed in Table 5.9 is that the large flow rates common in geothermal work offer a stumbling block because of the size of the purification plants necessary. All must remove a solid deposit of scale. In 1970 T. Yanagase and his associates described a retaining tank developed in Japan for this purpose.[64] Exhaust brines flow through this tank for long enough to permit silica deposition on the diaphragms, which are periodically cleaned. However, the kinetics of silica precipitation relate to the composition of the geothermal brine, which is obviously dependent upon its origins.

TABLE 5.9
Feasible techniques for downstream silica removed[57]

Treatment	Hydraulic load (m^3/m^2h)	Drawbacks
Sedimentation, i.e. thickening and clarification	1–2	Hydraulic and convective turbulence
Flotation	10–15	In part, affected by convective motion
Filtration	0·5–1·0	Expensive
Sand filtration	10–15	Not appropriate for high concentrations of suspended particles
Hydrocyclones	≥ 100	Low efficiencies and frequent breakdown of flocculates

Finally, a case of scaling in a geothermal reservoir was cited by K. Ölçenoglu in 1986.[65] He investigated the Kizildere field, where the solubility of carbon dioxide in the 200°C waters is given by the expression

$$\frac{[CO_2\,(mol)]}{[H_2O\,(mol)]} = 15\cdot6 \times 10^{-5} \times P \tag{5.27}$$

which is derived from J. Ellis and R. M. Golding in 1963 and in which the pressure is in kg/cm^2.[66] If a 20 wt% gas ratio be assumed, the saturation pressure of the carbon dioxide is $52\cdot4\,kg/cm^2$. If pressure falls below this level in the reservoir, then free gas will occur and this phenomenon has been observed in the field. The wells actually produce a mixture of steam and hot water. The solubility ratio of CO_2 in water and vapour phases is given by the expression

$$A = \frac{\text{Mass ratio (gas/water) in liquid}}{\text{Mass ratio (gas/water) in vapour}} \tag{5.28}$$

and the values of $-\log A$ were as listed in Table 5.10.

TABLE 5.10
Values for $-\log A$

Temperature, °C	100	125	150	175	200	225	250	275	300	325
$-\log A$	3·71	3·41	3·13	2·86	2·58	2·31	2·05	1·78	1·51	1·21

If eqn (5.28) be applied to the equilibrium separation of steam, then

$$A = \frac{(g_1 - Yg_1n)/(1 - X)}{Xg_1n/X} \tag{5.29}$$

and

$$n = \frac{1}{A(1 - X) + X} \tag{5.30}$$

in which g_1 is the initial gas percentage by weight in the reservoir fluid, g_2 is the amount of carbon dioxide entering the steam phase and X is a dryness factor.

Commencing with a 200°C production temperature, results at various separation temperatures were given by K. Ölçenoglu and are listed in Table 5.11.[65] This demonstrates that, when there is a small percentage of steam separation, most of the gas occurring in the vapour phase passes into the liquid one. As more steam separates, the quantity of gas remains practically constant so that the percentage of it in the vapour phase decreases.

In the Kizildere reservoir, the calculated saturation pressure for the gas was $52.4 \, kg/cm^2$. The master valve was shut at well KD-14 and, after 17 h, wellhead pressure reached $49 \, kg/cm^2$, thereafter staying constant. To obviate deposition in the reservoir, the gas must be retained in the fluid phase. Under static conditions, it is only necessary to maintain it above the gas saturation pressure of $52 \, kg/cm^2$, but, during flow, pressure decreases in the well and gas starts separating out. Because of this gas phase, the steam evaporates with its partial pressure corresponding to its temperature. The two phenomena take place spontaneously and cause an increased gas pressure as a result of the partial pressure of steam. Consequently, flashing

TABLE 5.11
Values for X, n and g_2 at the stated separation temperatures

Temperature ($^\circ C$)	X (%)	n	g_2
200	0	0	0
190	2·32	39·63	$0.92g_1$
180	4·44	21·63	$0.967g_1$
160	8·50	11·64	$0.99g_1$
150	10·42	9·53	$0.993g_1$

will take place at a deeper level. The pressure in the well will be the sum of the wellhead pressure and the pressure of the fluid column within the well. In theory, flashing takes place at approximately 700 mm of hydrostatic pressure above the reservoir. At Kizildere, it was recorded that, if the wellhead pressure is retained at $15 \, kg/cm^2$ gauge in the wells having ≥ 600 m of production casing, no encrustation took place in the reservoir. New production wells are sited where cap rock exceeds 600 m in thickness and then, under flow conditions, the pressure at the bottom of the production casing, i.e. at the top of the reservoir, will be above the saturation pressure of the gas.

5.8.7 Geothermal Power Plants
Obviously, it is necessary to ensure that materials to be used in construction of geothermal power plants are corrosion-resistant, particularly in respect of the turbines, which may encounter aggressive constituents of the relevant hot fluids. Naturally, appropriate design and proper installation

are mandatory. The fluid characteristics govern the selection of the turbine type, i.e. condensate or atmospheric exhaust; also operational flexibility is necessary to take account of the alteration of steam conditions and non-condensable gas contents with time. The turbines involved will use several different cycles. They may obtain clean steam through the heat exchangers, a practice falling into disuse because of its low level of efficiency. A new indirect approach utilizes organic fluids, but entails both new turbine and new power plant designs. A simpler cycle involves supply of heat directly from the production well, but there are the associated problems of scaling and corrosion as well as erosion of the equipment. In addition, the steam which reaches the turbine may be superheated or saturated and its temperature can reach 250°C. Finally, the turbine may be charged with flashed steam which is always wet, the operating cycle falling midway between the two cases cited above.

In California, Italy and Japan, direct steam turbines operate in vapour-dominated systems and the major drawbacks arise from corrosion and scaling. These may be reduced by initial washing by injecting an aqueous solution of suitable compounds with cyclone separation of solids and soluble substances. The process will remove chlorides and other deleterious chemical compounds as well as chlorine and other elements, but may cause thermodynamic degradation. Thermodynamically, non-condensable gases may influence the properties of the geothermal power plant. In condenser turbines, they have to be extracted from the condenser: either a compressor or a system of steam ejectors may be employed. The former absorbs up to 10% of the power produced and the latter up to 5% of the steam. If the gas content exceeds 15%, exhausting-to-atmosphere units are preferable. However, the power then generated is only half of that obtained from a condensing operation, but the latter necessitates more expense and complexity.

A condensing power plant uses a turbine with large adiabatic energy, a compressor, a condenser, a cooling tower, a water circuit with circulation pump, a water accumulation pool and an associated control system.[66] Most of this equipment is lost in the event that the geothermal production well becomes exhausted, whereas, with a condensing power plant, almost all of the associated components can be shifted to another site in such a case. Often an exhaust-to-atmosphere turbine is utilized initially because gas is present in the field being newly exploited, a condensing type being substituted later. Exhaust-to-atmosphere turbines have a power range of 1–15 MW as compared with condensing types which can reach 55 MW with a double-flow casing and 110 MW with two casings.

Turning now to single- or multi-flash turbines in liquid-dominated systems, the steam so derived contains less corrosive constituents; these are usually soluble in water and hence accumulate in the condensate. The steam also contains a lower percentage of non-condensable gases and they may be extracted using ejectors instead of compressors. There is also a lower percentage of condensable gases, and therefore water circulating in the condenser has lower acidity. In addition, there will be only a small entrainment of solids in the geothermal fluid. All of the above factors are advantages with regard to steam derived from flashing, but there are some disadvantages as well. The most important is that, commencing from a saturated wet condition, the steam is wetter at the last turbine stage so that it needs more careful moisture extraction, grooved buckets and low velocities at the tips of the blades. In multi-flash models, the turbines are more complicated and, to facilitate exploitation, a minimum of two cascade flashes is necessary, the second operating with the hot water left by the first and hence at a pressure of about 20% of the initial value. Obviously, an infinite number of cascade stages would produce total efficiency, but, for each one, the turbine would have to have an extra entry for steam, a physical and economic impossibility. Using geothermal turbines with flash steam is difficult and, if well exhaustion occurs, the plant may have to be written off.

Consequently, for small capacities, binary-cycle plants operated with organic fluids are utilized in order to exploit hot geothermal waters even at relatively low temperatures. A cycle combination employing flash turbines, superheating and binary cycle was considered excellent by S. Siena in 1986.[67] He mentioned that turbines made up until the mid-1970s were usually fed with low-pressure steam and operated by throttling. One reason was that they normally functioned at full load so that operation in a partial arc would not yield any advantages. That is to say, the steam flow needs to be controlled only during start-up and loading, turbine valves being left fully open after full load is reached. Consequently, the loss of efficiency during the initial transient stage is negligible. While functioning through throttling, steam is admitted to the turbine through emergency stop and control valves and reaches the first stage directly, i.e. by a total-arc admission process. Pressure upstream of this varies according to the load. If this is low, so is the pressure, the large drop between the pressure of the main steam and that of the first stage occurring through the control valves with no power yield; but when the full load is attained, the first-stage upstream pressure reaches a value comparable with that of the main steam. The reaction stages cannot be partialized and thus comply quite well with this operational method. As their construction is simple, they have been widely utilized. If there are minor differences in pressure between wells feeding different turbines of the same plant, these are adjustable by adding or subtracting one or two stages at admission. Through years, steam pressures in new wells may rise and vary from one well to another. Special turbine inlet heads may be necessary to deal with this situation.[67]

Reaction-type turbines have been commonly used, but lighter impulse ones are sometimes selected. They have a stage pressure drop across stationary parts, in contrast with reaction-type turbines in which the pressure drop takes place by division between stationary and rotating blades. As regards their construction, the impulse type employs a wheel-and-diaphragm arrangement whereas the reaction turbine has a drum rotor set-up. In the latter case, the design is restricted by the large pressure drop acting across the rotating blades to provide a large thrust load on the rotor. If such blades were to be mounted on a separate wheel, an extra thrust would develop in proportion to the area of the wheel and this would make it quite impractical to employ a wheel-and-diaphragm construction for the reaction turbine. The latter has more stages since it needs less energy per stage to reach maximum stage efficiency, i.e. in a given element less energy per stage requires more stages. On the other hand, fewer stages in the impulse design permit more space for packing teeth and a rugged construction. In the impulse case, there are fewer leakages of steam bypassing the actual working parts, particularly around the tips of the moving blades and the shaft packing. This is because there is a markedly lower drop in pressure across the tip of the impulse bucket, smaller diaphragm packing diameters and more teeth (due to the wider packing area), meaning fewer stages. Leakages around the working parts are reduced also by supporting the diaphragm on the centre line, resulting in less seal wear. This constitutes another reason why the impulse design possesses an inherently better sustained heat rate than the reaction one.

S. Siena considered the associated thermodynamics of the situation, alluding to the fact that the major difference between steam expansion and that occurring with a mixture of steam with non-condensable gases, e.g. carbon dioxide, is that the adiabatic expansion of steam between two pressures is an isentropic process whilst the other is not.[67] By definition, an adiabatic expansion of this type is a transformation taking place with no external heat exchange, an event which may be expressed as $dQ = 0$. Since the infinitesimal entropy variation is given by $dS = dQ/T$ and $dQ = 0$, dS is also zero. In fact, the adiabatic expansion of a steam and gas mixture is not isentropic, both constituents attempting to attain two different temperatures with the result that an irreversible heat exchange takes place and is accompanied by an increase in entropy. A 10% gas content produces a loss of efficiency of 0·6% as a consequence of this internal heat exchange between the steam and the gas. The product of the internal efficiency of the turbine and its adiabatic efficiency is the internal efficiency referred to the isentropic expansion, namely the isentropic internal efficiency.

Draining plays an important part in geothermal turbines. Humidity arising during expansion in a stage is extracted and this causes a decrease in the humidity of the downstream steam, which undergoes an isobaric enthalpy increase. The enthalpy drop available and the internal efficiency increase because of the removal of water drops and reduction in humidity at discharge. Clearly, this reduces erosion of the blades. In practice, to counter the effects of moisture on both rotating and stationary blades, each stage is drained by means of suitably calibrated holes so as to lessen the steam losses to the condenser. The removal of moisture is effected using a set of smooth or grooved buckets plus a collection chamber in each diaphragm. That steam lost through drainage is compensated for by the better quality of steam working within the turbine.

The Ansaldo component company in Genoa (which has experience in turbine remote-control regulators for smallish machines installed in various parts of large geothermal fields) used stainless martensitic steel in constructing geothermal rotors for use at Larderello with a 10-MW turbine and larger rotors were built for 55-MW turbines in California. The steel in question (13% chromium, 6% nickel) was chosen for its resistance to corrosion. There are two aspects of this corrosion, namely chemical (mostly when steam crosses the saturation line and the first droplets of liquid form, these comprising concentrated solutions of aggressive substances) and a later erosion and abrasion of oxide protective layers through the action of solid particles suspended in steam and travelling at the same velocity as that steam. On the last-stage blades of fossil-fuel turbines, stellite shields have been applied by brazing so as to protect them against erosion through impact with water droplets contained in steam. In the case of geothermal turbines, this is inapplicable because of the liability of the brazing material to corrosion. A new approach called Plasma Transferred Arc (PTA) permits automatic application of stellite powder and may be more suitable.

Incidentally, the remote-control regulators made by Ansaldo now have electronic regulation, the smallest turbine of 0·9 MW being radio-monitored and the largest one of 15 MW being monitored by analogue electronic control. They are claimed to have high stability and to need little maintenance.[67] This company mainly concentrates on skid-mounted small- and medium-sized units of 5–15 MW, either back-pressure or condensing in type.

REFERENCES

1. DONALDSON, I. G. and GRANT, M. A., 1981. Heat extraction from geothermal reservoirs. In: *Geothermal Systems: Principles and Case Histories*, ed. L. Rybach and L. J. P. Muffler, pp. 145–79. John Wiley & Sons, Chichester, 359 pp.
2. MCNABB, A., 1975. *A Model of the Wairakei Geothermal Field*. Unpublished report, Applied Mathematics Division, Department of Scientific and Industrial Research, Wellington, New Zealand.
3. GRANT, M. A. and SOREY, M. L., 1979. The compressibility and hydraulic diffusivity of a water–steam flow. *Bull. Water Resources Res.*, 15(3), 684–86.
4. MOENCH, A. P. and ATKINSON, P., 1978. Transient-pressure analysis in geothermal steam reservoirs with an immobile vaporizing phase. *Geothermics*, 7(2–4), 253–64.
5. BENSON, S. M., GORANSON, C. B., McEDWARDS, D. and SCHROEDER, R. C., 1978. *Well Tests*. Geothermal Resource and Reservoir Investigation of US Bureau of Reclamation Leaseholds at East Mesa, Imperial Valley, California. Report No. LBL-7994.
6. SCHROEDER, R. C., O'SULLIVAN, M. J., PRUESS, K., CELATI, R. and RUFFILLI, C., 1982. Reinjection studies of vapor-dominated systems. *Geothermics*, 11(2), 93–119.
7. BODVARSSON, G., 1969. On the temperature of the water flowing through fractures. *J. Geophys. Res.*, 74, 1987–92.
8. LIPPMANN, M. J., TSANG, C. F. and WITHERSPOON, P. A., 1977. Analysis of the response of geothermal reservoirs under injection and production procedures. SPE Paper 6537, presented at the 47th Regional Meeting of the SPE, Bakersfield, CA, USA.
9. MANGOLD, D., TSANG, C. F., LIPPMANN, M. J. and WITHERSPOON, P. A., 1979. A study of thermal effects in well test analysis. SPE Paper 8232, presented at the 54th Annual Fall Meeting SPE, Las Vegas, NV, USA.
10. WITHERSPOON, P. A., WANG, J. S. Y., IWAI, K. and GALE, J. E., 1979. *Validity of the Cubic Law for Fluid Flow in Deformable Rock Fracture*. US Bureau of Reclamation, Rept No. LBL-9557.
11. BODVARSSON, G., 1972. Thermal problems in the siting of reinjection wells. *Geothermics*, 1(2), 63–6.
12. O'SULLIVAN, M. J. and PRUESS, K., 1980. Analysis of injection testing of geothermal reservoirs. *Geothermal Resources Council Trans.*, 4, 401–4.
13. GARG, S. K., 1978. Pressure transient analysis for two-phase (liquid water/ steam) geothermal reservoirs. SPE Paper 7479, presented at the 53rd Annual Fall Technical Conference and Exhibition of the Society of Petroleum Engineers, Houston, TX, USA.
14. D'AMORE, F., FANCELLI, R. and PANICHI, C., 1987. Stable isotope study of reinjection processes in the Larderello geothermal field. *Geochim. Cosmochim. Acta*, 51, 857–67.

15. HENLEY, R. W., TRUESDELL, A. H. and BARTON, P. B., 1984. Fluid–mineral equilibria in hydrothermal systems. Contribution by A. J. WHITNEY. *Reviews in Economic Geology*, Vol. 1. Soc. Econ. Geologists.

16. ELLIS, A. J. and MAHON, W. A. J., 1977. *Chemistry and Geothermal Systems.* Academic Press, New York, 392 pp.

17. D'AMORE, F. and TRUESDELL, A. H., 1979. Models for steam chemistry at Larderello and The Geysers. *Proc. 5th Stanford Geothermal Reservoir Engineering Workshop*, pp. 283–97.

18. TRUESDELL, A. H., NATHENSON, M. and RYE, R. O., 1977. The effects of subsurface boiling and dilution on the isotopic compositions of Yellowstone thermal waters. *J. Geophys. Res.*, **82**, 3694–704.

19. AUSTIN, C. F., AUSTIN, W. H. JR and LEONARD, G. W., 1971. *Geothermal Science and Technology: A National Program.* Tech. Ser. 45-029-72. US Naval Weapons Center, China Lake, CA, USA.

20. GOLDSMITH, W., AUSTIN, C. F., WANG, H. C. and FINNEGAN, S., 1968. Stress waves in igneous rocks. *J. Geophys. Res.*, **71**, 8.

21. AUSTIN, C. F. and LEONARD, G. W., 1973. Chemical explosive stimulation of geothermal wells. In: *Geothermal Energy*, ed. P. Kruger & C. Otte, pp. 269–92. Stanford University Press, Stanford, CA, USA.

22. AMERICAN OIL SHALE CORPORATION–US ATOMIC ENERGY COMMISSION, 1971. *A Feasibility Study of Geothermal Power Plants.* PNE-1550.

23. BURNHAM, J. B. and STEWART, D. H., 1973. Recovery of geothermal energy from hot dry rock with nuclear explosives. In: *Geothermal Energy*, ed. P. Kruger and C. Otte, pp. 223–30. Stanford University Press, Stanford, CA, USA.

24. CHARLOT, L. A., 1971. *Plowshare Geothermal Steam Chemistry.* BNWL-1614.

25. BLUME, J. A., 1969. Ground motion effects. In: *Proc. Symp. Public Health Aspects of Peaceful Uses of Nuclear Explosives.* USAEC Tech. Rept No. SWRHL-82.

26. SANDQUIST, G. M. and WHAN, G. A., 1973. Environmental aspects of nuclear stimulation. In: *Geothermal Energy*, ed. P. Kruger and C. Otte, pp. 293–313. Stanford University Press, Stanford, CA, USA.

27. LESSLER, R. M., 1970. Reduction of radioactivity produced by nuclear explosives. In: *Symp. Engineering with Nuclear Explosives, Las Vegas, Nevada, 14 to 16 January 1970*, CONF-700101 (2). Clearinghouse for Federal Scientific and Technical Information, US Dept of Commerce, Springfield, VA, USA.

28. RAGHAVAN, R., RAMEY, H. J. JR and KRUGER, P., 1971. Calculation of steam extraction from nuclear explosion fractured geothermal aquifers. *Trans. Am. Nucl. Soc.*, **14**, 695.

29. STRAUB, F. G., 1964. Steam turbine blade deposits. *Engineering Experimental Station Bull., Univ. of Illinois*, Ser. No. 364, 43–59.

30. KRIKORIAN, O. H., 1973. Corrosion and scaling in nuclear stimulated plants. In: *Geothermal Energy*, ed. P. Kruger and C. Otte, pp. 315–34. Stanford University Press, Stanford, CA, USA.

31. SMITH, M. C., 1975. The Los Alamos Scientific Laboratory dry hot rock geothermal project (LASL Group). *Geothermics*, **4**, 27–39.

32. DASH, Z. V., MURPHY, H. D., AAMODT, R. L., AGUILAR, R. G., BROWN, D. W., COUNCE, D. A., FISHER, H. N., GRIGSBY, C., KEPPLER, H., LAUGHLIN, A., POTTER, R., TESTER, J. W., TRUJILLO, P. E. and ZYVOLOSKI, G. A., 1983. Hot dry rock geothermal reservoir testing: 1978 to 1980. *J. Volcan. Geotherm. Res.*, **15**, 59–99.

33. HOUSE, L., KEPPLER, H. and KAIEDA, H., 1985. Studies of a massive hydraulic fracturing experiment. Presented at Geothermal Resources Council, Annual Meeting, Kona, Hawaii, 26 to 30 August 1985.
34. DREESEN, D. S. and NICHOLSON, R. W., 1985. Completion and operation for MFH of Fenton Hill HDR Well EE-2. Presented at Geothermal Resources Council, Annual Meeting, Kona, Hawaii, 26 to 30 August 1985.
35. FEHLER, M., 1984. Source properties of microearthquakes accompanying hydraulic fracturing. EOS, Trans. Am. Geophys. Union, 65, 1011.
36. PARKER, R. and HENDRON, R., 1987. The Hot Dry Rock geothermal energy projects at Fenton Hill, New Mexico, and Rosemanowes, Cornwall. Presented at IEE Future Energy Options Conference, University of Reading, England.
37. HENDRON, R., 1987. The US Hot Dry Rock Project. Presented at 12th Workshop on Geothermal Reservoir Engineering, 20 to 22 January 1987, Stanford University, Stanford, CA, USA.
37a. GOFF, F. and GARDNER, J. N., 1988. Valles Caldera Region, New Mexico, and the Emerging Continental Scientific Drilling. J. Geophys. Res., 93(B6), 5997–9.
37b. SASS, J. H. and MORGAN, P., 1988. Conductive heat flux in VC-1 and the thermal regime of Valles Caldera, Jemez Mountains, New Mexico. J. Geophys. Res., 93(B6), 6027–39.
37c. GOFF, F., SHEVENELL, L., GARDNER, J., VUATAZ, F.-D. and GRIGSBY, C. O., 1988. The hydrothermal outflow plume of Valles Caldera, New Mexico, and a comparison with other outflow plumes. J. Geophys. Res., 93(B6), 6041–58.
37d. HULEN, J. B. and NIELSON, D. L., 1988. Hydrothermal brecciation in the Jemez Fault Zone, Valles Caldera, New Mexico: results from Continental Scientific Drilling Program Core Hole VC-1. J. Geophys. Res., 93(B6), 6077–89.
37e. SASADA, M., 1988. Microthermometry of fluid inclusions from the VC-1 core hole in Valles Caldera, New Mexico. J. Geophys. Res., 93(B6), 6091–6.
37f. STURCHIO, N. C. and BINZ, C. M., 1988. J. Geophys. Res., 93(B6), 6098–102.
37g. HIGGINS, M. D., 1988. Light lithophile elements (boron, lithium) abundances in the Valles Caldera, New Mexico, VC-1 cor hole. J. Geophys. Res., 93(B6), 6103–7.
37h. DEY, T. N. and KRANZ, R. L., 1988. State of stress and relationship of mechanical properties to hydrothermal alteration at Valles Caldera Core Hole 1, New Mexico. J. Geophys. Res., 93(B6), 6108–12.
38. BATCHELOR, A. S., 1982. Creation of hot dry rock systems by combined explosive and hydraulic fracturing. BHRA Fluid Engineering International Conference on Geothermal Energy, Florence, Italy, pp. 321–42.
39. PINE, R. J., MERRIFIELD, C. M. and LEDINGHAM, P., 1983. In situ stress measurement in the Carnemenellis granite. 2-Hydro-fracture tests at Rosemanowes Quarry to depths of 2000 m. Int. J. Rock Mech. Mining Sci. Geomech. Abstr., 20, 2.
40. BATCHELOR, A. S., 1983. Hot dry rock reservoir stimulation in the UK—an extended summary. Presented at 3rd Int. Seminar on the results of EEC Geothermal Energy Research, November 1983, Munich, FRG.
41. BATCHELOR, A. S. and PEARSON, C. M., 1979. Preliminary studies of dry rock geothermal exploitation in South West England. Trans. Inst. Min. Metall., Sect. B, Appl. Earth Sci., 88, B51–6.
42. PINE, R. J., BARIA, R., PEARSON, R. A., KWAKWA, K. A. and McCARTNEY, R. A., 1986. A technical summary of Phase 2B of the Camborne School of Mines

HDR Project, 1983–1986. Presented at EEC/US Workshop on Hot Dry Rock Geothermal Energy, May 1986, Brussels, Belgium.

43. HICKMAN, S. H., HEALY, J. H. and ZOBACK, M. D., 1985. *In situ* stress, natural fracture distribution and borehole elongation in the Auburn geothermal well, Auburn, New York. *J. Geophys. Res.*, **90**(B7), 5497–512.

44. PLUMB, R. A. and HICKMAN, S. H., 1985. Stress-induced borehole elongation: a comparison between the four-arm dipmeter and the borehole televiewer in the Auburn geothermal well. *J. Geophys. Res.*, **90**(B7), 5513–21.

45. ANON., 1987. Prospects for UK renewable energy technologies. *UK Department of Energy Review*, **1**, 4–5.

46. CHIEF SCIENTIST'S GROUP, 1985. *Prospects for the Exploitation of the Renewable Energy Technologies in the United Kingdom.* CSG, HL85/1364, Energy Technology Support Unit, ETSU R30, AERE, Harwell, UK, 107 pp.

47. HARRISON, R., 1984. Economics of low enthalpy geothermal district heating developments, Northern France and the UK. *Proc. Conf. Energy Options, IEE Conf., April 1984*, Publication No. 233.

48. KARLSSON, T., 1982. *Geothermal District Heating: The Icelandic Experience.* UNU Geothermal Training Programme, Reykjavik, Rept 1982–4, 116.

49. STREETER, V. L. (ed.), 1961. *Handbook of Fluid Dynamics*, 1st edn. McGraw-Hill, New York.

50. BARBIER, E., 1986. Technical–economic aspects of the utilization of geothermal waters. *Geothermics*, Sp. issue **15**(5/6), 857–79.

51. GOERING, S. W., GARING, K. L., COURY, G. E. and FRITZLER, E. A., 1980. *Residential and Commercial Space Heating and Cooling with Possible Greenhouse Operation.* US Dept of Energy Rept DOE/ET/28455-3.

52. COURY AND ASSOCIATES, 1980. *A Feasibility Analysis of Geothermal District Heating for Lakeview, Oregon.* US Dept of Energy Rept DOE/ET/27229-TI, 85.

53. BARBIER, E. and FANELLI, M., 1977. Non-electrical uses of geothermal energy. *Prog. Energy Combust. Sci.*, **3**, 73–103.

54. OLIVET, J., 1981. *Solaire et Géothermie contre Pétrole*, p. 240. Editions du Moniteur, Paris.

55. REISTAD, G. M. and MEANS, P., 1980. *Heat Pumps for Geothermal Applications: Availability and Performance.* US Dept of Energy Rept DOE/ID/12020-TI, 67.

56. SMITH, C. S. and ELLIS II, P. F., 1983. *Addendum to Material Selection Guidelines for Geothermal Utilization Systems.* US Dept of Energy Rept DOE/ID/27026-2, 213.

57. CORSI, R., 1986. Scaling and corrosion in geothermal equipment: problems and preventive measures. *Geothermics*, Sp. issue **15**(5/6), 839–56.

58. ELLIS, P. F. and CONOVER, M. F., 1981. *Materials Selection Guidelines for Geothermal Energy Utilization Systems.* US Dept of Energy, Contract No. DE-ACO2-79ET27026.

59. ODDO, J. E., 1982. Simplified calculation of $CaCO_3$ saturation of high temperatures and pressures in brine solutions. *J. Petrol. Technol.*, **7**, 1583–90.

60. KUWADA, J. T., 1982. Field demonstration of the EFP system for carbonate scale control. *Geothermal Resources Council Bull.*, **11**, 3–9.

61. CORSI, R., CULIVICCHI, G. and SABATELLI, F., 1985. Laboratory and field testing of calcium carbonate scale inhibitors. *Trans. Symp. Geothermal Energy, Geothermal Research Council, Hawaii*, **9**, 239–44.

62. FOURNIER, R. O. and MARSHALL, W. L., 1983. Calculation of amorphous silica solubilities at 25–300°C and apparent cation hydration numbers in aqueous salt solutions using the concept of effective density of water. *Geochim. Cosmochim. Acta*, **47**, 587–96.

63. VETTER, O. J. and KANDARPA, V., 1982. Handling of scale in geothermal operations. *Proc. Int. Conf. on Geothermal Energy, BHRA, Florence, Italy*, pp. 355–72.

64. YANAGASE, T., SUGINOHARA, Y. and YANAGASE, K., 1970. The properties of scales and methods to prevent them. UN Symp., *Geothermics*, Sp. issue **2**(2), 1619–23.

65. ÖLÇENOGLU, K., 1986. Scaling in the reservoir in Kizildere geothermal field, Turkey. *Geothermics*, Sp. issue **15**(5/6), 731–4.

66. ELLIS, J. and GOLDING, R. M., 1963. The solubility of CO_2 above 100°C in water and sodium chloride solution. *Am. J. Sci.*, **261**, 47–60.

67. SIENA, S., 1986. Small–medium size geothermal power plants for electricity generation. *Geothermics*, Sp. issue **15**(5/6), 821–37.

CHAPTER 6

Environmental Impact

'Everything has its beauty, but not everyone sees it.'
KUNG FU TSE, 557–479 BC

6.1 REFLECTIONS ON CONSEQUENCES

The consequences of exploitation may affect the site of a geothermal resource, but also spread outside it, depending upon the type of field, how it is exploited, and the geological and ecological framework in which it occurs. The high-temperature areas from which electric power is produced usually possess adverse chemical components. Normally, these are less deleterious to the surroundings than other types of thermal, nuclear or hydroelectric plants. On the other hand, even less severe effects arise from low-temperature areas most useful for direct application for space heating or process heat and located near urban or industrial consumers. Practically always, such associated activities as laying roads, drilling wells, installing pipelines and erecting power plants have a great influence on land utilization in any particular region. Whilst hydroelectric plants entail enormous construction work and other types of thermal power plant require considerable industrial support such as mines, transportation facilities and processing plants, geothermal power production is characterized by having every phase of the fuel cycle situated at the site. Thus the area of environmental impact of hydroelectric power plants extends a long distance from the actual power-generating plant, in contrast with the zone of influence of geothermal plants. In addition, geothermal plants constitute relatively clean energy sources and cannot have the potential for catastrophic failure which is such an inevitable adjunct of nuclear power plants. In fact, unlike fossil-fuel and nuclear plants, geothermal energy sources do not pollute on generation. Very undesirable environmental effects associated with other operations, but not with geothermal development, also include refining, strip mining, industrial wastage, off-shore drilling and various types of radioactive hazard. This does not imply that there is no radioactivity present in some geothermal areas, but that its levels are within prescribed limits; for instance, at The Geysers there is an α-radiation level of 0.015×10^{-7} mCi/cm^3 which is well below the US Public Health Service permissible concentration for drinking water.

Nevertheless, it must be admitted that earlier exploitation of several geothermal areas entailed aspects unacceptable nowadays, and some were developed experimentally so that the parties concerned had little incentive or indeed money to spare for the environment. A case in point is a plant at Wairakei in New Zealand which started in the middle 1950s in connection with production of heavy water by distillation together with a relatively small quantity of power, work described by T. G. N. Haldane and H. C. H. Armstead in 1962.[1] Observations at this plant were made during a four-month time interval in 1974 and a report was made by R. C. Axtmann in the same year.[2] The most significant pollution from this source involved unacceptable levels of hydrogen sulphide in gaseous effluent, introduction of arsenic effluent at concentrations exceeding World Health Organization recommended levels during periods of reduced river flow and possible mercury contamination of the Waikato River. In addition, production of 143 MWe entailed discharge of some 850 MWt into the river.

Remedies available for some of these problems include re-injection of waste water from the wells, treating and/or recovering hydrogen sulphide from the gas extractors, and building cooling towers instead of running the waste thermal waters into the Waikato River. In defence of the project, it was indicated that the plant in question contributed 7% of national electric power with the highest load factor and almost the minimum cost. Waste water added to the associated river produced an extra 2·4 MWe in downstream hydroelectric plants and the whole area comprises a profitable tourist attraction.

Another interesting case if the Cerro Prieto geothermal field in a desert region of northern Mexico where liquid and gaseous effluents from a 75-MWe power plant are disposed of on the surface, a practice which would not be permitted elsewhere. It has been stated that waste waters from steam separators and blow-down from cooling towers are piped to an 8-km^2 evaporation pond, the waste water containing on average 20 000 mg/kg of total dissolved solids which include various chemicals.[3] The pond had over twice the total dissolved solids content of seawater by 1974 and the system of containment and evaporation employed has an advantage in that it provides a possible means for recovering the potassium and lithium constituents of the brines. The steam from this field has a high content of non-condensable gases. S. Mercado in 1977 stated that when it enters the turbine it contains 14000 mg/kg of carbon dioxide, 1500 mg/kg of hydrogen sulphide and 110 mg/kg of ammonia.[4]

Finally, at The Geysers in California, a discarded turbine from an earlier fossil-fuel plant was used initially and excess condensate was permitted to flow into the Big Sulphur Creek. Later, excess condensate was re-injected, as A. J. Chasteen pointed out in 1976.[5] The main environmental problem here was hydrogen sulphide emission in large quantities resulting from a content of approximately 225 mg/kg in the steam. Abatement techniques entail the installation of heat-exchanger type condensers which can increase the amount of gas removable by gas ejectors, that gas then being treated by the Stretford process.

Geysers and hot springs contribute abnormal quantities of heat energy, mineral matter and water to relatively quite localized areas of eco-systems that are otherwise balanced. Therefore these systems become anomalous,

TABLE 6.1
The chemical compositions (wt%) of some sinters[6,7]

Mineral	Near Pohutu Geyser, New Zealand	Near Daisy Geyser, Yellowstone National Park, Wyoming, USA
Silica	85·58	89·75
Alumina	0·74	0·38
Iron oxide	0·09	0·06
Magnesium oxide	0·01	0·14
Calcium oxide	0·64	0·56
Sodium oxide	0·58	0·26
Potassium oxide	0·28	0·12
Titanium dioxide	0·03	—
Sulphur	0·09	—
Sulphur trioxide	—	0·05
Barium oxide	0·03	—
Chlorine	—	0·03
Manganese oxide	0·02	—
Phosphorus pentoxide	0·01	—
Zirconium dioxide	Trace	—
Water/organic matter	6·17	—
Water lost at 104–105°C	5·50	9·36

both biologically and geologically, and may even modify the atmospheric environment. In fact, since bleached bones have been found at the bottom of hot pools, it is clear that animals have sometimes fallen in. Indeed, the apparently solid peripheries of geysers and springs are frequently friable and highly dangerous because they are underlain by scalding water. In addition, whilst some geysers erupt predictably, others do not, and observers who are taken unawares can be injured as a result. Interestingly, with changes in activities of these features, some areas will cool and others heat up, and this affects the local plant life. For instance, in the Yellowstone National Park in Wyoming, USA, the death of many trees followed the Hebgen Lake earthquake through the expansion of hot areas. White and petrified remnants still exist and some are partially or totally enclosed by siliceous sinter, siliceous material being deposited also in the wood as water with a high silica content is taken up and evaporated. Actually, the quantity of mineral matter in geyser waters is rather low, usually from 1 to 10 g/litre, but it can accumulate and even be transported as time passes. The river waters at Yellowstone transport approximately 352 000 kg of mineral matter every 24 h. Some appropriate data are listed in Table 6.1. The higher water content recorded from Yellowstone sinter compared with that from New Zealand probably relates to the colder and wetter climate of the former.[8]

6.2 AIR AND WATER QUALITY

The gaseous constituents of geothermal effluents are extremely variable, both as regards concentrations and compositions. The information in Table 6.2 is relevant to concentrations of non-condensable gases from different sources. From the table it is clear that hydrogen sulphide ranges through almost four orders of magnitude in the fields considered and certainly it is that gas which creates the most adverse environmental effects. However, the importance of these effects varies according to where they take place. For instance, at Cerro Prieto over 800 tonnes are emitted annually, but, since this geothermal field is in an arid region, they have little impact upon human beings. On the other hand, the smaller emissions of 4500–6000 tonnes/year at The Geysers are far more serious: this reflects the location of the site near wine grape growing areas and recreational woods and grasslands. State law air quality standards in California necessitate maximum levels for this gas in ambient air not exceeding 42 g/m^3 (0·03 mg/kg) averaged over an hour. R. E. Ruff and his colleagues in 1978 effected

TABLE 6.2

Non-condensable gas concentrations (wt%) in selected geothermal regions

Gas	The Geysers[9]	East Mesa[12]	Cerro Prieto[10]	Wairakei[11]
CO_2	0·326	0·089 3	0·689–0·897	0·164 5
H_2S	0·022 2	0·000 005	0·017–0·035	0·04
CH_4	0·019 4	0·001 1	0·015–0·032	0·000 44
NH_3	0·005 2	—	0·012–0·019	0·000 75
N_2	0·005 6	0·003 7	0·001 5–0·004 1	0·000 46
C_2H_6	—	0·000 1	—	0·000 18
H_2	—	0·000 005	0·016–0·041	0·000 1
H_3BO_3	—	—	—	0·000 03
HF	—	—	—	0·001 78
Ar	—	0·000 1	0·000 04–0·000 1	—

measurements at The Geysers for SRI International and showed that, at sites 2–10 km from the development centre, the requisite standard was exceeded in 0·05–4% of total hours sampled.[13] Therefore reduction in emission was mandatory. New heat exchanger condensers were expected to cut the pollution by approximately 50% in new plants and, in existing plants using direct-contact condensers, Pacific Gas and Electric Company found that adding an iron salt to circulating water could reduce hydrogen sulphide emission by as much as 60%.[9,14] This action produces a sludge of native sulphur, various compounds of iron and fine rock which accumulates at a rapid rate (several thousand litres daily), to be removed to a far-off disposal site. As this is an expensive process, the company concerned has attempted to try other avenues such as removing hydrogen sulphide from steam prior to its entering the power plants.

In this geothermal steam there may be other undesirable elements such as traces of arsenic, boron, mercury and radon, of which details are listed in Table 6.3 of S. L. Altshuler (1978).[15] The actual partitioning of these elements between the re-injected condensate, the atmosphere and the liquid spray drift arising from cooling towers is not known with any certainty. However, the mercury content of soils up to a kilometre away can exceed the normal level by a factor of ten, although radon increases, whilst detectable, are much less significant. On the other hand, the boron content of soils as far as 800 m downwind from the older plants is up to 1000-fold greater than that found in the surrounding region, vegetation being markedly affected. Of course, boron may constitute a hazard elsewhere as well, for instance in Turkey, near the area of geothermal field development and power generation at Kizildere in western Anatolia not too far from

TABLE 6.3
Trace elements in steam at The Geysers and the projected output from new-type
plants (Unit 17)

Element	Concentration in steam (mg/kg)			Projected cooling-tower emission (g/MW h)	
	Low	High	Average	Gas or vapour	Liquid drift
Arsenic	0·002	0·05	0·019t	0·15	0·001
Boron	2·1	39·0	16·0	—	0·7
Mercury	0·000 31	0·001 8	0·005	0·039	0·000 09
Radon	5·0^a	30·5^a	16·2^a	0·13^b	—

^{a} In nCi/kg.
^{b} In μCi/MW h.

Izmir. According to E. Okandan and T. Polat in 1986, although the potential capacity there was estimated to be at least 150 MW, a plant of only 20·4 MW was opened on 14 February 1984 because any larger discharge of waste waters from the power plant could cause serious environmental problems in the Menderes River and adjacent soil because of high boron content (approximately 25 ppm) in the relevant discharge water.[16] Additional data from this interesting new project are given in Tables 6.4 and 6.5.

It must be added that the radioactive problem mentioned in Section 6.1 also involves curative thermal waters which have been discussed by E. Chiostri in 1975,[17] among others. H. A. Wollenberg in 1976 demonstrated that uranium and its daughters tend to have highest concentrations in low-temperature waters rich in calcium carbonate, with lowest values in silica-rich high-temperature systems.[18] Radiation levels of 0·25–0·5 mR/h were recorded over enclosed pools at Kyle Hot Springs in northern Nevada and well-defined radioactive anomalies existed downwind from these which

TABLE 6.4
Composition of non-condensable gases at Kizildere[16]

Gas	Quantity (%)
CO_2	99·9
H_2S	0·1

TABLE 6.5
The variation of non-condensable gases in different geothermal fields[16]

Field	Non-condensable gases present (%)
Cerro Prieto	1·25
Kizildere	11–15
Larderello	4·5–5·0
Matsukawa	1·10
Monte Amiata	10–25
The Geysers	0·6–1·0
Wairakei	0·35–0·5

showed that the source was the gas radon-222. Usually, hot spring areas have values smaller by an order of magnitude, even in the case of calcareous regions. In addition, siliceous hot springs either exhibit no anomalous radioactivity or, if any is present, it is smaller again by a further order of magnitude. The various chemicals emitted in some geothermal fields were listed by A. J. Ellis in 1978 and are given in Table 6.6.[19]

TABLE 6.6
Approximate amounts of emitted chemicals
(tonnes/year per 100 MWe power generated)

Emission	Cerro Prieto[20]	Wairakei	Broadlands	Larderello	The Geysers
Li	320	350	300	—	—
Na	150 000	35 000	30 000	—	—
K	40 000	5 500	5 000	—	—
Ca	10 000	500	80	—	—
F	40	200	200	—	—
Cl	300 000	60 000	50 000	—	—
Br	500	150	150	—	—
SO_4	170	700	200	—	—
NH_4	800	50	500	1 300	1 700
B	400	750	1 100	200	200
SiO_2	40 000	20 000	20 000	—	—
As	30	100	100	—	—
Hg	—	0·025	0·035	—	0·04
CO_2	150 000	8 000	400 000	400 000	300 000
H_2S	11 500	300	6 000	5 000	1 200[a] 900[b]

[a] 1976.
[b] 1977.

Turning now to water quality, geothermal fluids produce water varying from potable to so highly mineralized that D. E. White and his associates in 1963 described some as practically ore-depositing fluids.[21] Therefore handling them varies from place to place. M. D. Crittenden Jr in 1981 mentioned that environmental problems are usually worse at higher temperatures of such waters.[22]

In cases where geothermal water effluent is below surface-water quality, it must not be discharged on the surface. This usually entails re-injection, an undertaking requiring great care in order to avoid plugging of wells or formations which receive the fluids. In such steam fields as The Geysers, the re-injection process can be carried out by gravity because the reservoir pressure is much lower than the equivalent hydrostatic head. In addition, only excess condensate and the blow-down from cooling towers are being re-injected, these being both clean and of low salinity. Consequently, settling alone is sufficient.[5]

Under less favourable circumstances, chemical pretreatment and filtration may be mandatory. Interestingly, however, rather saline waters were successfully re-injected at Ahuachapán, El Salvador, as S. S. Einarsson and his colleagues indicated in 1976.[23] However, hypersaline brines are impermissible because of their capacity to produce scale and chemical precipitates should the temperature drop, this even where the heat content is high.[21,24] Salinity and chemical components content are critical factors in determining applications of geothermal fluids. In this respect, trace elements are significant also; for example, citrus fruits are highly sensitive to boron contents as low as 0.5 mg/litre, as F. M. Eaton and L. V. Wilcox indicated in 1939.[25] Relevant information is given in Table 6.7.

Incidentally, it may be that re-injection could prolong the life of a geothermal resource. Thus R. C. Axtmann indicated in 1974 that the re-injection of waste water then flowing into the Waikato River at Wairakei might conserve the equivalent of approximately 450 MWe of heat, a sizeable quantity which might act to ensure such prolongation. In addition, at this site, it would have reduced subsidence due to the extraction of geothermal fluids from a recorded value of 4.5 to 1 m. This matter was discussed by T. M. Hunt in 1975 and subsidence seems to be continuing, as he and R. C. Allis indicated in 1986.[27,28] A detailed discussion of their findings is given in Chapter 3. It is germane to add that the phenomenon has not been recorded at The Geysers. This is because steam temperature and pressure are of the order of 240°C and 34 kg/cm^2, the near constancy of pressure at depths exceeding 2.5 km (where hydrostatic pressures would normally approach 280 kg/cm^2) indicating that the relevant rocks must be competent so that the dry-steam field can exist at all. Therefore the removal of vapour does not trigger subsidence. However, with hot liquid-dominated fields such as Wairakei, non-maintenance of subterranean pressure through omission of re-injection of fluid causes subsidence (cf. J. W. Hatton in 1970[29]).

Risks to groundwater aquifers may arise from saline waters. For instance, brines constitute a potential contaminant at the Salton Sea in

TABLE 6.7
Tolerances for trace elements (mg/litre) in irrigation waters[26]

Element	Water continuously in use on all soils	Water in short-term use on fine-textured soils
Aluminium	1·0	20·0
Arsenic	1·0	10·0
Beryllium	0·5	1·0
Boron	0·75	2·0
Cadmium	0·005	0·05
Chromium	5·0	20·0
Cobalt	0·2	10·0
Copper	0·2	5·0
Lead	5·0	20·0
Lithium	5·0	5·0
Manganese	2·0	20·0
Molybdenum	0·005	0·05
Nickel	0·5	2·0
Selenium	0·05	0·05
Vanadium	10·0	10·0
Zinc	5·0	10·0

California should they be permitted to mix with irrigation water in the region. Although geothermal waters are usually less saline than these, they are more so than non-thermal waters. However, sometimes they are sufficiently pure to allow them to be used in agriculture and industry, and, in this case, it may be inferred that their higher temperatures were not sufficient to promote enough dissolution of volatile chemicals in the host rocks to pollute them appreciably. N. V. Peterson and E. A. Groh in 1967 alluded to the direct utilization of geothermal waters for stock watering at Klamath Falls in Oregon.[30]

6.3 WATER FROM GEOTHERMAL RESOURCES

Water is in short supply in many places and various measures have been proposed in order to alleviate such situations, among them importation, reclamation and desalination. All of these need large amounts of power and may involve pollution. Where they occur, geothermal resources, in particular liquid-dominated ones, provide rather clean, relatively cheap power and large quantities of water may be derived from them; for example, fresh water may be obtained from steam turbine condensates. If saline water emerges from a geothermal well, it can be distilled to yield fresh water and the feasibility of large-scale geothermal water distillation has been studied. Alternatively, electricity produced in a geothermal power plant can be employed to effect desalination of either seawater or brackish water. In fact, membrane desalting has been proposed as more appropriate than distillation.[31] Dealing with brines requires complex procedures and Table 6.8 lists instances showing the effects of various system parameters.

For case 1, A. D. K. Laird gave a simplified flow diagram and a thermodynamic process diagram.[31] Most of the well fluid is converted into fresh water. The brine returned to the reservoir becomes 4·7 times more concentrated than the well fluid and the calculated power production rate is 17·7 MW, the water production rate being about two million gallons daily,

TABLE 6.8

Approximate calculated water and power production for six cases (1–6) under different assumed conditions

Production parameter	1	2	3	4	5	6
Reservoir temperature, °C	315	315	315	315	205	150
Wellhead temperature, °C	229·2	229·2	229·2	229·2	150	93·3
Wellhead pressure, psia	400	400	400	400	67	11·5
Lowest temperature, °C	100	71·1	71·1	71·1	71·1	71·1
Fresh water/well fluid ratio	0·79	0·83	0·79	0·87	0·43	0·26
Rate of production of water at 90% plane availability, lb/h of fluid supplied						
10^6 gal/day	2·04	2·15	2·05	2·26	1·13	0·62
10^3 acre/ft year	2·29	2·41	2·30	2·55	1·27	0·70
Power production rate, MW	17·7	24·3	24·3	24·1	6·7	1·2
Percentage of case 2						
Water rate	95	100	95	105	53	29
Power	73	100	100	99	28	5

i.e. 2290 acre ft annually (at 90% availability and operating for nine-tenths of the year). The fresh water produced is above the boiling point and must be cooled; this may be done either through evaporation or by mixing with cooler water, although there will be an associated loss of geothermal energy. Case 2 illustrates the effects of temperature reduction relative to rejection of thermal energy and case 3 is a simplified version of case 2. All three cases (1–3) have the disadvantage that non-condensable gases pass through much of the equipment and can promote corrosion. At high pressures, such gases can be removed by mixing them with some of the liquid being re-injected back into the reservoir. Case 4 is a variant of case 2, more efficient in that it incorporates removal of non-condensable gases. In case 5, it was assumed that the thermal content of the reservoir is equivalent to that of liquid water at 200°C, and also that the fluid enters the system at a wellhead temperature of 150°C. Lower temperatures are involved in case 6. Penalties are incurred for such lowered temperatures and, in addition, for using higher waste heat rejection temperatures. Even a small redeployment of the heat of condensation of the vapours produces a definite improvement in water production.

6.3.1 Distillation

Multiple-effect distillation is carried out so that each event takes place at a lower temperature than the one preceding it. It is the oldest method of desalination and modern plants utilize as many as 25 effects which can yield up to 20 times as much water from steam condensed to heat up the first step of the process. In 1972 the US Bureau of Reclamation set up a three-stage vertical tube evaporator operating at 15–16°C from the assumed wellhead temperature of 200°C.[32] The incoming mixture is taken as comprising 20% steam and the first effect resembles case 4 above and entails removal of non-condensable gases. Two other effects also parallel the cases mentioned above. The brine from effect 3 is five times more concentrated than the wellhead mixture, i.e. four-fifths of the wellhead fluid is converted to fresh water. The steam from the third effect is transmitted through a turbogenerator so as to produce approximately 10 MW of power.

Another desalination method suitable for geothermal brines is the multi-stage flash process in which hot saline waters pass through a set of chambers of which each is at a lower pressure than its predecessor. One step of the process occurs at each stage. Some of the brine is flashed as it enters each chamber. Physically, the chambers are each 1 m or more in length because, since evaporation is not instantaneous, time must be allowed for the vapour to separate as the brine flows along the bottom of each. In each chamber there are tubes in which cooler brine flows towards the hot end of the set of chambers. Vapour is produced and condenses on the tubes and drains off into a trough underneath them. The distillate is also passed through the chambers in the same way as the hot brine and its vapours, too, condense on the tubing, thereby adding extra heat available for the cool brine.

Industrial multi-stage flash seawater distillation plants for desalination processes have 30 or more stages and the spent thermal energy is rejected to the sea. The multi-stage flash distiller is the most common technique for seawater conversion and is used in about 90% of all desalination work. Its advantages include the fact that the hot brine channel can be constructed so that it will not get blocked even if solid materials such as silica precipitate into it; also deposits adhering to the walls can be scraped off cheaply. In addition, feed water does not require extra heating in geothermal applications, although this is not true with seawater where a heater must be placed between the discharge ends of the condenser tubes and the entrance to the series of flash chambers.

This advantage of the geothermal usage is counterbalanced by a disadvantage. Whereas in seawater desalination the cool brine in the condenser tubes is incoming feed water in the process of being heated, in geothermal desalination there is no substitute available for this excellent method of cooling the condenser (which otherwise would permit quite a large fraction of geothermal brine to be desalted). Consequently, the multiple-effect approach is more suitable than the multi-stage flash one.

A typical layout for a desalination cycle is shown in Fig. 6.1.

6.3.2 Other Techniques

Thermal energy cannot be stored and must be converted into other forms of energy, e.g. electrical or chemical. Unfortunately, some of the energy derived from a geothermal reservoir cannot be converted and must either be transferred to materials at lower temperatures or dissipated, i.e. lost. Such wasted heat may enter the atmosphere, often without noticeable effects. Sometimes it may be beneficial and, at other times, detrimental. In any case, its disposal is important.

The amount of thermal energy which can be transformed into chemical energy by removal of salt from brines by a perfect process would be approximately 4 kW h/1000 gallons of water converted from seawater, a very small quantity in comparison with the geothermal energy which is convertible to electric power or returned to the reservoir in reject form. The inference is that if water alone is being produced, i.e. if no attempt is made

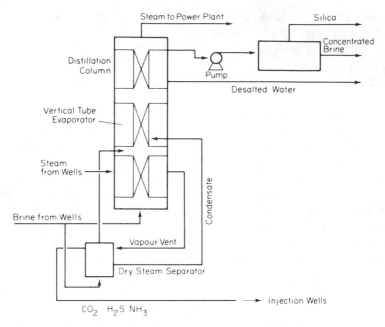

FIG. 6.1.   Schematic diagram of a desalination cycle.

to return thermal energy to the reservoir, almost all of the geothermal energy extracted constitutes waste heat. If electricity is exported from a geothermal field, desalination can also be decentralized, which can be valuable to a distant, more populated region. The major desalination processes operated by electrical power are vapour compression, reverse osmosis and electrodialysis.

6.3.2.1 Vapour Compression

The basis for this process is that the temperature of water vapour can be raised if a compressor is used to raise the pressure. A temperature difference is established and can promote condensation of compressed vapour on one side of a heat-transfer surface, thereby effecting evaporation of an equal quantity of water on the other. This latter is fed to the compressor. The actual temperature and pressure differences are kept small so as to minimize the power of the compressor and the staging principle can be incorporated into the vapour compression plant design; see the discussion by B. W. Tleimat in 1969 (Ref. 33). As the process is cheap, it is very competitive and particularly suitable to low-capacity seawater conversion of up to 500 000 gallons per day. But it is not appropriate for feed waters such as geothermal brines because these contain appreciable amounts of silica.

6.3.2.2 Reverse Osmosis

Sometimes termed hyperfiltration, this is based upon osmosis, i.e. the flow of water through a semi-permeable membrane, one permitting the passage of a solvent but not that of solutes. Desalination depends upon most of the included salts being retained by such a membrane when a pressure difference exceeding the osmotic pressure is maintained across it. The osmotic pressure rises with concentration difference across the semi-permeable membrane and hence both the requisite energy and cost of the fresh water produced increase with the salinity content of the feed water. Best results are given when this has a concentration range of 2000–5000 ppm.

6.3.2.3 Electrodialysis

Here two types of membrane are used and are selectively permeable to ions. One will pass cations better than the other which, in turn, passes anions preferentially. Alternating cation and anion membranes are arranged on the walls of parallel channels through which brine is allowed to flow. An electric current is passed across and carries ions of opposite sign through the membranes on opposite sides of each channel. The concentrations in every second channel are reduced and they increase in alternate channels.

Geometrically, the channels must be long enough to ensure that the necessary reduction is attained in concentration of salts in water produced. Theoretically, the electric current flowing across is proportional to the number of ions which it transports, and therefore to the required reduction in salinity. Clearly, the more concentrated the saline waters are, the greater the electrical energy necessary to achieve desalination.

Electrodialysis is thought to be most valuable in freshening brackish waters with salt concentrations up to 3000 ppm, a low level of concentration which keeps costs down.

6.3.3 Budgets

Where large-scale operations are involved, it may be cheaper to import water for pressure maintenance and cooling. Such water must be exported from elsewhere and, for the Colorado River, the Imperial Valley in California was suggested as a source. It lies in an arid region not very near the coastline. The exportation of as much as 2·5 million acre ft annually of desalted water was envisaged and the proposal made that seawater should be imported in order to compensate. This would have to be done from the Gulf of California some 160 km or more away.[34] Locally available water would be less expensive to obtain and one source here might be irrigation-drainage water. The arid basin of Imperial Valley is below sea level and has no drainage outlet, the drainage sump being the Salton Sea, which is sustained by such agricultural drainage water. The water inflow there is about 1 300 000 acre ft yearly and it is balanced by very high evaporation from the water surface covering over 900 km[2]. It was estimated that 150 000 acre ft of water could be withdrawn annually and used to maintain pressure in geothermal reservoirs until the evaporation reduction of 12% balanced the withdrawal.[34]

The US Bureau of Reclamation believed that 125 000 acre ft of water would have to be extracted from the Salton Sea to supply replacement and cooling water for their demonstration facility designed to produce about 200 000 acre ft annually of desalinated water and 420 MW of power.[32] Solids would have to be disposed of; salts can be removed by re-injecting brine into subterranean voids. Such work at the Salton Sea would produce stabilization of its level and salinity, reduce irrigation water salinity, provide cooling facilities for power plants and prevent subsidence as well as augmenting the flow of the Colorado River. Obviously, the benefits would exceed the cost, but the expense of any sort of desalination will always exceed that of large systems capable of supplying fresh water by just pumping it out of nearby rivers. Nevertheless, it is anticipated that desalination will get cheaper and its application to upgrading waste brines, saline groundwater and seawater remain competitive with long-distance importation schemes.

6.4 Geological Hazards

Among the hazards are landslides which, under conditions of adverse terrain, govern the siting of wells, power plants and other facilities in geothermal regions. One such area is The Geysers, which is in a location of steep slopes resulting from the erosion of rocks of marked structural instability. Almost half of it is underlain by landslides, according to R. J. McLaughlin in 1978.[35] Practically all were active during the Quaternary and a large number remain so. C. F. Bacon in 1976 reported that over 50% of the wells at The Geysers were drilled on such landslides, two of three well blow-outs in the region being ascribed to casing failure caused by this phenomenon.[36] In both cases, the wells were drilled prior to recognition of the dangers inherent in their siting.

Nowadays, developers place well pads on zones of stable bedrock and drill one or several inclined holes from a single site. In this manner, the bottom-hole spacing in producing horizons can be maintained so as to develop the resource optimally and at the same time ensure that wellheads are located in stable situations. Careful study of the site geology and soil stability entails expense, but this is far less than the cost incurred by losing a well through a blow-out. C. F. Bacon in 1976 stated that controlling, plugging and abandoning a well which blew out in 1975 almost equalled the cost of drilling a new well.[36] The same factors influence the siting of geothermal power plants and to an even greater degree because of the serious loss which would result from a landslide falling on to one. Other areas in the world in which these hazards may take place are some areas of Japan, the Philippines and Central America.

Seismicity is another geological problem and the link between silicic volcanism and geothermal resources has been established all over the world, as R. L. Smith and H. R. Shaw indicated in 1975.[37] It is based upon considerations of plate tectonics discussed in detail in Chapter 1, Section 1.3.3. A consequence is that much geothermal development occurs in places where there is a high regional seismicity. Thus earthquake-resistant construction of power plants and allied equipment must be coupled with avoidance of active or potentially active fault zones, into which no re-injection of fluids must take place. Protection against earthquakes entails designing and constructing such facilities in a manner whereby they can resist acceleration and vibration frequencies connected with the pheno-mena. The San Fernando earthquake of 9 February 1971 showed that the ground acceleration locally can be greater than $0.5g$, a higher figure than that provided for by many engineering designs of the past.[38]

In view of the foregoing, it is not surprising that many earthquakes have been recorded from geothermal fields, e.g. in the Imperial Valley–Mexicali part of the head of the Gulf of California, at the Salton Sea, Brawley, Heber, East Mesa and Cerro Prieto. The last five are situated close to the intersection of the San Andreas fault system with the East Pacific Rise, a point of great strain involving at least five active major faults covering a 100-km-wide geographical swath. In fact, this is probably the most active seismic region in the USA, earthquakes exceeding 6 in magnitude having taken place here in 1899, 1903, 1915, 1918, 1934, 1948 and 1968, according to W. Hamilton in 1976.[39]

Faults may produce surface rupturing which can damage geothermal facilities if they are located too near or across them.

Another variety of seismicity is that induced by man. It was first recognized in 1966 when liquid waste injected at moderate pressure in a deep well set off earthquakes near Denver, Colorado.[40] This implied that re-injection of waste geothermal fluids can produce the phenomenon elsewhere, as C. A. Swanberg indicated in 1976.[41] In 1976 it was determined that low-level seismicity can result from the withdrawal of fluid in oil fields, notably at Wilmington.[42] However, although cases of the phenomenon have been recorded from geothermal fields, it does not appear to constitute a particularly serious danger.

Well documented is the case of The Geysers, which is mentioned in Chapter 4. It was reported in 1976 that, after re-injection commenced in 1969, over 1.6×10^{10} litres had been returned to the geothermal reservoir at a rate sometimes as high as 170×10^6 litres daily.[5]

Monitoring of seismicity was effected by R. M. Hamilton and L. J. P. Muffler in 1972 using temporary stations.[43] Activity was discovered to centre along known faults in the area of steam production, although it was not possible to determine whether the recorded micro-earthquakes resulted from this together with fluid re-injection or from the liberation of built-up natural regional stress. Subsequent monitoring showed a large number of small quakes of magnitude <2 on the Richter scale which were extremely localized as regards both area and depth. Many epicentres are inside the producing area, i.e. the area of main pressure reduction.[44] Their hypocentres are shallow, occurring between 1 and 4 km, placing them in the area of steam production. The largest quakes take place near injection wells and have attained magnitudes as high as 3·8. There are now almost monthly earthquakes and high micro-earthquake activity in the region.

Some have concluded that the effects result directly from the geothermal operations and certainly this is partly true. Production-related continuous induced seismicity produces micro-quakes along fractures under 1 km in length which are mostly benign. Also, after the addition of four power units in March 1979, the production rose by 70% and formerly aseismic zones around three of them became active. However, as The Geysers and Clear Lake are situated quite near the San Andreas fault system, which passes only about 50 km southwest of them, it is not surprising that first-motion studies of quakes within the field and also in the surrounding region show that both respond to the same regional stress pattern.

The earthquakes taking place outside the producing field, which seem to belong to the normal regional tectonic regime (of non-geothermal character), are deeper (4–12 km) and less frequent. They lie along a major right-lateral shear zone sub-parallel to the San Andreas, passing within a few kilometres of the steam field.[45]

In the same state, micro-earthquakes were studied at East Mesa and interpreted to locate an active fault, the East Mesa fault, which turned out not to be their cause. They probably originate from the active Brawley fault, as T. V. McEvilly and his colleagues indicated in 1978.[46]

In Chapter 5 the Los Alamos Project is discussed: hydraulic fracturing experiments there induced seismicity in 1983. Section 5.5.1 gives full details.[47]

Interestingly, no seismic or micro-seismic connection was established with re-injection in the Viterbo region of Central Italy, as G. M. Camelli and G. Carabelli in 1976 indicated on the basis of an investigation into gravity flow into a well at a rate of 3×10^6 litres daily.[48] In the Otaki area in Japan, where geothermal hot water was being re-injected, K. Kubota and K. Aosaki in 1976 obtained similar results.[49] Flow under gravity into three wells at rates higher by an order of magnitude than that used in the Italian case produced no seismic events over a three-year period.[22]

To examine the matter in greater depth, the US Geological Survey started a series of appropriate experiments in 1969 at the Rangley oilfield in western Colorado.[50] Here the injection of water intended to stimulate the production of oil seemed to be responsible for earthquakes. Results demonstrated that the Hubbert–Rubey principle is valid; this relates the fluid pressure required to initiate fracture to the stress normal to a pre-existing fault plane and shear stress parallel to it. Initial measurements gave existing stress values and these were employed in calculation of the critical value above which fluid pressure might induce fault movement. During cyclic pumping tests, it was shown that earthquakes were initiated when the fluid pressure was elevated above this value and stopped when the applied pressure was reduced below it. Clearly, the circumstances under which fluid injection may trigger earthquakes include the presence of a fault already possessing a significant shear stress across it and the injection of such fluids at pressures above the critical value necessary to induce faulting. It may be added that reservoir-induced earthquakes may take place also; instances have been recorded in India and China (see Chapter 3, Section 3.2.2.6).[51,52]

If fluids are extracted from a geothermal reservoir for a long period of time, subsidence may result. The best known case of this is that of Wairakei, the first documentation on the matter being by W. B. Stillwell and his associates in 1976.[53] Measurements were commenced in 1956 after well-testing began and there was a maximum subsidence in the time interval 1964–74 some 1500 m east of the well field. There was only a minor amount of re-injection during the tests. Horizontal surveys showed a simultaneous inward movement towards the centre of vertical subsidence of amounts up to 0·5 m. Although the subsidence was quite considerable, there appears to have been practically no appreciable damage to the pipelines and power plant because these are peripheral to the subsidence centre. However, telescoping of as much as 0·8 m took place in concrete drains carrying hot water from the wellhead steam separators. The area of the maximum subsidence was associated with the thickness of the underlying Waiora Formation, a rhyolitic pumice with a porosity between 20 and 40% (see Fig. 2.11). In 1964 a maximum of 67 megatonnes/year of water and steam was discharged by the wells, and more wells were coupled to the power station, the quantity of electricity generated reaching 170 MW (1 megatonne, Mt = 1×10^9 kg). Most of the production wells were located in a 1-km^2 area called the bore field. Declining reservoir pressures thereafter limited the production of fluids and, by the early 1980s, the rate of withdrawal of mass decreased to approximately 47 Mt annually. The total mass withdrawn from the bore field to 1984 was 1350 Mt, equivalent to a volume of about 1·7 km^3 of water at 250°C. The extraction of such an enormous mass of fluid caused major hydrological changes.

The reservoir is thought to have consisted initially of a liquid-dominated two-phase zone over a deeper liquid zone.[54] Removal of the water caused the two-phase zone to expand both laterally and vertically, and steam formed at and above the level at which fluid was withdrawn (0·5–1 km). The two-phase zone now has two parts, an upper ('steam zone' or 'vapour-dominated zone' in which steam is the continuous pressure-controlling phase and a lower 'liquid-dominated zone' in which liquid water is the continuous phase, but boiling conditions (saturation pressure) exist and some steam may occur. By the early 1970s, the steam zone extended over an area of 15 km^2. Although the groundwater zone is now perched over much of the field, changes in groundwater level are so far restricted to a 2-km^2 area which contains much of the bore field and Geyser Valley, according to R. G. Allis in 1982.[55] Pressures in the deep, single-phase zone of hot water occurring beneath the two-phase zone fell by up to 2·5 MPa (1 MPa = 10 bar) in the 1950s and 1960s, but have remained quite stable since. This is taken as indicative of substantial recharge, but pressures in the steam zone went on declining at rates of up to 0·05 MPa annually. However, in spite of the pressure and temperature declines in the reservoir zone, the average discharge enthalpy of the geothermal fluids from the bore field stayed near 1100×50 kJ/kg during exploitation. This implies that the production wells are tapping a vast volume of hot water. However, since the late 1960s, the enthalpy has diminished at a rate of some 8 kJ/kg per year (equivalent to an annual drop in temperature of 1·6°C) and the falling temperature may be due to both a drop in reservoir pressure and an invasion of cold water.[54] At first, the creation and expansion of the steam zone produced increases in heat flow from *ca.* 400 to 800 MWt during the mid-1960s, but later declines to 600 MWt occurred in step with declining pressures. In addition, there has been natural steam loss through thermal areas on the surface. Transition in thermal activity from decreasing hot spring and geyser discharge in Geyser Valley to increased steam flows elsewhere caused the natural mass flow to drop from about 13 Mt annually in 1950 to approximately 5 Mt/year in the early 1980s.

Another consequence of exploitation was ground movement, i.e. ground surface subsidence, of which the maximum of over 9 m is recorded outside the bore field itself; see for instance the work of R. G. Allis and P. Barker in 1982.[56] Repeat triangulations demonstrated that horizontal movements of up to 1 m also took place. In general, the direction of movement has been towards the centre of subsidence.

It is apparent that these changes in temperature, pressure, heat flow, enthalpy and ground surface signify that exploitation produced important changes in mass distribution in the reservoir and these in turn promoted gravity changes. Corrected for subsidence, these have been up to -1000 $(\pm 300)\,\mu$Gal in the 1-km^2 area of the production bore field and smaller decreases cover a 50-km^2 surrounding region as well. Most took place during the 1960s, after which the net gravity change for the entire field has been zero (showing mass flow equilibrium). Gravity models discussed by R. G. Allis and T. M. Hunt in 1986 suggested that saturation of the steam zone was 0·7 ($\pm$0·1) in 1962, decreasing to 0·6 by 1972, and they attributed gravity increases in the northern and eastern bore field since the early 1970s to cool water invading the steam zone.[57]

Another example of centralized subsidence was recorded at The Geysers in California by B. E. Lofgren in 1978.[58] In spite of the much greater production here, the maximum subsidence noted from 1973 to 1977 was a mere 14 cm since the reservoir is situated in very strong rocks with low compressibility. Nevertheless, it is closely related to the area of maximum pressure drop and probably results from steam production. There are also localized horizontal movements of about the same magnitude, from which it may be inferred that the reservoir is contracting in both the vertical and horizontal senses as the internal pressure diminishes.

At both Wairakei and The Geysers there are regional tectonic strains, but these are almost negligible in comparison with the local effects of withdrawal of geothermal fluids. However, in the Imperial Valley, the reverse may apply.[22] Geothermal resources at the Salton Sea, Heber and

Brawley have been developed and Colorado River irrigation water is distributed to some 400 000 hectares through a 2700-km network of concrete-lined canals covering the Imperial Irrigation District. A local subsidence of 1 m annually would hinder operations and indeed there has been a regional subsidence of 5 cm/year near the edge of the Salton Sea between 1972 and 1977, as was recorded by the Department of Public Works of Imperial County. It showed a northeasterly to northerly inclination towards the sea in question. Such a large-scale phenomenon is easy to distinguish from the more localized results attributable to the extraction of geothermal fluids. There are governmental regulations for the region which necessitate precise first- or second-order surveys of wellhead benchmarks at least once annually in order that warning of such local subsidence can be given. The effect can be countered by injecting more fluid or reducing the extraction rate, or both.

Gravity and elevation changes were examined by G. Geri and his colleagues in the Travale geothermal field of Tuscany in Italy and reported in 1982.[59] This geothermal resource was discovered much later than Larderello and, after renewal of exploration there in 1973, a levelling network was organized. Five observations made during the period 1978–80 demonstrated subsidence of the central part of the area at an average rate of 20–25 mm/year. The geothermal gradient exceeds $1°C/10$ m over an area of over $250 \, km^2$ and can rise to as much as $3°C/10$ m, the heat flow at Larderello–Travale being higher than $3 \, \mu cal/cm^2 s$, according to R. Cataldi and others in 1978.[60] The temperatures in the geothermal reservoir exceed $200°C$ and, in a few wells, they rise to $280°C$. There is no volcanism in the region, but there may well be a granite pluton at depth. A granite porphyry dike at Boccheggiano just southeast of the Travale geothermal field is about 2·3 Ma old.

A precise gravity network was installed at Travale in 1979 so as to obtain data on any density variations or mass movements within the reservoir. Such information would allow removal of the gravity effect of differential subsidence, of regional tectonics not influenced by geothermic considerations, or both. The residue should be interpreted in terms of fluid mass displacements, compaction of porous rocks, phase changes in the fluid contents of the reservoir due to temperature or pressure variations, and other factors. The main micro-gravimetric network of eight well-constrained stations was measured using two Lacoste–Romberg gravimeters. All gravity stations were located on levelling benchmarks and three external bases were situated in places thought to be geologically stable and outside the geothermal field; thus they were intended to act as reference points. In 1981 an absolute gravity station was set up and incorporated within the grid to solve any ambiguities arising with respect to gravity reference data or variations in instrumental characteristics with time. The site chosen for this was Palazzo al Piano, Sovicille, Siena. The main gravity grid was extended using an auxiliary network of 27 bases in and around the main geothermal field.

The general form of the observation equation utilized in the data reduction is

$$g_i - g_j - k(r_i - r_j) - d(t_i - t_j) = \varepsilon_{ij} \tag{6.1}$$

in which g is the unknown gravity value, k the known scale i factor, d the unknown drift value, r the dial reading i corrected for dial tables and earth tides, ε the observation error and j indicates a gravimetric factor.

The data-processing system employed was that developed in the Earth Physics Branch of the Department of Mines and Resources in Canada.[61] To obviate non-linearities of scale, the information from the two gravimeters was handled separately and a calibration line covering the gravity range of the Travale networks was organized in the Pisan Mounts near Pisa, some 90 km from the Travale geothermal region. Absolute stations in Rome, Trieste and Palazzo al Piano constitute a valuable long-range calibration line. The absolute gravity station in the latter was measured by the symmetrical free-fall absolute gravimeter of the Istituto di Metrologia G. Colonnetti and the accuracy of the instrument established as ca. 10 μGal. This facilitates detection of long-term gravity variations resulting from geodynamic events.

The major results of the investigation are that, for elevations, variations exceeding 30 mm ($\approx 9 \mu$Gal) are detectable by gravity measurements if there have been no changes in density. The observations on the auxiliary network defined an area which may be affected by events related to geothermal exploitation. Clearly, gravity and geodetic surveys have proved to be important in the Travale geothermal field as a means of studying the environmental effects of such exploitation. Perhaps they could contribute to understanding the evolution of such a field during exploitation and permit distinction of preferential areas of fluid flow.

The system water–carbon dioxide–methane–hydrogen has been studied in the Travale–Radicondoli geothermal field by S. Nuti and his co-workers, who tentatively reported in 1980 that the carbon dioxide in the field originated mostly from marine carbonates, as is probably also true of carbon dioxide at Larderello.[62] Also D. Patella et al. in 1979 reported that a dipole electrical sounding approach had been employed there and proved to be capable of distinguishing, at least in the geological situation of the Travale area, various structural features of the relevant field such as cover, reservoir, substratum, uplifted structures and tectonic depressions.[63]

The following general remarks are germane. Geothermal energy, unlike that arising from fossil fuels, is used on site and hence has practically no important distant impact. Drilling a geothermal well requires disturbing approximately two hectares of the land surface, but once the operation is completed only about $100 \, m^2$ need be involved further. To obtain 100 MWe, 15 or so such wells are necessary, so a relatively small area is utilized. The most obvious environmental change lies in the network of production pipelines, but often this can be minimized by using appropriate colours to camouflage them. Sometimes the pipelines can be raised so as to permit continued farming or grazing underneath them: this has been done at Larderello for many years. Occasionally, the geothermal facilities constitute a tourist attraction, as is the case at Wairakei.

TABLE 6.9

The non-condensable gas production of a typical fossil-fuel plant and a geothermal plant (The Geysers) in mg/kg day MWe

	CO_2	H_2S (abated)	SO_2	S (content)
Coal (1% S)	18 000	—	127	64
The Geysers	780	25	—	23·5

The effects of geothermal energy on the atmosphere may range from practically none where closed-loop circulations or systems with very low gas contents are involved, e.g. at East Mesa, to moderate where there is a higher content of non-condensable gases, e.g. at Cerro Prieto, The Geysers, Larderello and Wairakei. The data in Table 6.9 from R. G. Bowen in 1973 are relevant.[64]

It may be noted that the geothermal plant yields less than 5% as much carbon dioxide and only a little over one-third as much sulphur as the equivalent steam plant burning coal. Where fluids are re-injected, there will remain almost no chemical impact on the surface and groundwater, thermal loading being the sole residual effect. Thus, if cooling towers are employed an effect upon the atmosphere will result, and where rivers are used in preference to evaporative cooling, for instance in New Zealand, the surface water will be affected. Where such extra heat causes eutrophication or plant growth, as it seems to do in New Zealand, cooling towers are preferable.[2]

Among other factors consequent upon geothermal energy development is that of noise, levels of 80 dB resulting from drilling in dry-steam areas. This is as much as the noise emitted when a jet aircraft takes off. Mufflers are used to reduce the sound level, but nevertheless residual sound can carry for kilometres and this could be a nuisance in populated areas or places utilized for recreation. One advantage is that the direct venting of geothermal wells occupies only a few days annually once production commences.

Geysers emit noises also. In this regard, it is interesting to quote from the description given in 1837 by Krug von Nidda of the ground noises of the Great Geyser in Iceland while he was standing about 60 paces away from it.[65] He wrote that

> We heard a dull thunder-like noise under our feet, which soon became louder, and was changed to sounds resembling a series of shots, and following each other in rapid succession. The earth experienced a trembling movement; I hastened out of my tent and saw great masses of steam bursting from the interior of the geyser, and the water of the spring thrown out to a height of 15 to 20 feet. This agitation of the geyser lasted scarcely a minute, and the usual perfect tranquillity was then restored... This... was one of the smaller eruptions... the roaring noise resounded from beneath... 12 to 15 formidable thundering reports followed, during which the ground was violently agitated by a vibrating movement. I hastened from the edge of the basin for it threatened to burst asunder under my feet.

At that time, the geyser erupted every two hours or so, but it is now dormant, although the ground surrounding it goes on reacting to events below the surface. A seismometer placed at the lip of its basin showed a slow undulatory heaving motion with a time constant of 10–15 s and a surface velocity of about 1×10^{-6} cm/s. This is most likely the result of forces associated with periodic upwelling of water in the geyser basin at the footings of the cone.

Investigations were made at Old Faithful, the famous geyser in the Yellowstone National Park in Wyoming, and here the same type of motion was recorded, each individual pulse lasting for about 10 s and the seismic signals comprising a short series of long-period movements enduring for approximately a minute. These relate to movements of the subterranean water which quickly fills the geyser within some 20 min or so. The first series is succeeded by another one and later the effect ceases. Sometimes the motion is accompanied by an audible booming sound. Then short (0·25 s) high-frequency (20–50 Hz) pulses of seismic energy start appearing at a repetition rate of *ca.* 50/min, some being generated by the collapse of steam bubbles forming in warmer lower waters and rising into cooler upper ones, and others arising from the actual explosion of steam bubbles as they attain the surface. It has been seen that the quiet venting of steam from the opening of this geyser after an eruption is replaced by a continuing set of intermittent and vigorous puffs of steam. The temperature measurements demonstrate that there is considerable circulation after the high-frequency seismic signals commence, especially at the lower levels.

Riverside Geyser in the Upper Geyser Basin in the same National Park is the most regular in Yellowstone. There are about 6 h between every eruption and this interval never varies by more than half an hour either way. It erupts as an inclined jet 25 m high from a small hole in a sinter pad lying on the east bank of the Firehole River some 2 km downstream from Old Faithful. It plays to maximum height for about 5 min and then dies down slowly, stopping entirely in another quarter of an hour. Like Old Faithful, it is a columnar geyser, i.e. one displaying a cone or protuberance above its narrow sub-surface tube, which is filled with water and emptied partly or completely during an eruption. It differs in that its pulses contain higher frequencies (50–60 Hz, occasionally 80–90 Hz) and the amplitude of the pulses varies characteristically by more than ten-fold throughout the time interval between eruptions.

The rate of occurrence of such pulses varies considerably from geyser to geyser and also during the time intervals between eruptions of a particular geyser. Usually, the seismic behaviour can be correlated with some specific and observable geyser action. However, Lone Star Geyser is an exception. Boiling and eruptions produce the signals at Bead, Jewel, Plume and Seismic; one-sided impulsive-type seismic pulses recorded at Old Faithful, Riverside, Castle, Plume and White Dome are indicative of ground heaving and breakage plus falling which are connected to mass movement of water. The most useful signals are derived from Old Faithful, where there is a strong correlation between the time of onset of seismic activity and the length of time to the next eruption. Some relevant information is shown in Table 6.10.

Reyholtshver Geyser in Iceland is smaller, but generates similar high-frequency seismic pulses to those occurring at Old Faithful. Almost every 9 min it erupts from its rather small standpipe and its seismic signature comprises a set of short (0·5 s) bursts of high-frequency 20-Hz pulses, spaced at *ca.* 3-s intervals. They start immediately after an eruption and abruptly cease after 2 min, commencing again after the next eruption 7 min later.

TABLE 6.10
Frequency content and magnitudes of seismic pulses generated by selected geysers in Yellowstone National Park, Wyoming, USA[66]

Geyser	Predominant frequency (Hz)	Magnitude of surface velocity (10^{-3} cm/s)
Old Faithful	30–50	2·5
Riverside	50–60	1·8
Castle	20–40	0·1
Bead	30–40	2·0
Plume	50–60	2·5
Jewel	6–10	1·3
White Dome	15	2·5
Seismic	50–60	5·0
Lone Star	30–40	1·3

Also in Iceland, Strokkur has a seismicity characterized by a series of four or five short 1–2-s bursts of oscillations following an eruption. The first one takes place *ca.* 9 s after the eruption and the rest follow at 2- or 3-s intervals. They are probably associated with the abrupt refilling of the subterranean reservoir by the plunging down of near-surface water to produce a hammer-like effect. The main ground motion is upwards. Other Icelandic geysers include South Geyser, Beowawe and Uxhaver in the northern part of the country.

South Geyser has an opening which is 30 cm in diameter and enters a shallow 4 m-diameter catch basin. During eruptions which take place every 4 min, the water is emitted to a height of ~2 m for about a minute with much overflow. Eruption is succeeded by a sudden upward movement of the ground lasting about a second, stops and recommences—the total upward movement endures for about three-quarters of a minute. The upwards displacement and the associated later partial recovery are perhaps the result of downwardly rushing water refilling emptied subterranean cavities, the first sudden motion being caused by the initial slap of returning water against the roof of cavities. Uxhaver is different in that, a few seconds after erupting, the ground starts to subside at a velocity of approximately 2×10^{-4} cm/s compared with the rise in the South of 8×10^{-6} cm/s. The subsidence goes on at this constant rate for a time interval ranging from 1 to 20 s, averaging *ca.* 8 s. After some eruptions, subsidence stops for a short while, sometimes more than once. Where it continues for a longer period of time than 8 s, a more prolonged stoppage takes place and may last as long as 5 s. In any case, the ground surface ultimately stops moving downwards and starts to rise, stabilizing after a few minutes. This subsidence is probably caused by collapsing of the reservoir emptied during the eruption and the recovery by the refilling of the reservoir.

Subterranean environments are affected by geothermal factors as well as those above ground; an excellent instance of this is the hydrothermal metamorphism recorded from the Larderello field in 1980 by G. Cavarretta and others.[67] Study of this phenomenon can give valuable information on the nature of the interactions between geothermal fluids and host rocks. Ultimately, permeability may be lost or at least reduced by mineral deposition, i.e. self-sealing, and, in New Zealand, a close relationship has been found to exist between hydrothermal minerals and permeability, for instance by A. Steiner in 1977.[68] Interestingly, although Larderello is generally accepted as belonging to the vapour-dominated geothermal system category, G. Marinelli in 1969 suggested, following a pioneering study of hydrothermal minerals there, that it was preceded by a hot-water system.[69] The hydrothermal mineral assemblages in rocks underneath the current productive horizon (the Triassic *Calcare Cavernoso* Breccia) may not be stable at the measured or inferred well temperatures. Indeed, E-an Zen in 1974 claimed that one may not assume that metamorphism is still active in geothermal areas.[70] Also temperature measurements for some of the wells were not available in 1980, e.g. that of Larderello Profondo.[67] However, there is no apparent discrepancy between the hydrothermal mineralogy and the thermal conditions of the field and the mineral assemblages demonstrate a temperature range close to measured values or those inferred from isotope geothermometry.

Geologically, the region comprises several tectonic units. At the top, there are Neogene sediments followed by Ligurian Nappes (flysch with occasional ophiolites) and Tuscan Nappe (Mesozoic limestone, Cretaceous marlstone or *Scaglia* and Oligocene sandstone or *Macigno*). The base of the last is made up of the gypsum–anhydrite-bearing *Anidriti di Burano* and/or the *Calcare Cavernoso* alluded to above. Most of the the steam production is from wells tapping this particular horizon and deriving from its high permeability caused by intense fracturing. Below there is the Triassic and beneath that a Palaeozoic Complex.

Rocks from the deepest geothermal wells show high-temperature, low-pressure parageneses developed in the course of a post-tectonic thermal event of uncertain age and origin. Good agreement was obtained between measured temperatures and those deduced from some hydrothermal parageneses, e.g. pyrite + cubanite + pyrrhotite + chalcopyrite. On the whole, the hydrothermal mineral assemblages can be related to high-temperature conditions. Hence the mass transfer process from which the hydrothermal phases originated was the last recorded metamorphic event in the rocks which bear the circulating fluids. From this it is taken that the hydrothermal mineral assemblages seen in the geothermal field today are still in equilibrium. It is presumed that the hydrothermal metamorphism in question is related to a convective hot-water circulation which occurred in the system before exploitation commenced. Nevertheless, there is also the possibility that water or brine could exist in a zone below the vapour-dominated reservoir. At present, water at Larderello is to be found in fractured basement around the major producing horizons near the peripheral charging areas and hydrothermal metamorphism is thought to be still active in those zones where appropriate temperatures are attained.

As many of the hydrothermal minerals contain iron, this element has to be supplied in both oxidation states (II and III) and, because of the abundance of ilmenite as a primary mineral in the metamorphic complex, G. Cavarretta et al. in 1980 hypothesized the following reaction:[67]

after ilmenite

$$TiO_2 + CaCO_3 + H_2SiO_4 \rightarrow CaTiSiO_3 + H_2CO_3 + H_2O \qquad (6.2)$$

leucoxene   calcite   sphene

$(H_2CO_3 = H_2O_{(1)} + CO_{2(1)})$. The inclusion of dissolved silica is based upon the presence of hydrothermal quartz and represented above as H_2SiO_4. Bivalent iron may derive from the hydrolysis of magnetite (after ilmenite), thus

$$Fe_3O_4 + 6H^+ \rightarrow 3Fe^{2+} + 3H_2O + \tfrac{1}{2}O_2 \qquad (6.3)$$

Some say that this reaction has a log K too low to be applicable to natural processes and they substitute solvation reactions which involve the presence of chlorine.[71] A typical one is as follows:

$$Fe_3O_4 + 6H^+ + 3Cl^- \rightarrow 3FeCl^+ + 3H_2O_{(1)} + \tfrac{1}{2}O_2 \qquad (6.4)$$

This may well fit the hydrothermal process at Larderello because up to 6000 ppm of Cl^- have been found in some undiluted deep (1000 m) waters there.

The existence of chlorine in the geothermal field probably relates to the evaporitic formations of Triassic and Palaeozoic ages such as the *Burano*, *Calcare Cavernoso* and *Boccheggiano*, although chlorides have been reported only rarely from them; for instance, sodium chloride was reported in 1955 by L. Trevisan.[72] Pyrite may have formed in the following manner:

(i) Up to 250°C:

$$FeCl^+ + 2H_2S_{(aq)} + \tfrac{1}{2}O_{2(g)} \rightarrow FeS_2 + 2H^+ + Cl^- + H_2O_{(l)} \qquad (6.5)$$

(ii) Up to 350°C:

$$FeCl_2 + 2H_2S_{(aq)} + \tfrac{1}{2}O_{2(g)} \rightarrow FeS_2 + 2H^+ + 2Cl^- + H_2O_{(l)} \qquad (6.6)$$

Of course, pyrrhotite and other sulphides or copper, cobalt, arsenic and zinc could have originated in an analogous fashion.

6.5 PLANT AND ANIMAL LIFE

Plant and animal life often luxuriates in geothermal areas, particularly around hot springs with nitrogen in their waters and geysers. On the other hand, where sulphated springs and fumaroles are concerned, the reverse is true. In the Yellowstone National Park in Wyoming, in the spring and early summer prolific yellow monkey flowers and the lily-like yellow fritillary abound, and indeed vegetation can almost cover vents where the water temperature in the spring is sufficiently low. Of course, where hotter water is involved, e.g. when a new spring forms or an existing hot spring overflows and changes the course of its channel, existing plants can be killed. A brilliant display of colours around hot pools and geyser run-off channels results from the activities of living organisms adapted to extreme temperatures and pH values of hot spring waters. Some of these temperatures are listed in Table 6.11 in reference to alkaline springs in the Yellowstone National Park.[8,73]

Although the pH of a hot spring is usually quite constant in value, its temperature is not: this is because of cooling as water flows out and down drainage channels or spreads towards shallower peripheries. Obviously, many organisms adapted to higher temperatures will grow near the water source whereas others, preferring lower temperatures, will flourish in the cooler areas around the spring. As the centre of a drainage channel of a hot spring or geyser can be at maximum temperature for life, i.e. around 73°C, it is often white in colour because of bacteria living there. Numerous springs are rimmed with blue–green algae and it may, rightly, be inferred that the different colours represent defined temperature zones. Higher-temperature organisms are white and, as temperatures decline, colours change through yellow, orange, red to dark green. Colour masses are caused by aggregations of unicellular bacteria and blue–green algae, pigments in each individual cell contributing the colour. By far the most common colour is

TABLE 6.11
Biologically significant temperatures

	Temperature (°C)
Maximum temperature for bacteria	≥ 100
Temperature at which water boils at sea level	100
Temperature at which water boils at Yellowstone National Park	93
Maximum temperature limits for	
Blue–green algae	70–73
Fungi	60–62
True algae	55–60
Protozoa	56
Crustacea	49–50
Mosses	50
Hot-spring flies	45–50
Higher plants	45
Vertebrata (frogs and fish)	38

green due to chlorophyll which exists in all algae and many bacteria. Some algae also contain phycocyanin which absorbs sunlight at a different wavelength from chlorophyll. The combination of chlorophyll and phycocyanin produces the characteristic colour of blue–green algae. Other pigments which may be found include carotenoids, yellow, orange and red in colour, present in all blue–green algae and some bacteria. If there is a deficiency of chlorophyll, they will give their colour to the relevant algae and bacteria. There is one type of bacterium, a carotenoid orange–red one found in alkaline springs, which possesses a special variety of chlorophyll and this demonstrates that it derives part of its energy from sunlight.

Practically every neutral or alkaline spring contains bacteria which sometimes grow in boiling water and can double their numbers in a couple of hours. These organisms, unlike algae, do not form mats, but normally cling on to sinter walls and pebbles. Between 50 and 60°C they flourish with algae and can create a network entrapping the former and holding them in position in mats which can attain thicknesses as great as 5 cm in channels even where there is constant flow.

Animals found around hot springs at Yellowstone include the buffalo and the elk. They feed on the algae and small plants when their usual source areas are snow-covered in winter, and no doubt they enjoy the warmth as well. Extremely numerous are the non-biting flies *Ephydra bruesii* which occur all over the world and inhabit cooler waters of alkaline hot-spring drainage channels. They lay vast numbers of orange–pink eggs above algal mats on pebbles and stones; these eggs are very colourful during winter, spring and early summer seasons. These flies eat algae and bacteria. The similar but larger colichopodid fly eats the ephydrid ones as adults and also as eggs and larvae. Other predators of the ephydrids are dragonflies, spiders, tiger beetles, wasps and killdeer. Another indigenous fly at Yellowstone is *Parocoenia turbita*, which buries white eggs in algal mats where they almost disappear from view. These flies swarm on the warm algal mats during cool summer nights and in the early morning. They have an interesting habit. They overcome the restrictions of their maximum survival temperature of 43°C by enclosing themselves in an insulating air bubble so that they can look for food in waters hotter than this. Their larvae can be very abundant, reaching numbers as great as $100\,000/m^2$ in optimum temperature conditions (around 30°C). The egg hatches within a day and the emergent larva consumes part of the algal mat. After a week it becomes a pupa, and an adult fly a few days later.

There is a great difference in plant and animal life found around acid springs: it is much less abundant and cannot sustain such high temperatures. The algae have a more complex cellular structure including a distinct nucleus and can exist in highly acidic waters. However, their upper temperature limit is only 56°C.

6.6 ENVIRONMENTAL ISSUES

Interest in what may be termed the integrative energy sources, namely geothermal together with solar, wind, biomass, bituminous shales, etc., was stimulated by the Arab oil embargo of 1974, the accompanying energy crisis of the early 1970s and the Three Mile Island disaster of 1979, but subsequently diminished when the price of oil fell drastically thereafter. Although these sources might have been expected to benefit again in the aftermath of the Chernobyl catastrophe of 1987, no significant positive effects have been observed. On the other hand, as M. Fanelli indicated in 1986, the geothermal sector has not suffered any important reverses to that time either.[74] Of course, the volatility of the international oil market could involve geothermal resources at any time and cause projects in their early phases to be cancelled and much accumulated experience to be lost. All human activities entail risk and this is true of both the renewable and conventional energy technologies. Harm may be caused to human health, property can be damaged, there may be an impact on industry or on society as a whole, aesthetic values may be offended and the eco-system may be disrupted.

Particular aspects of environmental impact that affect renewable energy technologies include the possibilities that environmental issues will curtail potential large-scale technological development and that they will increase the associated costs. Actually, the renewable energy sources concerned have relatively low energy density compared with fossil fuels and nuclear power. The inference is that they require large-scale operations in the form of, for example, numerous geothermal stations. But fossils-fuel combustion and nuclear power release far more pollutants (some radioactive) to the environment, necessitate mining and fuels extraction, and entail high concentrations of energy with concomitant possibilities of serious accidents such as gas explosions, oil fires and radioactive hazards. However, renewable energy technologies often require significantly larger tracts of land and may involve using large quantities of materials in their construction.

Consequences of geothermal energy resources development such as landslides, seismicity and subsidence induced using well-established methods have been discussed above in Section 6.4. At present, it is difficult to define the environmental results of geothermal HDR because it is still in the teething stage. There might be a problem with visual intrusion of cooling towers in the case of geothermal HDR electricity stations, but heat-only ones seem to be unobjectionable. The factors which could restrict exploitation of geothermal HDR fall into two categories. One relates to the relevant doublets, the hot-water gathering mains joining doublets and the mains distributing hot water. The second is connected with the number and location of power stations. There is also the potential noise problem during constriction and here the much longer drilling times required play a key role. The difficulty can be solved by soundproofing drilling rigs and double-glazing nearby properties. In the event that HDR were really to 'take off', many cooling towers would be vital and the degree to which they would intrude upon landscapes could become a bone of contention. In addition, the water consumption at each such plant could be great, and major losses could stem from evaporation in the towers. Actually, the cooling water needs for a geothermal HDR station could be as much as seven-fold those for an equivalent conventional station; hence such facilities will probably have to be sited near appropriate sources such as the sea, estuaries, rivers and lakes. There is an alternative, namely the dry cooling tower system, but this is not considered economically viable. However, if the geothermal HDR resource were to be utilized for directly supplying heat to cities, the problems would not arise and the operation of this approach would be just as benign as geothermal aquifers.

Another associated hazard that may be mentioned is heat rejection to the atmosphere. Should the ten 1000-MWe geothermal electricity-generating plants contemplated at Imperial Valley in California actually be installed, then the total of such heat over the envisaged 2500 km^2 which would be occupied might reach as much as 5% of the solar energy impinging upon the region. Naturally, such an appreciable increment of heat might well influence significantly the local pattern of weather.[8]

Public and occupational health and safety risks include injuries and illnesses, and no energy source can be considered to be free of such potential hazards. In the case of renewable energy sources, some risks have been both identified and quantified by studies in the USA. Naturally, the objective work involved sometimes provides a perception differing from that of the general public. In practice, occupational accidents are the most easily assessed and, in all cases studied to date, the risks from renewable energy sources fall within a range applicable to conventional energy technologies. Indeed, they may well be modest compared with those associated with coal and oil extraction and processing. Usually, they can be minimized by known technical and managerial measures. It is a pity that most of the evidence regarding renewables comes from the USA because a corollary is that some of their data may not be applicable elsewhere in places with different circumstances.

It is appropriate to consider some legislative aspects as these may affect exploitation of geothermal energy. Curiously, in the UK there is no overall legal framework, nor are there statutory provisions regulating the search for such energy. This causes uncertainty over actions permissible in terms of public law such as Town and Country Planning Acts and private law such as property rights. It is merely agreed that heat is a 'fugitive substance' and therefore it cannot be owned. Since there is no defined ownership of exploration rights, this inhibits a normal drilling industry involvement, entailing funding subsequent work on the basis of successful initial wells.

J. D. Garnish and others in 1986 suggested that, since there is no specific legislation on geothermal energy in the UK, commercial exploitation may be affected by existing laws relating to other substances from which energy is derived, to mineral exploitation or to the extraction and disposal of water.[75]

The relevant features of these laws seem to include the following important aspects. As regards other energy-producing substances, coal and petroleum products are state-owned under the Coal Acts 1937 and 1946 and the Petroleum (Production) Act 1934, while 'atomic' minerals such as uranium ores can be purchased compulsorily by the state under the Atomic Energy Acts 1946 and 1954. Other mineral substances, except gold and silver, usually belong to the owner of the land under which they occur, although over the years these rights may have been transferred to third parties. It is illegal to tunnel to derive minerals lying under the land of a neighbour unless the tunneller has been granted specific rights to the said minerals. There are additional constraints resulting from pieces of legislation such as the Town and Country Planning Acts, and there are also controls on methods used in running such operations.

It may be of interest to mention too that, in the UK, it is lawful for an owner to extract water from boreholes on his own land even if this causes water to be withdrawn from under the land of a neighbour. However, since 1963, it has been necessary to obtain a licence from the appropriate water authority granted under the Water Resources Act 1963.

One problem which may well arise with extraction of geothermal fluids is that they often contain mineral salts of which some will derive from under the land of a neighbour. The ensuing legal position remains undefined.

Disposal of cooled geothermal fluid is covered by the Rivers (Prevention of Pollution) Acts 1950–1961, which are in the process of replacement by Part II of the Control of Pollution Act 1974. The latter prohibits the discharge of noxious or poisonous substances to rivers and streams in England and Wales, and there are similar provisions which apply in Scotland and Northern Ireland. If resort was taken to dumping on land, this would necessitate acquiring a licence under Part I of the Control of Pollution Act 1974. There is no legislation against re-injection so long as the receiving subterranean strata are not involved in the weather/water supply cycle. In practice, the Regional Water Authority would doubtless require to be satisfied that other aquifers would not be placed in hazard.

It is possible that some parts of existing mining legislation might apply to geothermal work, e.g. that necessitating notification to the British Geological Survey prior to any shaft or borehole being drilled in connection with any mineral exploration or exploitation. Strangely, other mineral legislation is apparently inapplicable; for instance, mining inspectors do not seem to enjoy jurisdiction over geothermal sites.

It has been proposed that legislation be introduced in regard to potential subsidence resulting from removal of subterranean hot water. This is because existing laws regarding brine pumping [Brine Pumping (Compensation for Subsidence) Act 1981] and coal mining subsidence [Coal-Mining (Subsidence) Act 1957] do not appear to be applicable to geothermal energy development. But the Health and Safety at Work, etc., Act 1974 does pertain to such work and is probably widely enough drawn up to include provision for inspectors.

There is absolutely no legislation at present which controls spacing of individual schemes or requires developers to utilize re-injection in order to maintain reservoir pressure. Consequently, these would have to be covered by some sort of licensing procedure, perhaps including other aspects, which in itself will necessitate a definition of and a legal framework for geothermal energy.

As for the related question of taxation, the UK Inland Revenue Department published in July 1985 a consultation document referring to *Mines and Oil Well Allowances*. The major purpose of this is to simplify the existing code and integrate it better with the general capital allowance system. Another intention is to broaden the code so as to include expenditure on 'searching for, testing or developing a source of geothermal energy'. Clearly, the proposals embodied could be extremely helpful as regards the commercialization of geothermal energy in the UK. The legal situation elsewhere is discussed below in Chapter 7, Section 7.1.

REFERENCES

1. HALDANE, T. G. N. and ARMSTEAD, H. C. H., 1962. *Proceedings of a Joint Meeting of the Institutions of Civil Engineers, Mechanical Engineers and Electrical Engineers, London*, Vol. 176 (23), 603–34. Reprinted in *Hearings before Subcommittee on Energy of Committee on Science and Astronautics, House of Representatives, 93rd Congress, September 1973*, US Government Printing Office, Washington, DC, USA.
2. AXTMANN, R. C., 1974. *An Environmental Study of the Wairakei Power Plant*. Physics and Engineering Laboratory, New Zealand Department of Scientific and Industrial Research, Lower Hutt, New Zealand, Rept No. 445, 38 pp.
3. MERCADO, S., 1976. Cerro Prieto geothermoelectric project: pollution and basic protection. In: *Proc. 2nd UN Symp. on Development and Use of Geothermal Resources, San Francisco, California, USA, 1975*, pp. 1385–98.
4. MERCADO, S., 1977. Disposiciones de desechos geotermicos. In: *Simposio International Sobre Energia Geothermica en America Latina, Guatemala City, 1976*, pp. 777–99. Instituto Italo–Latino Americano, Rome, Italy.
5. CHASTEEN, A. J., 1976. Geothermal steam condensate reinjection. In: *Proc. 2nd UN Symp. on Development and Use of Geothermal Resources, San Francisco, California, USA, 1975*, pp. 1335–6.
6. ALLEN, E. T. and DAY, A. L., 1935. *Hot Springs of the Yellowstone National Park*. Carnegie Institution of Washington, Washington, DC, USA, Pubn 466, 525 pp.
7. LLOYD, E. F., 1975. *Geology of Whakarewarewa Hot Springs*. New Zealand Dept

of Scientific and Industrial Research, Wellington, New Zealand, Informn Ser. No. 50.

8. RINEHART, J. S., 1980. *Geysers and Geothermal Energy*. Springer-Verlag, New York, 223 pp.

9. ALLEN, G. W. and MCCLUER, H. K., 1976. Abatement of hydrogen sulfide emissions from The Geysers geothermal power plant. In: *Proc. 2nd UN Symp. on Development and Use of Geothermal Resources, San Francisco, California, USA, 1975*, pp. 1313–15.

10. NEHRING, N. L. and FAUSTO, J., 1979. Gases in steam from Cerro Prieto, Mexico, geothermal wells with a discussion of steam/gas ratio measurements. *Proc. Symp. Cerro Prieto, San Diego, California, USA*, Lawrence Berkeley Laboratory, LBL-7098, pp. 127–9.

11. AXTMANN, R. C., 1975. Environmental impact of a geothermal power plant. *Science*, **187**(4179), 795–803.

12. REPUBLIC GEOTHERMAL INC., 1978. Unpublished data.

13. RUFF, R. E., CAVANAUGH, L. A. and CARR, J. D., 1978. *1977 Executive Summary, Specialized Monitoring Services*. SRI International, Menlo Park, CA, USA.

14. DORIGHI, G. P., 1978. Hydrogen sulfide emissions inventory for The Geysers power plant January, 1976–December, 1977. Pacific Gas and Electric Co., San Francisco, CA, USA, Report 420-77.119.

15. ALTSHULER, S. L., 1978. Studies of cooling tower emissions at The Geysers power plant. Pacific Gas and Electric Co., San Francisco, CA, USA, Report 420-78.104.

16. OKANDAN, E. and POLAT, T., 1986. Field development and power generation in Kizildere, Turkey. *Geothermics*, Sp. issue **15**(5/6), 725–30.

17. CHIOSTRI, E., 1975. Geothermal resources for heat treatments. In: *Proc. 2nd UN Symp. on Development and Use of Geothermal Resources, San Francisco, California, USA*, pp. 2094–7.

18. WOLLENBERG, H. A., 1976. Radioactivity of geothermal systems. In: *Proc. 2nd UN Symp. on Development and Use of Geothermal Resources, San Francisco, California, USA*, pp. 1283–92.

19. ELLIS, A. J., 1978. Geothermal fluid chemistry and human health. *Geothermics*, **6**, 175–82.

20. WHITE, D. E., 1973. Characteristics of geothermal resources. In *Geothermal Energy*, ed. P. Kruger & C. Otte, pp. 69–94. Stanford University Press, Stanford, CA, USA.

21. WHITE, D. E., ANDERSON, E. T. and GRUBBS, D. K., 1963. Geothermal brine well: mile-deep drill hole may tap ore-bearing magmatic water and rocks undergoing metamorphism. *Science*, **139**(3558), 919–22.

22. CRITTENDEN, M. D. JR, 1981. Environmental aspects of geothermal development. In *Geothermal Systems: Principles and Case Histories*, ed. L. Rybach and L. J. P. Muffler, pp. 199–217. John Wiley & Sons, Chichester, 359 pp.

23. EINARSSON, S. S., VIDES, R. A. and CUÉLLAR, G., 1976. Disposal of geothermal waste water by reinjection. In: *Proc. 2nd UN Symp. on Development and Use of Geothermal Resources, San Francisco, California, USA*, pp. 1349–63.

24. MUFFLER, L. J. P. (ed.), 1979. *Assessment of Geothermal Resources of the United States—1978*. US Geol. Surv. Circular 790, 163 pp.

25. EATON, F. M. and WILCOX, L. V., 1939. *The Behavior of Boron in Soils*. US Dept of Agriculture, Tech. Bull. 696.
26. US NATIONAL ADVISORY TECHNICAL COMMITTEE, 1968. *Report on Water Quality Criteria*. Federal Water Pollution Control Administration, 234 pp.
27. HUNT, T. M., 1975. *Repeat Gravity Measurements at Wairakei, 1961–74*. New Zealand Department of Scientific and Industrial Research, Geophysics Division Rept 111, 27 pp.
28. ALLIS, R. C. and HUNT, T. M., 1986. Analysis of exploitation-induced gravity changes at Wairakei geothermal field. *Geophysics*, **51**(8), 1647–60.
29. HATTON, J. W., 1970. Ground subsidence of a geothermal field during exploitation. In: *UN Symp. on Utilization of Geothermal Resources, Pisa, Italy*. Maxwell Scientific International, New York.
30. PETERSON, N. V. and GROH, E. A., 1967. Geothermal potential of the Klamath Falls area, Oregon: a preliminary study. *Ore Bin*, **29**(11), 209–31.
31. LAIRD, A. D. K., 1973. Water from geothermal resources. In: *Geothermal Energy*, ed. P. Kruger and C. Otte, pp. 177–96. Stanford University Press, Stanford, CA, USA.
32. US BUREAU OF RECLAMATION, 1972. *Geothermal Resource Investigations, Imperial Valley, California*. US Dept of Interior.
33. TLEIMAT, B. W., 1969. Novel approach to desalination by vapor-compression distillation. *Annual Meeting of the American Society of Mechanical Engineers*, ASME Publn 69-WA/PID-1.
34. REX, R. W., 1970. *Investigation of Geothermal Resources in the Imperial Valley and their Potential Value for Desalination of Water and Electricity Production*. Inst. Geophysics and Planetary Physics, University of California, Riverside, CA, USA.
35. MCLAUGHLIN, R. J., 1978. Preliminary geologic map and structural section of the central Mayacamas Mountains and The Geysers Field, Sonoma, Lake and Mendocino Counties, California. US Geol. Surv. Open-file Rept 78-389, 2 sheets on a scale of 1:24000.
36. BACON, C. F., 1976. Blowout of a geothermal well. *California Geology*, **29**, 13–17.
37. SMITH, R. L. and SHAW, H. R., 1975. Igneous-related geothermal systems. In: *Assessment of Geothermal Resources of the United States—1975*, ed. D. E. White and D. L. Williams, pp. 58–83. US Geol. Surv. Circular 726.
38. MALEY, R. P. and CLOUD, W. K., 1971. Preliminary strong motion results from the San Fernando earthquake of February 9 1971. *US Geol. Surv. Prof. Paper 733*, pp. 163–76.
39. HAMILTON, W., 1976. Plate tectonics and man. *US Geol. Surv., Ann. Rept for Fiscal Year 1976*, pp. 39–53.
40. EVANS, D., 1966. The Denver area earthquakes and the Rocky Mountain Arsenal disposal well. *Mountain Geologist*, **3**, 23–6.
41. SWANBERG, C. A., 1976. Physical aspects of pollution related to geothermal energy development. In: *Proc. 2nd UN Symp. on Development and Use of Geothermal Resources, San Francisco, California, USA*, pp. 1435–43.
42. YERKES, R. F. and CASTLE, R. O., 1976. Seismicity and faulting attributable to fluid extraction. *Engineering Geology*, **10**, 151–67.
43. HAMILTON, R. M. and MUFFLER, L. J. P., 1972. Microearthquakes at The Geysers geothermal area, California. *J. Geophys. Res.*, **77**, 2081–6.

44. LIPMAN, S. C., STROEBEL, C. J. and GULATI, M. S., 1978. Reservoir performance at The Geysers Field. *Geothermics*, **7**(2/4), 209–19.

45. HERD, D. G., 1978. Intracontinental plate boundary east of Cape Mendocino, California. *Geology*, **6**, 721–5.

46. MCEVILLY, T. V., SCHLECHTER, B. and MAYER, E. L., 1978. East Mesa seismic study. In: *Geothermal Exploration Technology: Annual Report 1978*, pp. 23–5. Lawrence Berkeley Laboratory, University of California, USA.

47. PARKER, R. and HENDRON, R., 1987. The Hot Dry Rock geothermal energy projects at Fenton Hill, New Mexico, and Rosemanowes, Cornwall. IEE Future Energy Options Conference, University of Reading, April 1987.

48. CAMELLI, G. M. and CARABELLI, E., 1976. Seismic control during a reinjection experiment in the Viterbo Region (Central Italy). In: *Proc. 2nd UN Symp. on Development and Use of Geothermal Resources, San Francisco, California, USA, 1975*, pp. 1329–34.

49. KUBOTA, K. and AOSAKI, K., 1976. Reinjection of geothermal hot water at the Otake Geothermal Field. In: *Proc. 2nd UN Symp. on Development and Use of Geothermal Resources, San Francisco, California, USA, 1975*, pp. 1379–83.

50. RALEIGH, C. B., HEALY, J. H. and BROEDEHOEFT, J. D., 1976. An experiment in earthquake control at Rangley, Colorado. *Science*, **191**, 1230–7.

51. PATIL, D. N., BHOSALE, V. N., GUHA, S. K. and POWAR, K. B., 1986. Reservoir-induced seismicity in the vicinity of Lake Bhatsa, Maharashtra, India. *Physics of the Earth and Planet. Interiors*, **44**, 73–81.

52. YUANZHANG, D., ANYU, X. and YIMING, C., 1986. Induced earthquakes in Zhelin reservoir, China. *Physics of the Earth and Planet. Interiors*, **44**, 107–14.

53. STILLWELL, W. B., HALL, W. K. and TAWHAI, J., 1976. Ground movement in New Zealand geothermal fields. *Proc. 2nd UN Symp. on Development and Use of Geothermal Resources, San Francisco, California, USA, 1975*, pp. 1427–34.

54. GRANT, M. A. and HORNE, R. N., 1980. The initial state and response to exploitation of Wairakei geothermal field. *Trans. Geothermal Resources Council*, **4**, 333–6.

55. ALLIS, R. G., 1982. Geologic controls on shallow hydrologic changes at Wairakei field. *Proc. 4th New Zealand Geothermal Workshop*, pp. 139–44.

56. ALLIS, R. G. and BARKER, P., 1982. Update on subsidence at Wairakei. *Proc. 4th New Zealand Geothermal Workshop*, pp. 365–70.

57. ALLIS, R. G. and HUNT, T. M., 1986. Analysis of exploitation-induced gravity changes at Wairakei geothermal field. *Geophysics*, **51**(8), 1647–60.

58. LOFGREN, B. E., 1958. *Monitoring Crustal Deformation in The Geysers–Clear Lake Geothermal Area, California*. US Geol. Surv. Open-file Rept 78-597, 99 pp.

59. GERI, G., MARSON, I., ROSSI, A. and TORO, B., 1982. Gravity and elevation changes in the Travale geothermal field (Tuscany), Italy. *Geothermics*, **11**(3), 153–61.

60. CATALDI, R., LAZZAROTTO, A., MUFFLER, P., SQUARCI, P. and STEFANI, G., 1978. Assessment of geothermal potential of Central and Southern Tuscany. *Geothermics*, **7**, 91–131.

61. MORELLI, C., GANTAR, C., HONKASALO, T., MCCONNELL, R. K., TANNER, J. G., SZABO, B., UOTILA, U. and WARREN, C. T., 1971. *The International Gravity Standardization Net—1971*. Int. Assoc. Geodesy, Special Publication No. 4.

62. NUTI, S., NOTO, P. and FERRARA, G. C., 1980. The system $H_2O–CO_2–CH_4–H_2$ at Travale, Italy: tentative interpretation. *Geothermics*, **9**, 287–95.
63. PATELLA, D., ROSSI, A. and TRAMACERE, A., 1979. First results of the application of the dipole electrical sounding method in the geothermal area of Travale–Radicondoli (Tuscany). *Geothermics*, **8**, 111–34.
64. BOWEN, R. G., 1973. Environmental impact of geothermal development. In: *Geothermal Energy, Resources, Production, Stimulation*, ed. P. Kruger and C. Otte, pp. 197–215. Stanford University Press, Stanford, CA, USA.
65. VON NIDDA, K., 1837. On the mineral springs of Iceland. *Edinburgh Phil. Mag. J. Sci.*, **22**, 90–110, 220–6.
66. NICHOLLS, H. R. and RINEHART, J. S., 1967. Geophysical study of geyser action in Yellowstone National Park. *J. Geophys. Res.*, **72**, 4651–63.
67. CAVARRETTA, G., GIANELLI, G. and OUXEDDU, M., 1980. Hydrothermal metamorphism in the Larderello geothermal field. *Geothermics*, **9**, 297–314.
68. STEINER, A., 1977. The Wairakei geothermal area, North Island, New Zealand. *New Zealand Geol. Surv. Bull.*, **90**, 1–136.
69. MARINELLI, G., 1969. Some geological data on the geothermal areas of Tuscany. *Bull. Volc.*, **33**(1), 319–33.
70. ZEN, E-AN, 1974. Burial metamorphism. *Canad. Min.*, 444–55.
71. CRERAR, D. A., SUSAK, N. J., BORCSIK, M. and SCHWARTZ, S., 1978. Solubility of the buffer assemblage pyrite + pyrrhotite + magnetite in NaCl solutions from 200 to 300°C. *Geochim. Cosmochim. Acta*, **42**, 1427–37.
72. TREVISAN, L., 1955. It Trias della Toscana e i problemia der Verrucano triassico. *Atti. Soc. Tosc. Sci. Nat. Mem.*, **A62**, 1–30.
73. BROCK, T. D., 1978. *The Geysers of Yellowstone*. Colorado Associated University Press, Boulder, CO, USA, 225 pp.
74. FANELLI, M., 1986. Foreword, Proc. 1st Afro–Asian Geothermal Seminar, Chiang Mai, Thailand, 4–9 November 1985, ed. E. Barbier. *Geothermics*, **15**(5/6), 565.
75. GARNISH, J., VAUX, R. and FULLER, R. W. E., 1986. *Geothermal Aquifers. Department of Energy R&D Programme 1976–1986*, ed. R. Vaux, pp. 99–100. Energy Technology Support Unit, Atomic Energy Research Establishment, Harwell, Oxfordshire OX11 0RA, UK, for the Department of Energy, 109 pp., 6 Appendices.

CHAPTER 7

Uses of Geothermal Energy

'To see, open then thine eyes, And behold...'

SELVARAJAN YESUDIAN, 1953

7.1 WORLD ENERGY

In 1985 E. Barbier[1] gave a summary of world energy consumption from primary sources and oil equivalent tons (OET) for three years, namely 1962, 1972 and 1982 (Table 7.1). Expanding these data, details of the consumption of energy on the planet between 1925 and 1984, expressed as a percentage of primary energy sources as given by E. Barbier in 1986, are listed in Table 7.2.[4]

TABLE 7.1
Global energy consumption (million OET)

Year Energy source	1962	1972	1982
Oil	1 275	2 655	2 900
Coal	1 280	1 410	1 775
Natural gas	430	980	1 300
Hydropower	220	315	450
Nuclear	—[a]	40	225
Other[b]	65	65	50
Total	3 270	5 465	6 700

[a] Negligible.
[b] Industrial usage of wood, geothermal energy, alcohol from biomass.[2,3]

TABLE 7.2
Energy consumption on Earth (percentage of primary energy sources)

Year Energy source	1925	1938	1950	1955	1960	1963
Oil	13·0	20·3	26·5	31·7	34·6	39·7
Coal	81·3	69·7	58·5	52·1	45·7	38·1
Natural gas	3·2	5·5	9·2	10·0	12·6	14·2
Hydropower	2·5	4·5	5·8	6·2	7·1	6·2
Nuclear	—	—	—	—	—	—
Other[a]	—	—	—	—	—	1·8

	1970	1973	1974	1980	1983	1984
Oil	45·7	49·2	48·8	46·5	41·9	42·3
Coal	30·3	25·7	25·8	26·4	27·2	27·5
Natural gas	17·3	17·6	17·8	18·3	19·3	19·0
Hydropower	6·2	5·7	5·8	6·2	7·0	7·0
Nuclear	0·5	0·8	0·9	2·6	3·6 ⎫	4·2
Other[a]	NA[b]	1·0	0·9	NA[b]	1·0 ⎭	

[a] Industrial usage of wood, alcohol from biomass.[5–9]
[b] NA, not available.

Some remarks on these primary sources are appropriate. From 1963 onwards oil became the major energy source for the world, accounting for up to 40% of total energy consumption. More than 50% of known reserves are in the Middle East, but the political power accruing from these diminished after the energy crisis of 1973, consumption dropping from 49·2% then to 42·3% in 1984. No doubt oil will continue in importance at least until 2000.

However, coal represented the origin of over half of the world's energy consumption until the 1950s and was indeed the first choice both for generation of power and raising of industrial steam as well as constituting a main fuel for domestic heating and providing town gas.

Approximately half of the planetary reserves of coal are situated in Eastern Europe, the USSR and China. Coal has played a great part in transport in the form of steam-driven ships and railway engines. Its consumption steadily rises and, over the last one-third of a century, its consumption has practically doubled. However, that of other fuels increased far faster so that the percentage contribution of coal declined from over 80% in 1925 to a mere 25% or so in 1973. Attempts to revive the coal industry have not proved particularly successful and countries mining it tend to use almost all of it themselves, only 10% of annual production being traded on an international market. Among major producers, the USA gets only approximately half of its electric power from coal and in Japan the figure is as low as 4·7%. The drawback with coal is the vast investment required to exploit new mines and to continue to exploit old mines, coupled with higher transportation costs and the greater capital costs of coal-fired power stations.

Natural gas, from being relatively insignificant as an energy source in 1925 (3·2%), contributes today no less than almost one-fifth of total world energy. Nevertheless, it has until now been mostly a domestic energy source, used mainly in industrialized countries with their own reserves. The largest producer and consumer is the USSR, consumption there being approximately one-third of that of the whole world in 1984 (although the USA utilized almost as much). Eastern Europe and China are also major producers. The USSR, with The Netherlands, Norway and Canada, dominates the exportation of pipeline natural gas, liquid natural gas being exported by Algeria, Indonesia and Brunei.

Hydropower has been used for millenia in the form of the mechanical energy arising from either falling or flowing water or both, and towards the close of the nineteenth century it was applied to the generation of electricity. In 1984 it contributed 7% of the consumption needs of the world: but it may be added that in some countries it has a much greater importance. For instance, in Norway it yields 99% and in South America 55% of requirements. Although huge hydropower stations exist, mini-hydro schemes have been considered as well, these involving plants with generating capacities of only 1 MW or so. These are aimed at installation in rural areas in developing countries and, like their bigger brothers, have the advantage that they are indigenous sources of both clean and relatively inexpensive energy.

Nuclear energy is a newcomer on the world scene, having started to contribute to energy needs at the end of the 1960s and by 1983 accounting for 3·6% of global consumption. As much energy is produced from 1 kg of uranium-235 as that obtained by combustion of 2000 tons of oil (two million times more per unit weight of radioactive nuclide). In 1981 uranium production in the Free World was some 44 000 tons and reserves were estimated to be 1·7 million tons (this relating to the uranium content of ores at a production cost of under US$30/lb of U_3O_8, at 1981 prices).[10] Since the expansion of nuclear capacity after Three Mile Island and Chernobyl will be severely curtailed, as is evidenced by the fact that the number of plants ordered is, at 87, much lower than the number being constructed in 1984, namely 180, this will more than suffice. On 1 January 1984, 300 nuclear reactors were in operation throughout the world and these had a total capacity of 193 000 MWe. The 180 referred to would add another 167 000 MWe to that figure and the extra 80 a further 87 000 MWe.[11]

On the question of the lifetimes of proven fossil-fuel reserves, let it be assumed that their extraction rates remain approximately those of 1982, as listed in Table 7.3, together with known reserves. It is a simple calculation to determine that oil should last another 31 years, coal an additional 250 years and natural gas a further 60 years. As for oil production from tar sands and oil shales, the current production rate is experimental and low, but proven reserves comprise 41×10^9 OET for the former and $27\cdot4 \times 10^9$ OET for the latter. However, they are often located in unfavourable environments, and in any case the costs involved in exploiting them would be perhaps as much

TABLE 7.3
Fossil-fuel extraction rates in 1982 and known reserves

Fossil fuel	1982 production (OET)	Reserves (OET)
Oil	$2\cdot9 \times 10^9$	$93\cdot3 \times 10^9$
Coal	$1\cdot8 \times 10^9$	452×10^9
Natural gas	$1\cdot3 \times 10^9$	$79\cdot5 \times 10^9$

as an order of magnitude higher than those entailed with oil from conventional sources. It is interesting that geothermal energy average world cost has been estimated, as of 1987, as between US$50 and US$140 for the energy of 1 barrel of oil equivalent (OEBBL) on a thermal basis (1 barrel of oil $= 137$ kg $\times 10\,000$ kcal/kg $= 1\,370\,000$ kcal).[4] The greater figure is approximately the actual cost of geothermal power production in the Japanese power stations, as M. Kaneko indicated in 1983.[12] Cost of energy in developing countries is critical because, whilst in 1972 only 8% of the export earnings of oil-importing developing countries had to be expended on this, by 1978 the percentage has risen to 20%.

Table 7.4 summarizes world energy consumption in million oil equivalent tons. E. Barbier in 1986 indicated that these data demonstrate an annual increase in world energy consumption during the period 1960–73 of approximately 6% yearly, but, owing to the energy crisis, this dropped to about 3% annually between 1974 and 1980, and indeed there was no increase at all, but rather a slight decrease, between 1980 and 1983.[4] Also,

TABLE 7.4

Consumption of energy in selected years throughout the world (million OET)[2,5-9]

Energy source \ Year	1960	1961	1962	1963	1970	1971	1972
Oil	1 100	1 175	1 275	1 345	2 375	2 530	2 655
Coal	1 450	1 200	1 280	1 290	1 575	1 430	1 410
Gas	400	400	430	480	900	910	980
Hydropower	225	215	220	210	325	205	315
Nuclear	—	—	—	—	25	40	40
Other[a]	NA[b]	65	65	60	NA	65	65
Total	3 175	3 055	3 270	3 385	5 200	5 280	5 465

	1973	1974	1980	1981	1982	1983	1984
Oil	2 835	2 800	3 175	3 000	2 900	2 810	3 000
Coal	1 480	1 480	1 800	1 750	1 775	1 820	1 950
Gas	1 015	1 020	1 250	1 275	1 300	1 290	1 350
Hydropower	330	330	425	425	450	470	500
Nuclear	50	50	175	200	225	240 ⎱	300
Other[a]	55	55	NA[b]	75	50	70 ⎰	
Total	5 765	5 735	6 825	6 725	6 700	6 700	7 100

[a] Industrial usage of wood, alcohol from biomass.
[b] NA, not available.

for the decade 1963–73 energy consumption rose by a total of 70%, but for the succeeding ten years 1973–83 the increase was only 16%. The consumption of oil between 1973 and 1983 stayed the same, whereas there was a 110% increase between 1963 and 1973.

Energy consumption in industrialized and developing countries is shown in Table 7.5. From this information it is apparent that, in 1962, the OECD countries together with the USSR and Eastern Europe used 81% of total energy, a figure which rose to 82% by 1972. Interestingly, for 1982 the group utilized less (77%), so that the share of the developing countries then reached 23%. This was despite the fact that 68% of the population of the world lived in them. However, it must be borne in mind that most of the relevant countries also have much more favourable climatic conditions than the industrialized lands. On the other hand, only about one-seventh of their total energy needs are applied domestically and this is a parameter significant in terms of their standard of living. For comparison, the equivalent quantities in the USA, OECD and Europe, and Japan are 23, 27

TABLE 7.5
Consumption of energy for selected groups of countries and years
(million OET)

Countries	1962	1972	1982
OECD[a]	2 105	3 445	3 550
USSR and Eastern Europe	550	1 060	1 625
OPEC	40	100	225
Developing countries	575	860	1 300
Totals	3 270	5 465	6 705

[a] EEC and industrialized countries.

and 21%, respectively. The sectors in which the energy was consumed are given in Table 7.6.

As regards the contribution of geothermally generated electricity to world needs, the information in Table 7.7 is germane.

The globally installed geothermal electrical power resources in MW, as of the end of 1985, are listed together with the data for 1982 and an estimate for 1992 in Table 7.8.[14,15] If the information in Table 7.8 is compared with that in Table 7.7, it is obvious that geothermal energy contributes only 0·2% of global electrical power. Nevertheless, in developing countries with limited electrical consumption, its importance is much greater than this statistic would imply; for example, in the Philippines it makes up no less than 18% of the needs of that country. Apart from generation of electricity, there are few uses for geothermal energy. Therefore it is not surprising that J. S. Gudmundsson in 1985 stated that, of the 7100 million OET consumed around the world in 1984 (see Table 7.4), low-enthalpy geothermal energy is thought to have engendered savings of 2·8 million OET in total, i.e. only 0·04%.[16]

TABLE 7.6
Usage (million OET) in various sectors

Sector	USA	OECD and Europe	Japan	Developing countries
Domestic	365	305	65	115
Energy industry use and loss	510	325	95	285
Industry	255	255	90	225
Transport	435	230	65	210
Total	1 565	1 115	315	835

TABLE 7.7
Global electrical power for 1981[13]

Countries	Power (MW)
North America	700 000
Western Europe	450 000
USSR	250 000
Japan	150 000
Asia (excluding USSR and Japan)	130 000
Eastern Europe (excluding USSR)	100 000
South America	60 000
Australia and Oceania	25 000
South Africa	20 000
Africa (excluding South Africa)	20 000
Total	1 910 000
Distribution	
Developed countries	1 695 000 (88·75%)
Developing countries	215 000 (11·25%)
Total	1 910 000

In 1984 P. R. Donovan reported that, as of that year, a minimum of 40 countries were actively engaged in geothermal exploration and at least 15 were exploiting geothermal resources for the generation of electricity.[17] Another 11 were substituting it for more expensive heat sources in the fields of agriculture, domestic heating and industrial processes. Some interesting facts were revealed in the UN Seminar on the Uses of Geothermal Energy held in Florence, Italy, in 1984. For instance, the huge financial commitment of the Italian government in geothermal development, particularly at Larderello, has hardly compensated for the depletion of the relevant wells. The maximum anticipated total production over the next half century probably cannot exceed 1000 MW and only 700 MW is envisaged for 1992 (see Table 7.8). In Greece, wet-steam reservoirs have been identified on the islands of Milos (see Chapter 4, Section 4.5.1.2) and Nisyros in the Aegean Sea. There may be a potential as high as 100 MW for the former and an initial exploratory well on the latter demonstrated that a deep, high-salinity reservoir exists there and has temperatures around 350°C. In Spain, there is a National Geothermal Programme funded by the government to the tune of some US$8 million annually.[1]

The above work has stimulated international cooperation and the United Nations recognized the potential significance of geothermal energy

TABLE 7.8
The geothermal electrical power (MW) of the world

Country	1982	1985	1992
Azores (Portuguese)	3	3	?
Chile	—	—	15
China	4	14	30
El Salvador	95	95	150
Ethiopia	—	—	5
Greece	—	2	10
Guatemala	—	—	15
Iceland	41	41	71
India	—	—	5
Indonesia	30	32	144
Italy	440	520[a]	700
Japan	215	215	325
Kenya	30	45	120
Mexico	180	645	2 440
New Zealand	202	167	302
Nicaragua	35	35	180
Philippines	570	894	1 042
Turkey	0·5	20	130
USA	936	2 022	4 370
USSR	11	11	241
West Indies (French)	—	4	?
Totals	2 792·5	4 766	10 300[b]

[a] Compare also Table 7.16 below.
[b] Historically, there has been a tremendous development of the resource from the relatively minute 0·3 MW produced at Larderello in 1913, only three-quarters of a century ago.

over two decades ago, initiating the Revolving Fund for Natural Resources Exploration which is active in the field of mineral exploration projects as well as in regard to geothermal evaluation studies. It is concentrated on exploratory drilling projects at present, but it funds projects up to the discovery stage.

The member countries of the European Economic Community also undertake such work and have begun funding demonstration projects. Apart from France, they do not have specific geothermal legislation and the situation in the UK was alluded to in some detail in Chapter 6 (see Section 6.6). This is indeed curious, especially in the case of Italy, with its long history of involvement with geothermal energy. In fact, only six of the 40 countries mentioned by P. R. Donovan in 1984 have such laws.[17] Besides France, these are USA, Canada, Iceland, Philippines and New Zealand.

Some reference to the subject may be made. In the USA, a Geothermal Steam Act was enacted in 1970 to regulate geothermal work on Federal land, the individual states regulating within the non-Federal land. The first and most detailed legislation was established by California, but J. W. Aidlin in 1975 claimed that fear in America inhibits the development of the resource.[18] He emphasized that governmental institutions there were too slow in setting up appropriate criteria, probably because of potential, but unlikely, anticipated dangers.

In the case of Canada, a Geothermal Resources Act was passed in 1973 (see A. M. Jessop in 1975),[19] but the country still does not have geothermal objectives.

In Iceland, the 1968 Energy Act incorporates a 1940 Act on the Right of Ownership and Use of Geothermal Resources and, in its ninth article, states that every unit of landed property carries with it the right to possess and exploit geothermal heat from the property. However, under Article 14, the Icelandic Government can expropriate geothermal heat in the public interest, as H. Torfason indicated in 1975.[20] The National Energy Authority was established as the body responsible for supervising the national interest in exploring for geothermal resources. Electricity generation is a state operation, but geothermal district heating is a municipal one; for example, such heating is utilized by 80% of the population of Reykjavik and the town alone runs it. In fact, the 1968 Energy Act specifies that the development of geothermal resources for the space heating of communities should be in the hands only of the municipalities.

In the Philippines, Law 5092 was enacted in 1967 in order to regulate geothermal energy and its exploitation. Such energy is regarded as comprising steam and hot water and it is managed by the Director of Mines (see A. Alcaraz in 1975[21]).

Turning to New Zealand, the Geothermal Steam Act of 1953 vests jurisdiction over geothermal resources in the government and excludes waters at temperatures up to 70°C, emphasizes safety and does not impose unnecessary restraints on small-scale domestic usage. The disregarding of low-enthalpy waters is acceptable in a country where well temperatures can easily reach 300°C. Sites with such waters are quite secure against hydrothermal eruptions.[22] While on this subject, it is interesting to record that mankind put natural heat to work ever since New Zealand was first inhabited, and the Maoris have poetic legends explaining its origins. They used and still use it for cooking, washing clothes and bathing. The later European arrivals soon discovered that, even with primitive heat exchangers, they could obtain clean hot water for all purposes. Today, especially in the Rotorua–Wairakei region, there is widespread employment of geothermal heat for homes, hospitals and public buildings as well as for various agricultural industries such as pig farming and the hothouse production of plants and vegetables. A major industrial user in the Tasman Pulp and Paper Company factory at Kawerau. The volcanic regions of the central part of the North Island are highly active; in fact Mount Tarawera erupted in 1886.

It was stated earlier that France is the only EEC country with proper geothermal legislation and, indeed, during the past 10 or 15 years rising interest has been demonstrated in the subject there, particularly with regard to the hot waters of the Paris and Aquitaine Basins. The government created the Comité Technique de Géothermie in 1982 as a part of the Agence Française pour la Maîtrise de l'Energie, a body responsible for stimulating the involvement in geothermal development of local administrations, with whom it also organizes joint ventures aimed at geothermal exploitation and management.

7.2 NON-ELECTRICAL USES OF GEOTHERMAL ENERGY

Of the non-electrical geothermal energy applications listed in Table 7.9 the most important are space and process heating. Although they are examined separately below (Sections 7.2.2 and 7.2.3), it is feasible to integrate them and indeed to link both with the geothermal generation of electricity into one system. Thus, where the main objective of utilizing geothermal resources is (say) space heating, secondary electrical power generation may be feasible. Where process heating is required, depending upon the geothermal source, some associated generation of electrical power may prove to be possible and there are usually opportunities to develop various heating applications.

Some places where such applications are proceeding are cited in Table 7.10, and important additional information on planned uses in Thailand is given in Table 7.11.

The economic implications of Table 7.11 are especially important in an isolated community such as Phrao, which is entirely dependent on energy and fuel imported from Chiang Mai and Li. As regards the electrical uses, according to available geochemical data, a flash system or binary system

TABLE 7.9
Electrical and non-electrical geothermal energy applications

Temperature ($°C$)^a	Uses
Saturated steam	Process heating
180	Evaporation of highly concentrated solutions
	Refrigeration by ammonia absorption
	Digestion in paper pulp, kraft
170	Heavy water through the hydrogen sulphide process
	Drying of diatomaceous earth
160	Drying of timber
	Drying of fish meal
150	Alumina through the Bayer process
140	High-speed drying of farm products
	Food canning
130	Evaporation in sugar refining
	Evaporation and crystallization of salts
Saturated steam/water overlap	
120	Fresh water by distillation
	Most multiple-effect evaporations
	Concentration of saline solutions
	Refrigeration by medium temperatures
110	Drying and curing light-aggregate cement slabs
Water	
90	Drying of stock fish
	Space heating
80	Greenhouses
70	Refrigeration at low temperature
60	Animal husbandry
	Greenhouses serviced by combined space and hotbed heating
50	Mushroom growing
	Therapeutic baths
40	Soil warming
30	Swimming pools
	Fermentations
	De-icing
	Warm water for all-year mining in cold climates
20	Fish hatching
	Fish farming

^{a} The temperature range of conventional electrical power production is 140–180°C.

TABLE 7.10
Existing and planned process uses

Product	Country	Applications	Type of geothermal energy
Pulp, paper	New Zealand	Evaporating and digesting	Primary and secondary steam
Timber drying and seasoning	Iceland and New Zealand	Drying and seasoning	Steam and hot water
Diatomite processing	Iceland	Drying, heating and de-icing	Steam
Seaweed drying	Iceland	Drying	Hot water
Wool washing	Iceland, USSR	Heating and drying	Steam
Curing and drying building materials	Iceland	Heating and drying	Steam and hot water
Stock fish drying	Iceland	Drying	Hot water
Hay drying	Iceland	Drying	Hot water
Salts from geothermal brine	Iceland	Evaporation	Steam
Salt recovery from seawater	Japan	Evaporation	Steam
Brewing and distillation	Japan	Heating and evaporation	Steam
Boric acid recovery	Italy	Evaporation	Steam

could be employed. The latter is based upon the transfer of heat from the geothermal fluid to a working fluid with lower flashing temperature. In spite of its low efficiency (7–10%), the geothermal power plant has few operational costs and a 200–2000 kW capacity would suffice. Its output could entail using two geothermal sites which produce *ca.* 200 t/h of hot water and steam from 1000 m. Residual power could sustain centralized air conditioning which, together with food conservation, would consume large amounts of electricity, particularly in April to July with a 41·3°C maximum and an average temperature of 31·5°C. Geothermal fluid from a nearby hot spring or from the subterranean reservoir below Amphoe Phrao with temperatures exceeding 80°C could substitute cheaper means of space cooling than those using ammonia, bromide, etc., as working fluids. In fact, the centre of Phrao is near the shallow geothermal wells and this could prove valuable in developing a space-cooling facility which might include mushroom cultivation. Northeastern Thailand is famous for its mushrooms and produces over 600 tonnes annually, a figure which, although it rises yearly, is insufficient to permit exporting the fungi. The geothermal fluid of

TABLE 7.11

Geothermal fluid temperature ranges in the Phrao Basin, Thailand, usable for the generation of electricity and general servicing[23]

Temperature range (°C)	Application
30–50	Aquaculture and mushroom growing, fermentation
30–125	Animal husbandry
30–90	Washing
70–90	Pasteurization
30–80	Soft drinks
40–75	Meat
40–120	Milk and cheese
50–140	Sugar-alcohol
70–150	Vegetable oil
75–150	Beer-malt
90–150	Preserved fruit and vegetables
60–130	Heat exchange for space cooling
70–180	Electric flash
70–175	Electric binary cycle

the Phrao Basin could be applied to cultivation of raw mushrooms for the domestic market and dried ones for export. After rice, tobacco curing is the most important activity in the Phrao Basin and a number of tobacco factories are situated near the geothermal springs. The average production of each is about 80 tonnes annually and processing is effected using hot-air circulation (conventional system) and forced-air circulation (bulk system). The requisite heat energy could be supplied by hot water and switching to geothermal resources would save wood, a solution to the problem of forest preservation in the region.

Almost three-quarters of the cooperative agricultural land in the southern Phrao Basin is devoted to sugar-cane production; the vegetable is cultivated in terrace deposits with good drainage. Nevertheless, the total output is not enough to warrant a single refinery, the production having to be sent away to Chiang Mai and Lampang, 110 and 200 km distant respectively. Obviously, geothermal applications would help here. Finally, the temperature ranges of geothermal fluids can be used in aquaculture, which can involve algae, eutrophic bacteria, crustaceans and fish farms.

From this information it may be seen that space heating and greenhouses fall into the relatively low-temperature category, process heating belonging to the higher-temperature one.

7.2.1 Transmission of Geothermal Heat

Where natural steam is available, normally at pressures ranging from 5–20 atm, exploitation near the source is desirable because, owing to the high volume of steam, pipelines are expensive to construct. In fact, it may be that the economics of process-heating applications justify transporting raw materials to the steam source rather than conveying the steam to them. Steam pipelines can be made of low-carbon steel and should incorporate means of expansion. They are usually insulated with glass wool or other suitable material and may be covered with an aluminium jacket. This

TABLE 7.12

Hot-water supply pipelines for district heating systems in Iceland[24]

Characteristics	Reykir–Reykjavik		Hveravellir–Husavik supply conduit
	Main supply conduit	Older supply conduit	
Type of water	Geothermal	Geothermal	Geothermal
Length, km	13	16	19
Water temperature, °C	86	86	98
Pipeline diameter, mm	700	350 (two)	250
Capacity, tonnes/h	5 000	2 000	200
Pumping power, kW	1 800	700	Self-flowing
Material of pipe	Steel	Steel	Asbestos-cement
Insulation	Rock-wool	Turf	Earth
Temperature loss, °C	1·5	5	18
Protection method	Concrete conduit	Concrete conduit	Earth cover

makes them rather costly, another reason to keep them short. A detailed discussion of piping has been given in Chapter 5, Section 5.8.2, to which reference may be made if required.

Water is separated when the steam is wet, usually by means of special cyclones situated at the borehole, each phase often being piped separately (although two-phase flow in pipelines is possible). If long-distance transport is mandatory, only the water is sent. In Iceland, this has been done for decades, and distances up to 16 km have been involved, as B. Lindal indicated in 1977.[24] Details of some Icelandic pipelines are given in Table 7.12. It was found that dispatching heat energy as water may be cheaper than transmitting electric power, especially where high-energy loads are concerned.

Although the temperature of water in pipelines is frequently less than 100°C, it can be as high as 125°C. G. C. Bodvarsson and J. Zoëga suggested in 1964 that, under appropriate circumstances, the optimum temperature from a cost point of view may be between 169 and 180°C.[25] Such very hot-water supply pipelines may be made out of ordinary insulated steel and major lines are protected by concrete conduits. Multi-stage centrifugal pumps maintain the flow rate of the water.

TABLE 7.13
Process design features for application to geothermal steam and water[24]

Operation	Geothermal steam		Geothermal water	
	Type	Examples	Type	Examples
Drying	Indirect heating	Steam tube and drum dryers	Indirect heating	Multideck conveyer
Evaporation	Primary heat exchangers	Forced circulation evaporators	Countercurrent heaters	Preheaters
Distillation	Steam distillation	General	—	—
Refrigeration	Freezing	Ammonia absorption	Comfort cooling	Lithium and bromine absorption
De-icing	—	—	Direct application, indirect heating	Dredging aid, pavement de-icing

7.2.2 Process Heating

On this there are three basic questions, namely what are the products which can be treated using heat from geothermal fluids, what advantages does geothermal heat offer as against conventional methods and, if there is a disadvantage as regards site location, can this be offset by the lower cost of the geothermal energy involved? Table 7.10 deals with the first issue; as regards the other two, B. Lindal in 1977 was of the opinion that, since existing technology is overwhelmingly orientated towards fossil fuels, no conclusive answers can be given from current engineering practices and economics.[24] He also summarized process design features for geothermal steam and hot water, and these are given in Table 7.13.

7.2.3 Space Heating

Space heating is probably most widely achieved geothermally in Iceland, where the majority of the population have houses warmed in this way and there is a continual extension of the service. The fluids used range in temperature from 60°C to as high as 150°C. Bearing in mind the thermal characteristics of arctic, temperate and tropical parts of the world shown in Fig. 7.1, such heating is possible in many regions in the first two categories. It is acceptable to environmental agencies also because it entails no emission of smoke, and warm effluents are widely dispersed within the sewage system.

Where hot water is the medium, single-pipe systems are normally used and water is later discharged into sewage. Preferably, the temperature of the distribution water is in the 80–90°C range and cools to approximately 40°C in use. Supply mains to the distribution system usually discharge into storage tanks which aid in coping with diurnal fluctuations in hot-water load. So as to maintain enough pressure in the distribution system, booster pumping is often necessary. In towns, the distribution network may be installed beneath the streets and those mains with diameters exceeding 7·5 cm or so are emplaced in concrete channels and insulated by some suitable material such as glass wool or aerated concrete. The channels are themselves embedded in gravel together with concrete drainpipes. The minimum inclination of such channels is kept to 0·5%. At street intersections, the channels may meet in concrete chambers where valves, fastening bolts and expansion joints are located. Appropriate ventilation is

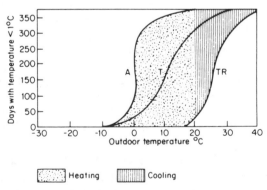

FIG. 7.1.   The thermal characteristics of arctic (A), temperate (T) and tropical (TR) climates. From: S. S. Einarsson, 1973. Geothermal district heating. In: *Geothermal Energy*, p. 129. UNESCO, Paris.

provided and drainage effected from the bottom. Where this is not feasible, a pump is used. Smaller street mains and house connections from them can be insulated with polyurethane foam and protected with a jacket of high-density, water-tight polyethylene.

The heating system in a district must be suitable for the prevailing climate, of which the most significant element is probably the daily variation to be expected in temperature. From time to time there will be an over-capacity in the heating supply, because this must be adequate to cover the coldest period of the year. However, the cost of geothermal energy for space heating is usually closely proportional to the maximum requisite capacity. Methods may be used to increase the annual load factor and optimize the heat use. For instance, the system may be designed for an outdoor temperature somewhat above that of the coldest day of the year, or the system may be constructed so as to incorporate a fossil fuel booster which can raise the water temperature during very cold spells. Another possibility is to include a local geothermal reservoir where deep pumps are installed in the boreholes. This arrangement can yield increased production over the short time intervals of coldness by pumping when the drawdown of the water reaches a prearranged level.

7.2.4 Group Heating Schemes
From the above it is clear that the exploitation of geothermal aquifer heat will depend upon local conditions, upon which the aquifer characteristics and the proposed heat load will have an important bearing. This makes it apparent that generalizations are less valuable than studies of various cases into which economic parameters and constraints can be introduced.

7.2.4.1 United Kingdom
In the UK, such work has been done in the context of the geothermal aquifer programme; the main case studies were for a group heating scheme in Southampton associated with a deep aquifer, and a hypothetical scheme at Grimsby associated with an intermediate-depth aquifer and heat pumps. British investigations showed that a typical deep-aquifer single well or doublet might be capable of supplying approximately 4 MW; this, ideally, could supply part of the needs of a heating scheme with a total load of 10 MW.

Recently, district heating in the UK was the subject of major studies and the Marshall Committee issued several Energy Papers (20, 34 and 35 cited in Ref. 26) which concentrated on combined heat and power (CHP). EP35 recommended that a lead city scheme ought to be initiated as soon as possible, suggesting also that this could require government support. Later work by W. S. Atkins and Partners reviewed the prospects for large-scale CHP district heating schemes in nine city areas, namely London, Belfast, Edinburgh, Glasgow, Leicester, Liverpool, Manchester, Sheffield and Tyneside. Using CHP would involve higher utilization of fuel than if heat production and electricity generation were to be effected separately. The 1982 Atkins Report considered all nine schemes feasible and potentially commercially viable. A significant parameter is the heat load density and a minimum peak density of 20 MW/km^2 was proposed in EP34. Below such a density the distribution costs would render a scheme uneconomical. Of course, the studies in question relate to large-scale schemes, for instance that at Leicester would entail 372 MW.

On a smaller scale, there are many group heating schemes which apply to a number of dwellings and small commercial premises. For heating homes, central-heating systems can be used: the hot water can be admitted directly to them with discharge to sewage after use. Inferential water meters with magnetic coupling between the flow sensor and the registering mechanism may be utilized. The maximum flow of hot water can be controlled by means of a sealed maximum-flow regulator, although sometimes minimum-flow ones are substituted. If the hot water cannot be supplied directly, e.g. because of high and adverse mineral contents which could promote scaling, then heat exchangers may be applied between the hot water supply and water which circulates in the central-heating system. Of course, public buildings and industrial installations may be heated similarly.

However, the group approach offers many advantages such as the higher boiler efficiency attainable by better management and control and the lower capital costs of the boilers (economy of scale); also it might be possible to employ a cheaper fuel. Nevertheless, there are some disadvantages which include the capital costs of the supply and distribution mains, heat losses from them and the expenses of heat metering (necessary to avoid heat wastage). According to J. D. Garnish and his colleagues in 1986, the Combined Heat and Power Association estimated that there are approximately 1000 group and district heating schemes currently in operation in the UK, most of them on a small scale (with the notable exception of Nottingham).[26]

7.2.4.2 Eire
Researches are going on in the Republic of Ireland and initially are concerned with making an inventory of hot springs.

7.2.4.3 France

In France, group and district heating schemes are popular, for instance at Villeneuve-la-Garenne. This followed an extensive exploration programme for hydrocarbons in the 1950s and 1960s; although it was unsuccessful in this respect, it identified vast hot-water aquifers through the 5000 deep wells drilled. A doublet scheme was commissioned at Melun in 1970, but no further progress was made until after the tremendous oil price increases of 1974 and 1975. About 74 other schemes had commenced by the end of 1984 and the work has gone on apace ever since. Thus in 1981 two more were added, seven more in 1982, 14 in 1983 and so on to the present time. The objective is to save a million tonnes of oil equivalent (Mtoe) annually by 1990, a target which seems likely to be achieved.

The French Government encourages such activities because of the low level of indigenous reserves of fossil fuels in France. Consequently, geothermal schemes receive a direct grant of 30% of the cost of the initial borehole and, if this fails, an extra 50% (so that the developer risks only 20% of the cost of this first borehole). There is an insurance coverage against possible subsequent failure of the wells, financed by an initial governmental contribution of five million francs and supported by levies on every scheme. Finally, there is also a cash grant of Fr. 400 per toe saved annually. As regards the associated surfacer work involved in such geothermal development, there are low interest rate funds available associated with HLM housing (low rent housing association).

Most exploitation has been from the Dogger Limestone of the Paris Basin, with 51 successful schemes and only two failures. After Paris, Aquitaine is next in importance and has 12 successful schemes with only a single failure. In other parts of the country, the exploration was less successful, 12 wells having been abandoned in southeast France, Alsace and Limagne. At Cretail in Paris, recently completed drilling provided the following data: depth, 1658 m; artesian flow, 210 m^3/h (58 litre/s); transmissivity, 30 D m (Darcy m); temperature, 79°C. This is average for the Dogger Formation, 36 boreholes delivering water from it at temperatures exceeding 60°C with an average transmissivity of 42 D m, most free-flowing at rates of 28–67 litre/s. This is in marked contrast with the situation in the Permotriassic sandstones there and in the UK. These are characterized by intergranular flow, whilst the Dogger aquifer is fissure-dominated.

A large-scale scheme in the Paris Basin is at Meaux, where four doublets have been drilled and there are three separate heating loads, namely a hospital, a swimming pool and domestic buildings equivalent to 14 800 dwellings. The total peak demand is 135 MW and the scheme saves some

20 000 toe annually. The Collinet doublet in the Pierre Collinet District was drilled initially in conjunction with a vertical production well so as to facilitate both logging and coring, the re-injection well being located about 150 m away on the ground surface. All the other three doublets had both wells deviated from the same drilling pads and the extraction temperature is 78–79°C. The flows are pump-boosted and amount to 250 m³/h at Collinet and the hospital, and 600 m³/h (170 litre/s) total at the two Beauval doublets. Artesian flows were 240, 202 and 190 m³/h at Collinet, the hospital and Beauval, respectively. Well depths are 1750–1760 m into the Dogger Formation and the overall cost of the scheme was Fr. 300 million with a payback period of seven years.

During 1984 and 1985, Meaux hosted a demonstration of a new French turbine-driven down-hole pump which may be capable of offering much longer maintenance-free periods of operation than the average nine months experienced with electrically driven down-hole pumps in the Paris Basin. An efficiency of 40–45% is claimed, somewhat lower than the range of 60–65% for an electrically driven pump. The new turbine-driven down-hole

TABLE 7.14

Comparison of various French 10-MW heat-capacity group heating schemes

Fuel base	Capital cost kFr	Operating cost[a] kFr			Capex/15 kFr	Total cost (kFr)
		P1	P2	P3		
Oil	4 000	4 000	165	400	266	4 831
Coal	10 000	2 000	520	1 000	666	4 186
Geothermal	22 000	1 000	700	1 000	1 466	4 166

Capital cost is divided by 15 to give an annual cost (if DCF methods had been utilized, then the corresponding figure for 5% discount rate and 20-year life would be 12·46, and 15 would correspond to a 3% discount rate, but the table is intended to provide an indication of relative values). It is noteworthy that annualized geothermal costs are close to those of coal, but it is French energy policy to lower dependence upon imported fuel, which explains the support given to possibly peripheral geothermal schemes (once these exist they benefit from later increases in the real prices of competitive fossil fuels).

[a] The French normally categorize operating costs into three sub-divisions, P1, P2, P3: P1 refers to fuel and electricity costs, in the case of a geothermal system being the cost of the electricity required to drive the extraction and re-injection pumps; P2 refers to minor maintenance, day-to-day operating staff and so on; P3 refers to major maintenance and renewals such as down-hole pumps, borehole servicing, etc. Since such contingencies are sporadic, the P3 sector accumulates a fund to prepare for them.

pump seemed good over its first year, but it is an open question whether the longer maintenance-free time interval will suffice to compensate for the lower efficiency.

Meaux exemplifies a scheme in which the aquifer temperature is high enough for no heat pumps to be necessary to boost temperatures. Where this is not the case such boosting is mandatory, for instance in some schemes in the Paris Basin and those at Aquitaine. This is so in the Paris Basin at Beauvais, where three gas-engine-driven heat pumps each of 600 kW shaft power are used and the heat load includes a hospital and is equivalent to 1500 dwellings. There the geothermal resource temperature is approximately 40°C and the peak load 10·5 MW. In the Aquitaine Basin, at Pessac not far from Bordeaux, the wellhead temperature is 47°C and the flow rate is 200 m³/h (56 litre/s).[27] The depth to the surface of the aquifer in the Cenomanian (Late Cretaceous) Formation is 825 m. This particular scheme covers 1526 dwellings with a peak heat load of 6·8 MW. Actually, 56% of the annual heat supply is given by direct exchange with the geothermal fluid, 35% from the heat pumps and 9% from peak-lopping boilers. As noted, both electrically and gas-driven heat pumps are used in France, but there is a trend towards the employment of electric drive because of the favourable electricity costs.

The economics of French schemes are summarized in Table 7.14.[27]

7.2.4.4 Belgium and The Netherlands

Both of these countries have a deep karstic-type limestone formation of Early Carboniferous age which occasionally demonstrates geothermal potential. There were two attempts to use this, one at Hoogstraten and the second at Spijkenisse, but they had to be abandoned. This was because the reservoir characteristics were unsatisfactory and also because of competing fuel costs. Recently, evidence has been presented to show that there is a widespread and interconnected karstic system in the Turnhout region. A doublet has been drilled at Merksplas and this yields more than 156 m³/h (42 litre/s) of 75°C water with a salinity of 132 g/litre. The objective is to heat a prison complex and a few glasshouses. At Turnhout, approximately 10 km away, a shallower Late Cretaceous aquifer has been tapped through a well 800 m deep and gives water at 37°C. A heating system using heat pumps was scheduled here. Across the Franco-Belgian border there is more karstic Early Carboniferous limestone and wells at Ghlin and Ghislain produced water at 71–73°C from depths of 1570 and 2500 m, respectively. Unlike the northern reservoir at Merksplas, here the salinity is low (2 g/litre). A space heating system was mooted for St Ghislain.

7.2.4.5 Denmark

There have been exploratory drillings at Aars and Farso, but they have not been successful. However, another at Thisted in the Triassic Gassum Formation appears to be more promising at some 1300 m, where the water temperature is 46°C.

7.2.4.6 Federal Republic of Germany

There are operating geothermal schemes in seven West German cities and they have a total thermal power of 6 MW. Eight more were in the planning stage in 1986.[26]

7.2.4.7 Italy

The world-famous high-enthalpy resources of this country, most prominently at Larderello, have been discussed exhaustively earlier in this book and mention has been made of low-enthalpy resources in the Po Valley. However, there are other very small-scale schemes at Metanopoli near Milan, Vicenza and Ferraro as well as in the Rome area, but there has been little operating experience gained as yet.

7.2.4.8 Hungary

As noted in Chapter 4, Hungary, located in the Pannonian Basin, has huge low-enthalpy geothermal resources and the average temperature gradient is 50°C/km with an associated heat flow of 90–100 mW/m^2, well above the world average. Limestones and sandstones are exploited, permeability in the former being fissure-controlled and in the latter resulting from intergranular connections. Carbonate reservoirs produce a number of natural springs bordering mountainous areas and indeed recharged from these features. Sand and sandstone aquifers lie beneath the Great and Little Hungarian Plains. The boundary of the Upper and Lower Pannonian is at the bottom of the useful reservoir at depths reaching 2500 m, and the relevant permeable sequences are thick. At least 1000 wells operate (100 or so of them from unsuccessful oil exploration) within the temperature range 30°C to >90°C. Some 220 are at temperatures exceeding 60°C. The low-temperature wells are applied to balneological–therapeutic purposes in the shape of numerous indoor and outdoor baths and swimming pools. Warmer waters are used mostly in agriculture and horticulture—totalling about 1250 GJ/year (equivalent to heat loads 15-fold greater than that considered at Southampton). Most deep wells are artesian with wellhead pressures of 2–3 bar and artesian flows reaching as much as 33 litre/s. There has been very little re-injection work.

7.2.4.9 Sweden

There are apparently no deep aquifers in Sweden, although the exploitation of shallow aquifers is under consideration. At Lund, there are four wells of 700 m depth and they yield water at 21°C. A 20-MW heat pump was constructed and integrated with a district heating scheme that already existed. Discharge from the heat pumps is at 4°C. It is claimed that the payback period is three years. Two comments are appropriate. One is that the heat pumps substitute for electrically heated boilers and hence it might be expected that they are more favourable economically. The other is that hydroelectrically generated electricity is extremely cheap in Sweden. In fact, this country is famous for development of very large heat pumps to be employed in district heating systems. One specification comprises 25 large-scale heat pump installations within the range 6–100 MW, with 30 MW as the largest unit size. The heat sources include lake water, treated sewage, seawater and industrial cooling water.

7.2.5 Horticulture and Fish Farming

As well as space heating and supplying domestic and service hot water to buildings, geothermal heat may be used in both horticulture and fish farming.

Glasshouse complexes up to 8 hectares in area need large quantities of heat related both to the crop and its growing season. As peak loads of as much as 0·75 MW/acre (1 acre = 0·4 ha) are involved, a complex of such a size corresponds to the output of a single geothermal well or doublet system supplying 40% of the peak load. Of course, the growing season may extend from February through to October or even longer and, whilst the interior temperature relates to the crop, it is normally around 21°C. Where exotic plants are concerned, it will be higher. In contrast to usual domestic heating, the night load will exceed the day load, so that diurnal heat storage could be cost-effective. The annual load factor of a glasshouse heat load will be approximately 30% and the load factor of the geothermal constituent may be much greater if the system is suitably sized and diurnal storage is employed.

It is possible to use a non-geothermal system. This entails a number of plain steel tubes of 50 mm diameter to which hot water or steam is supplied at around 100°C, radiation offering an important part of the heat transfer. Where the geothermal resource possesses a temperature exceeding 70°C, this approach would still be appropriate.

Attempts have been made to exploit lower-temperature geothermal resources in glasshouses, one being to employ forced-draught convector heaters. This was done at a glasshouse near the UK Central Electricity Generating Board power station at Drax. However, strong air currents of this sort may be adverse because they might transmit diseases. In France, synthetic rubber piping has been used and can provide a large heat transfer surface more cheaply than if steel piping were employed. An alternative, namely to utilize heat pumps to increase the circulating water temperature, does not appear to be economic. This is because of the cost of the equipment and the employment of dearer electricity or diesel oil as compared with coal and heavy fuel oil.

Geothermal heat is also applicable to fish farming (pisciculture), for instance with brown trout and eel. These have growth rates which decrease with rise in temperature above an optimum for each. Indeed, the fish can lose weight near the upper extremes of their respective temperature ranges, i.e. around 22°C for brown trout and 31°C for eel. The cause of this is probably increased activity and respiratory rates, but it could relate also to reduced rates of feeding at higher temperatures. Optimum growth temperatures are maintained at fish farms near power stations by a controlled mixing of suitable quantities of warm cooling water and water at ambient temperatures.

7.2.6 Heating Equipment Appropriate to Geothermal Systems

If a heating system is to be used in conjunction with a geothermal resource, it must be designed for lower flow temperatures than conventional ones, these relating to the supply temperature from the geothermal aquifer, and for maximum heat extraction from the geothermal fluid. Several approaches are feasible. Thus convector radiators of larger size or more sophisticated nature can be applied, or underfloor heating can be used. The latter has a bad reputation in the UK due to leaks in wet systems. Another possibility is warm-air heating, which is recommended by W. S. Atkins and Partners, and may offer the best means of ensuring a low return temperature. Heat pumps may be utilized in order to elevate the flow temperature and/or reduce the return temperature.[26]

J. Garnish and his associates in 1986 gave an excellent illustration regarding convector radiators, considering one supplied with water at a temperature T_F and returning it at a temperature T_R, after heating a room with a temperature T_A.[26] Even where the radiator involves two corrugated steel plates welded together, only part of the heat transfer will be by radiation, actually some 40% with an 82°C/71°C system, the rest taking place by convection. Radiators which use enhanced natural convection due to the selected geometry are available, as well as forced convector heaters, but they are somewhat noisy for domestic usage. If lower flow/return temperatures be employed, the radiation constituent declines rapidly because radiative heat transfer is proportional to the difference of the fourth power of temperature, whereas natural convection is proportional to temperature difference. Over the appropriate temperature range, the following empirical correlation pertains:

$$\text{Heat output} = \text{constant} \times (\text{temperature difference})^n \qquad (7.1)$$

in which n is usually 1·2 or 1·3.

In radiator tests, the use of the log mean temperature difference (LMTD) rather than the simple temperature difference produced a better correlation, where

$$\text{LMTD} = \frac{T_F - T_R}{\ln\left(T_F - T_A\right)/(T_R - T_A)} \qquad (7.2)$$

As well as following the relationship expressed in eqn (7.1), the heat output of the radiator is equal to the heat extracted from the water:

$$\text{Heat output} = \text{specific heat} \times \text{mass flow rate} \times (T_F - T_R) \qquad (7.3)$$

In fact, eqns (7.1), (7.2) and (7.3) comprise the basis for radiator control.

First studies of convector radiators were aimed at using water hotter than 60°C and conventional systems are designed with flow/return temperatures of 90°C/70°C or 82°C/71°C. Hoare, Lee and Partners had higher-performance radiators tested at the Building Services Research and Information Association (BSRIA) in 1981 at 69°C/30°C flow/return temperature (in comparison with a simple radiator at 82°C/71°C). BSRIA proposed that advantage should be taken of the fact that the geothermal supply was intended to cover some 40% of the peak load, suggesting that a return temperature should be chosen and the convector radiators sized so as to provide full output under flow/return conditions of, say, 90°C/30°C. If geothermal heat is intended to yield 40% of peak load, then radiators are adequately sized down to a flow temperature under 60°C. Taking the domestic hot supply into consideration complicates matters, but designing on this basis necessitates convector radiators approximately 2·5 times greater than the size for conventional heating. Using such larger radiators together with lower flow temperatures needs lower flow rates through individual radiators, and this could promote problems of dirt blocking the small apertures which would be involved. Most of the geothermal schemes installed in the UK would entail retrofitting and relevant heating equipment would have to be site-specific.

7.3 GEOTHERMAL PROGNOSIS

The development of the geothermal resource has been considerable in the last few years. For instance, M. Fanelli in 1986 indicated that, whilst in 1970 geothermoelectric power installed in the world amounted to just over 680 MW of which 385 MW was in Italy, by 1985 the figure had reached 4766 MW, a 600% increase in 15 years.[28] Progress is most impressive in the developing countries: in 1970 there were just two plants, a 3·5-MW one in Mexico and a small one of only a few kilowatts in China. By the middle 1980s no less than 1568 MW of geothermoelectric power installations exist in them, 61% of these being in Asia and 3% in Africa. The rapid increase in research and development programmes in the field of geothermal energy stimulated the demand for expertise so that an estimate made at the end of the 1960s that approximately 25 new geothermal specialists every year would be necessary has been proved inadequate. At the present time, there are four international centres where appropriate training is being given: these are at Pisa in Italy, in Iceland, in Japan and in New Zealand. The output from them is approximately 70 to 80 experts annually. In addition, there is a certain amount of training going on locally in a number of countries, but the numbers required still exceed the supply available. There is an ever-growing urgency in the matter, as may be seen from the discussion below.

7.3.1 India

A vast country such as India will need many specialists optimally to develop its geothermal resources. In this particular case, no suitable geothermal environments for the generation of electricity have been identified as yet.

However, although there are no post-Tertiary volcanic regions on the sub-continent, except for the Bay Islands, middle-enthalpy geothermal systems may be connected with anomalous thermals in the crust which occur in young granites in the higher Himalayas, in granitic associations in the lesser Himalayas, in tectonic features and lineaments as in the West Coast area of Maharashtra and at Tattapani, and in the rifts and grabens of Gondwana sedimentary basins. A suture zone in the Indus Basin of Ladakh, where there are geothermal fields such as Puga–Chumathang, has been interpreted as a plate tectonic boundary and constitutes the most favourable possibility for eventual generation of electricity. This area is a prime candidate for intensive expert investigations. Another one is the Cambay Basin in which almost 2000 m of Quaternary and Tertiary sediments occur in a graben, below which a rather young pluton lies at shallow depth (between 6 and 20 km), the former possibly constituting a reservoir and the latter a heat source. Binary-cycle power generation could be attempted here and also at Tattapani, and is planned according to S. K. Guha in 1986.[29] A drawback of the approach is that it necessitates a very large flow rate and greater supply of cooling process water. However, at least in the Cambay Basin, this would not be a great problem. The situation is different in hard rock terrains such as Puga and Tattapani. There are other places possessing reasonably high temperatures ($>75°C$) which might be utilized as pre-heater media for the economic production of process steam by conventional means.

For secondary applications of low- and medium-enthalpy geothermal fluids, the demand would have to be assessed. For instance, whilst space heating in dwellings and glasshouses may be required in the higher Himalayas, this would not be the case in the tropical peninsula region of the country. Nevertheless, in isolated parts of it, refrigeration together with other light industrial and agricultural processes could be supported locally by geothermal heat. Interestingly, it seems that the geothermal fluids in India do not possess a high mineral content and hence present no difficulties as regards corrosion and scaling. Regarding direct heat use, it was estimated in 1982 by V. S. Krishnaswamy and R. Shankar that, from 46 hydrothermal systems through a binary process, a total of 1840 MWe power could be anticipated and, from all of the 113 systems identified, a total of 10 000 MW could be extracted from such low- to medium-enthalpy fluids.[30] In addition, it is apparent that the many hundreds of Indian hot springs might be utilized in balneotherapy and tourism, an aspect so far rather neglected except in Bihar, Himachal Pradesh and the West Coast area.

7.3.2 Thailand

On the other side of the world in Thailand, geothermal development in the San Kampaeng and Fang areas proceeds apace. The former is underlain by a Carboniferous sedimentary sequence of hard white massive sandstone and thin-bedded shale with numerous quartz veinlets. This formation is folded and constitutes the basement, contacting Permian rocks centrally. The latter comprise lower sedimentary and upper volcanic rocks. Triassic porphyritic granites are exposed in the northeast. At Fang, there are complex tectonics with accompanying metamorphism and faulting with a stratigraphic sequence extending from the Precambrian to the Devonian with Quaternary alluvium. Carboniferous and Triassic granites are intrusives in the region. Thermal waters seem to arise from the same source as cold ones and the sodium bicarbonate type prevails, their salinity being low as well. The sodium–potassium–calcium and silica geothermometry indicates temperatures of 180–200°C in the main reservoir. A geophysical survey revealed an anomaly and exploration drilling was undertaken. The sub-surface thermal anomaly in the San Kampaeng area may derive from a granitic body and meteoric water is thought to penetrate to depth, where it becomes heated to 200°C or so. In 1985 a geothermal well 1227 m deep was drilled. Unfortunately, it had a bottom-hole temperature of only 99°C and could not be considered economically viable.

Future work includes a further geochemical survey (fingerprint method) to confirm the presence of faults or fracture zones, and also shallow hole drilling to measure the actual heat flow. At Fang, the shallow reservoir is believed to be restricted to the narrow zone along a fault plane and the French Bureau des Recherches Géologiques et Minières decided to use a down-hole pump for producing hot water from the available wells. An initial geothermal power plant was envisaged. More work will be necessary, for instance additional deep drilling, so as to gain a better understanding of the subterranean geology and hence to modify reservoir models of San Kampaeng, as S. Praserdvigai indicated in 1986.[31]

7.3.3 Kenya

Here, in the East Olkaria geothermal region, a detailed three-dimensional well-by-well model was developed by G. S. Bodvarsson and others in 1987, and this has been fully discussed in Chapter 2 (see Section 2.6.3).[32] It matched the flow rate and enthalpy information from all wells as well as the pressure decline in the reservoir and provided enlightenment as regards the spatial distribution of various hydrological parameters such as porosities and permeabilities within the reservoir system.

Interesting follow-up work is continuing into the capability of the model in question actually to forecast reservoir performance and that of a number of individual wells for differing scenarios of future exploitation. The researchers involved effected two simulation studies in the early 1980s: the first showed that excessive production from the steam zone would promote both localized boiling and accompanying pressure decline, so restricting the productive lifespan of the system. The second ignored the steam zone, investigating only a liquid-dominated region 550 m thick. It was demonstrated that a long-term power-generating output of 45 MWe is feasible at this geothermal field. However, there is a caveat: if the most pessimistic estimates of horizontal permeability and negligible vertical permeability turn out to be correct, then the production area will have to be expanded by approximately 2.3 km^2 by future replacement drilling. Of course, this assumes that the geothermal field is large enough.

There are three main geothermal investigation interests currently. One is to assess the response of the reservoir to power production of 45 or 105 MWe, a second to look into the effects of injection, and the third to estimate the optimum well spacing for drilling in the future. In 1987 G. S. Bodvarsson and the same group discussed aspects of the matter.[33] They recorded an existing well density of 20 wells/km^2, an average spacing of only 225 m, and also interference between wells expected to worsen through the estimated 30 years' duration of the geothermal field. Probably the well density is too high and should drop to, say, 11 or 5/km^2 (i.e. well spacings of 300 and 450 m respectively). Various performance predictions are made and the major results of the study include the observation that a well density of 5/km^2 would be too low to achieve a satisfactory recovery efficiency and would also increase demand on well field area, with a mere 25% savings in the number of development wells. G. S. Bodvarsson and the rest suggest that the optimum well density at Olkaria is somewhere around 11/km^2 and, for this value, 24 extra wells would have to be drilled over the next 30 years in order to maintain a power production of 45 MWe. Approximately 4 km^2 are required for this. To obtain 105 MWe, 9.5 km^2 would be mandatory.

An appropriate injection programme under which more than half of the fluids produced are returned to the reservoir would probably reduce the areal requirements appreciably and thus reduce the number of development wells required too. Naturally, injection wells will exert great influence on flow rates and enthalpies in adjacent wells, but they will not increase the steam rates from wells over the short term. In fact, injection will keep steam rates up over quite considerable time intervals, again reducing the number of development wells which will have to be drilled. If there are no preferential flow paths capable of causing premature thermal breakthrough, then the higher the injection percentage, the more beneficial will be the consequences of the exercise. Full, i.e. 100%, injection would cause a semi-steady-state flow field to develop quite quickly and, in this situation, very few additional wells would be needed for 45 MWe power production.

7.3.4 Santa Lucia
Unlike India, say, the tiny country of St Lucia, a Caribbean island, may become self-sufficient in energy should current testing for geothermal energy prove positive. There an exploratory well at the site of an extinct

volcano produced considerable quantities of hot water at 300°C at a depth of only 1500 m. This result is so encouraging that an electricity-generating power plant with an output of 8–10 MW might be feasible and this would entail input of appropriate expertise.

7.3.5 A Developed Country—Italy

To counterbalance this picture from four developing countries, one enormous, one relatively minute and the others medium-sized, the status of geothermal development in Italy is relevant. There have already been some 85 years of electricity generation from geothermal steam there; hence it may be regarded as among the most highly developed lands in the world from the standpoint of exploration and exploitation of geothermal energy.[34]

The first experiment took place on 4 July 1904 at Larderello, a little south of Florence, when Prince Piero Ginori Conti lit five bulbs with electricity generated by a small dynamo driven by geothermal steam. As E. Barbier indicated, those few watts have risen to about 3000 MWe worldwide today, a truly astounding development from the minute cradle village of Larderello, which was originally involved only in producing boric acid, although around it there were hot-water pools and springs used for centuries as therapeutic baths.[34] One of the most famous of these was the Morbo bath near the Montecerboli pools, now the site of Larderello, and in it many famous people took the waters, among them Lorenzo the Magnificent, the Lord of Florence.

The state electricity corporation ENEL is heavily involved in geothermal work. Using geothermal resources for industrial purposes actually commenced over 150 years ago and before 1913 geothermal fluids were employed for heat and the manufacture of chemical products. During the period from 1913 to 1965, combined chemical and electrical production was attained, and from 1965 onwards power generation predominated. Chemical production was then abandoned altogether. Between 1975 and 1984 geothermal energy was mainly devoted to generation of electricity, although some thermal projects were begun as well. B. Billi and others in 1986 gave an excellent survey of geothermal development in Italy from 1975 to the beginning of 1985 and data from this are given in Table 7.15.[35] In 1975 electricity production was as follows: installed capacity, 417·6 MWe; generated electricity, 2483 GW h/year (1·7% of Italian electric power). In 1984 the corresponding figures were 459·2 MWe and 2840 GW h/year (1·6% of Italian electric power).

Electrical power is produced from Larderello, Travale, Radicondoli and Monte Amiata in Tuscany, and Latera in Latium; some appropriate comments may be made about each of them.

TABLE 7.15

Drilling in Italy from January 1975 to January 1985[35]

Area	Jan. 1975		Jan. 1975–Jan. 1985		Jan. 1985	
	No. of wells drilled	Average depth (m)	No. of wells drilled	Average depth (m)	No. of wells drilled	Average depth (m)
Larderello	539	656	39	1 800	578	735
Travale–Radicondoli	34	644	25	1 874	59	1 165
Mt Amiata (Bagnore, Piancastagnaio, Poggio Nibbio)	60	780	12	2 796	72	1 115
Latera	—	—	11	2 130	11	2 130
Cesano	—	—	12	2 520	12	2 520
Torre Alfina	5	749	4	1 770	9	1 200
Phlegraean Fields[a]	—	—	12	2 380	12	2 380
Others[a]	12	951	13	1 860	25	1 420
Total	650	—	126	—	776	—

[a] Excludes shallow wells drilled in the 1940s in the Phlegraean Fields and those drilled in the 1950s on Ischia and near Viterbo in Latium.

At Larderello, the total area explored is approximately 250 km² and, of 578 wells drilled to December 1984, some 200 currently produce. During the time interval from 1975 to January 1985, 39 wells were drilled and 25 are productive. Geothermal fluid used here totals *ca.* 2800 t/h, usually superheated steam at pressures of 1–11 bar and temperatures between 130 and 260°C.

Some interesting recent work on helium in these fluids was published by F. D'Amore and A. Truesdell in 1984.[36] Their results showed that there has been a strong decrease in helium concentrations in Larderello steam since earlier surveys and the gas is concentrated in central zones of steam upflow. The helium decline and existing distribution are ascribed to the release of stored helium produced by production-induced micro-fracturing.

On the elementary and isotopic compositions of noble gases in the geothermal fluids of Tuscany, S. Nuti in 1984 analysed not only helium but also radon obtained from fluids from over 90 geothermal wells.[37] His results gave good correlations between constituents of atmospheric origin and those of deep origin, which implies that there is a mixing between cold waters and hot fluids at considerable depth. The helium isotope ratios indicated that a radiogenic crustal component is present. Enrichment in radiogenic argon and excess nucleogenic neon-21 was recorded as well.

Water is believed to reach the deep parts of the Larderello geothermal system so that its included atmospheric gases can mix with radiogenic ones produced in the crust and also with the carbon dioxide released by the hot rocks and the deep gases from the mantle. Excellent correlation between constituents of atmospheric and deep origins suggest that the mixing between cold water and the hot geothermal fluids cannot take place at shallow depths and the process itself is confirmed by extremely strong correlations between helium and nitrogen, and helium and methane. High He/Ne ratios exclude the existence of a large contribution of atmospheric helium. Helium isotopic ratios reveal a crustal radiogenic component and also a ^{3}He-enriched component which is probably of primordial origin.

At Larderello, the installed capacity is 384·7 MW. After 1975, drilling has been mostly peripheral; this aids in defining the limit of the main reservoir in carbonate rocks. Through the years in question, average depths of wells have increased because of efforts to locate permeable horizons in the metamorphic basement down to about 4 km. Re-injection is proceeding also and, by 1981, about 600 t/h of condensate was being disposed below ground. In addition, old generator units were modernized.

Very interesting work on the age and duration of geothermal anomalies involved the Kawerau geothermal field in New Zealand and similar studies have been done at Larderello as well (see P. R. L. Browne in 1979[38] and A. Del Moro and his co-workers in 1982[39]). In the latter case, rubidium–strontium and potassium–argon ages were obtained on six biotites, two muscovites and a hornblende, metamorphic minerals sampled from cores of micaschist, gneiss and amphibolite of Early Palaeozoic to Precambrian age from a depth greater than 2 km in basement from the Larderello–Travale geothermal region. Mostly the data cluster within the range 2·5–3·7 Ma and reveal the occurrence of a Pliocene thermal event which probably originated the geothermal field. Consequently, the duration of the Larderello field is longer than was anticipated. In the basement levels of the two wells examined, unstabilized minimum temperatures of 290 and 380°C were recorded and at these all biotites demonstrate practically complete argon-40 and strontium-87 retention. Actually, the petrological evidence (stilpnomelane stability) and empirical information from activation energies and diffusion coefficients also favour a closure temperature exceeding 400°C for rubidium–strontium and potassium–argon in biotites, this according with recent direct experimental assessments. For the past 3 Ma the mean geothermal gradients of 120–150°C/km have been evaluated in the first 2 or 3 km and 60–65°C/km in the underlying 2 km. An approximate evaluation of the total cooling in the last 3 Ma gave a value of 120°C at 2·5 km depth and 50°C at 4 km depth in Sasso 22 well. Most of the samples cored derived from this well, although two came from the

Sperimentale Serrazzano well. A mean rate of uplift of *ca.* 0·2 mm annually was calculated independently.[39]

As regards mineralogy, the formation of authigenic minerals and their uses as indicators of the physico-chemical parameters of the geothermal fluids in the Larderello–Travale geothermal field were investigated by G. Cavarretta and others in 1982.[40] Various assemblages of authigenic minerals with differing compositions were obtained from a number of areas of the Larderello–Travale geothermal field, the most common in the peripheral area being quartz, calcite, chlorite and sulphides. They are believed to have formed in a temperature range of 150–250°C and with a P_{CO_2} range of 0·1–1 atm. In the hotter parts of the field in question, at depths between 200 and 1300 m, i.e. in the zone of maximum productive fractures, the most important authigenic minerals are quartz, chlorite, calcite, potassium feldspar and potassium-bearing mica with which wairakite, haematite, anhydrite and barite are occasionally associated. The deposition of all these minerals takes place in a temperature range of 200–300°C under P_{CO_2} of approximately 1 atm and at relatively high P_{CO_2} in geothermal fluids; this is sometimes characterized by silica over-saturation. Beneath the 1300 m depth in the hottest parts of the field, potassium feldspar, chlorite, epidote, sulphides with minor sphene, prehinte, clinopyroxene and clinoamphibole occur. These minerals form within a temperature range of 250–350°C under a P_{CO_2} under 1 atm and a P_{CO_2} which ranges from 10^{-40} to 10^{-33} atm and reasonably reflect the physico-chemical parameters of the geothermal fluids prior to exploitation.[40] Relic wollastonite and andradite provide evidence of an earlier hydrothermal stage marked by higher temperatures.

Travale–Radicondoli lies some 15 km east-southeast of Larderello. The area explored here covers about 30 km² and 14 wells produce superheated steam while others yield a hot water and steam mixture unsuitable for the economical generation of power. Now the installed capacity is 48 MW and the main reservoir has characteristics similar to those of that at Larderello. Deep exploration demonstrated that permeable horizons exist also in the metamorphic basement at depths of 2·5–3·5 km. Well testing here was reported by P. Atkinson and others in 1978.[41] They pointed out that the reservoir is geologically and hydrologically complex, lying on the edge of a graben near an extensive outcrop of the reservoir rocks which constitutes a recharge area.

The rocks in question are a 'basement' Palaeozoic Complex composed of phyllites and quartzites with interbedded saccharoidal limestones and anhydrites, and a mainly carbonate complex including anhydrites, dolomites and stratified limestones (Noric to Rhaetic in age), massive limestones, cherty limestones, marls and radiolarites of Jurassic age. The explored region is divisible into three zones. In one, that nearest the recharge area, a few non-commercial wells produce two-phase water/steam mixtures. In the second, the wells yield superheated steam, whilst a well drilled within the graben gives a geothermal fluid with a non-condensable gas content of some 80%.

The performance history before the arrival of pressure transient studies relates mostly to the 'old' Travale reservoir which is situated southwest of the 'new' Travale–Radicondoli reservoir in which the newer wells are drilled and in which modern well test analysis technology has been utilized. The most important well in the 'new' reservoir is Travale Well 22, which has been thoroughly tested. However, almost all of the other wells in the 'new' reservoir became involved through well-interference tests. In these, those around Travale Well 22 are shut in and their pressure responses for different Travale Well 22 production rates measured. Well-interference tests show the characteristics of fluid flow in the reservoir between well tests, and qualitatively illustrate the heterogeneity of the said reservoir. The main results of the research demonstrate fracturing at various levels, the minimum distance from the plane of fracture to any boundary being at least 2 km (although this may have to be modified); the direction of the fracture is N 73° W and the whole reservoir could be faulted in the Apennine direction.

In 1979 the dipole electrical sounding method was applied by D. Patella and his associates.[42] This approach was based upon the dipole technique which allows investigations to be effected satisfactorily at very great depths. The impermeable cap rocks were found to propagate heat by conduction, with temperatures ranging from normal ambient ones at the surface to a maximum of 150°C at 1·5 km depth, assuming a geothermal gradient of up to 10°C/100 m. The zones corresponding to the 'old' Travale part are different in having a geothermal gradient as high as 30°C/100 m. Here cover rocks can have temperatures as high as 200°C. The temperature distribution within the carbonate formations is governed by a more complex mechanism, namely the circulation pattern of hot saline geothermal fluids and cold meteoric waters. The resistivity values exceeded 1000 Ω m for cold, outcropping carbonate rocks, ranged from 20 to 100 Ω m for permeable hot rocks including intense circulation of hot saline geothermal fluids, and probably vary from 100 to 1000 Ω m for the deep carbonate rocks with high temperatures, low fluid contents and low permeability. The metamorphic complex is characterized by low permeability and low fluid contents, and the geothermal gradient is different. The average resistivity is much higher than the values found in overlying rocks.

The nature and tectonic setting of the Travale–Radicondoli field was discussed in 1983 by P. Castellucci and his fellow-workers.[43] They identified two different tectonic units in the basement, namely the Boccheggiano Formation and micaschists. The latter always occur in other areas of the Larderello field, but the former never. Four lithostratigraphic and/or tectonic sequences were recognized. These are carbonatic quartzite, dolostone, minor anhydrite, graphite schists with sulphide ore-bodies at the top; metagraywacke, dolostone, anhydrite, basic metatuffite and sulphide ore-bodies; metagraywacke with minor graphitic schists and anhydrite; and, basally, metatuffites.

Crustal deformation and gravity changes during the initial decade of exploitation of the new Travale–Radicondoli geothermal field were recorded in 1985 by G. Geri and his associates.[44] Their precise levelling measurements on a specially constructed network of benchmarks revealed that the central portion of the region is subsiding with an average rate during 1973–83 of between 20 and 25 mm annually. A set of gravity measurements was effected every year between 1979 and 1982 on a network of gravity benchmarks. The g variations observed attained a maximum of 40 μGal.

At Monte Amiata, there are the Bagnore, Piancastagnaio and Poggio Nebbio geothermal fields and the explored area occupies some 50 km^2. Now only the first two fields are exploited and yield steam from a shallow reservoir at pressures varying from 2·5 to 10 bar. The thermodynamic behaviour of the Bagnore field was discussed by P. Atkinson and his co-workers in 1978 and again in 1980.[45,46] Production here began in 1959 and the reservoir gas was originally at a pressure of 23 atm. The non-condensable gas content was over 80% by weight and most of this was carbon dioxide. During the initial years, this and the reservoir pressure declined to about 10% by weight and 7 atm, respectively. Since then the change has decelerated considerably. The initial reservoir temperature was estimated as between 170 and 180°C.

A mathematical model accounting for thermodynamic and chemical equilibria between vapour, liquid and solid carbonate phases in the reservoir was developed and applied to studying initial conditions in the latter. A lumped-parameter model of a two-phase two-component system was developed and this CO_2–H_2O liquid–vapour model was used so as to calculate history of pressure and composition within the reservoir. The work confirmed that there was at first a large accumulation of non-condensable gases within the reservoir and that it was withdrawn during the early years of exploitation. Model calculations for the initial state of the reservoir demonstrate that the carbon dioxide initially present could not have derived from the local carbonate rocks. Since additional calculations from the producing-state lumped-parameter model showed that the long-term producing concentration of carbon dioxide cannot be explained on the assumption of reasonable quantities of carbon dioxide-saturated liquid–water influx into the reservoir, the work demonstrated the need for further investigations into the nature of carbon dioxide and water influx into the reservoir.

It is believed that in the Bagnore steam field the early behaviour of the reservoir was controlled by a blowdown of carbon dioxide in the gas cap, the long-term behaviour there being controlled by water influx. At Piancastagnaio, a deep drilling programme showed large permeable layers within the metamorphic basement at 2·5–3·5 km depth, actually more than 2 km below the shallow reservoir under exploitation. The temperature at the deeper layers is 330–350°C and the pressure approximately 200 bar. There is now 22 MW installed, but an extra 30 MW could be obtained by using additional geothermal fluid.

In Latium, there are several geothermal fields and one, Latera, was mentioned earlier and is located within the Vulsini volcanic complex. It is one of the geothermal areas of the peri-Tyrrhenian belt along which a regional high geothermal anomaly exists. Nine deep wells have been drilled in the Latera Caldera and four have been productive. In 1980 a joint venture between ENEL and AGIP discovered a new liquid-dominated type and five of the 11 wells drilled are productive. With the exception of one which gives a gaseous mixture (more than 90% carbon dioxide), all yield a two-phase fluid at 210–220°C with a total dissolved solids content of 10–12 g/litre (mostly alkaline chlorides and alkaline earth bicarbonates) together with a gas content of 5 wt% (98% carbon dioxide). The fluid available is anything from 1000 to 1200 t/h and roughly corresponds to the amount requisite for the supply of 30 MW.

The reservoir comprises fractured carbonate rocks of the Tuscan Nappe. Overlying volcanic units are sealed by hydrothermal minerals, mainly calcite and anhydrite, and act as an impervious cap rock. The evolution of the geothermal system has been examined by G. Cavarretta and others in 1985[47] from a deep aquifer (1000–1500 m depth), now not connected with the shallow aquifer in the volcanoclastic units. These aquifers comprise a sodium chloride and water solution with significant quantities of sulphate and carbonate. The fluid temperatures range between 200 and 230°C and in-hole temperatures as great as 343°C at 2775 m have been measured in dry wells.

The newly formed mineral assemblages from volcanic and sedimentary units were studied so as to reconstruct the thermal evolution of this field. These high-temperature hydrothermal minerals may be the results of ongoing active metamorphism and some are in equilibrium with the existing geothermal fluids. They can be divided into 'contact metasomatic' calcite, etc., 'high-temperature hydrothermal' calcite, etc., and 'low-temperature hydrothermal' calcite, etc. The first group is now regarded as relic and parts of the remaining two are still in equilibrium with the existing conditions in various parts of the geothermal system. Their appearance may signify the early beginning of the Latera geothermal system where the generation of an aqueous fluid phase by secondary boiling initiated extensive hydrofracturing of the enclosing rocks.[48] The intrusion of a syenitic melt-up to a depth of approximately 2 km and dated as 0·86 Ma represents the initial stage in the geothermal evolution of the Latera system.

A second geothermal field in Latium is Cesano, a hot brine one discovered at depths of 1·5–3 km, corresponding to fractured carbonate formations. The salinity is 300–350 g/litre of chlorides, sodium and potassium sulphates, and the temperature is about 250°C. The highest levels of total dissolved solids occur in the fluid coming from the centre of the field, where the production of electricity has only a marginal value. A possible integrated use of the brine is under consideration.

A third geothermal field in Latium is Torre Alfina, which is liquid-dominated with the reservoir located in fractured carbonate rocks at shallow depth (500–1000 m). Over 500 t/h of fluid is produced and has a temperature of 140–150°C with a salinity of some 7 g/litre comprising mostly alkaline earth bicarbonates. It is saturated with carbon dioxide and consequently there is plentiful carbonate scaling when the fluid flashes during production.

In Campania, the Phlegraean Fields (see Fig. 4.18) contain 12 wells at depths between 900 and 3000 m. They are concentrated in two areas, namely Mofete and San Vito. At the former, three permeable layers were located, of which the shallowest (600–1300 m depth) corresponds to fractured volcanics and possesses water with a total dissolved solids content of approximately 38 g/litre. The deepest layer relates to fractured sedimentary rocks at 2500–3000 m depth and has a hypersaline fluid with a total dissolved solids content of over 500 g/litre. The reservoir temperatures are 200, 250 and 350°C, respectively, for the three layers. At San Vito, the bottom temperature in one well 3000 m deep is more than 420°C.

Until 1970 the generation of electricity was the sole industrial aim in Italy and this concentrated interest in the key areas of Larderello, Travale, Monte Amiata and, to a lesser extent, in some other zones of the pre-Apennine belt in Tuscany, Latium and Campania considered promising for finding steam or fluids with temperatures in excess of 180°C. The oil crisis of 1973 widened the scope of exploration to find geothermal resources applicable also to non-electrical uses. The installed capacity of Italian geothermal power plants in January 1975 and December 1984 is shown in Table 7.16. The situation regarding direct heating in Italy using geothermal energy, as of 1986, is summarized in Table 7.17.

Other useful information is shown in Table 7.18.

In 1975 a deep exploration programme was launched in an attempt to discover whether the basement metamorphic complex lying between 2·5 and 5 km below the surface included permeable horizons containing geothermal fluids of higher temperature and pressure than in the reservoirs exploited until then. The work continued for a decade or more and produced encouraging results in the Larderello, Piancastagnaio and Travale areas (see G. Cappetti and others in 1985[49]). Fractured horizons were revealed in the basement at Larderello to the investigated depth of 4 km and a vapour-dominated system was shown to exist down to ca. 3 km. At Travale and Monte Amiata, also in outlying areas at Larderello, very thick impermeable formations were found beneath the shallow exploited reservoirs so that deeper productive horizons could be attained without drilling problems. Deep well work confirmed that high-temperature zones in the upper portions of reservoirs are restricted in size by circulation of meteoric waters and a geothermal anomaly 3 km deep is much more extensive for the Larderello and Travale fields. At Piancastagnaio near Monte Amiata, deep drilling showed an almost impermeable basement to exist, this being mostly composed of phyllites, at depths from 2·6 to 3·5 km beneath the exploited shallow reservoir. At these depths a liquid-dominated reservoir occurs and has temperatures between 300 and 350°C with pressures of 200–260 bar. There is only a limited transmissivity in the deep productive layer (0·1–0·5 D m), but the wells are commercially viable because of the high enthalpies and pressures. Flashing takes place in the formation after a short period of production. In fact, ten wells were drilled to 1986 and all are productive.

A re-injection programme commenced in the 1950s in order to discharge waste condensate from the Larderello power plant back below ground and

TABLE 7.16
The installed capacity of Italian geothermal power plants in 1975 and 1984[35]

Region	Power plant	Jan. 1975			Dec. 1984[b]		
		No. of units	Unit rating (MW)	Total installed capacity (MW)	No. of units	Unit rating (MW)	Total installed capacity (MW)
Larderello	L2	4	14·5	69	4	14·5	58
		1	11	—	—	—	—
	L3	3	26	120	4	26	113
		1	24	—	1	9	—
		2	9	—	—	—	—
	Gabbro	1	15	15	1	15	15
	Castelnuovo	2	11	50	2	11	50
		1	26	—	1	26	—
		1	2	—	1	2	—
	Serrazzano	2	12·5	47	2	12·5	47
		2	3·5	—	2	3·5	—
		1	15	—	1	15	—
	Sasso Pisano	1	12·5	22·7	1	12·5	19·2
		1[a]	3·2	—	1	3·2	—
		2	3·5	—	1[a]	3·5	—
	Lago	1	12·5	33·5	1	12·5	33·5
		1	6·5	—	1	6·5	—
		1	14·5	—	1	14·5	—
	Monterotondo	1	12·5	12·5	1	12·5	12·5
	Vallonsordo	1[a]	0·9	0·9	—	—	—
	Molinetto	1[a]	3·5	3·5	—	—	—
	Lagoni Rossi 1	1[a]	3·5	3·5	1[a]	3·5	3·5
	Lagoni Rossi 2	1[a]	3	3	—	—	—
	Lagoni Rossi 3	—	—	—	1	8	8
	San Martino	—	—	—	1	9	9
	La Leccia	—	—	—	1	8	8
	Molinetto 2	—	—	—	1	8	8
	Subtotals	33	—	380·6	31	—	384·7
Travale–Radicondoli	Travale 2	1[a]	15	15	1[a]	15	18
		—	—	—	1	—	—
	Radicondoli	—	—	—	2	15	30
	Subtotals	1	—	15	4	—	48
Mt. Amiata	Bagnore 1	1[a]	3·5	3·5	1[a]	3·5	3·5
	Bagnore 2	1[a]	3·5	3·5	1[a]	3·5	3·5
	Piancastagnaio	1[a]	15	15	1[a]	15	15
	Subtotals	3	—	22	3	—	22
Latium	Latera	—	—	—	1[a]	4·5	4·5
Total		37	—	417·6	39	—	459·2

[a] Discharging-to-atmosphere units.
[b] Compare Table 7.8.

TABLE 7.17
Direct heating utilization of geothermal energy in Italy[35]

Site	Type of use[a]	Rated use					Average use annually				
		Flow rate (kg/s)	Temperature (°C) I[b]	Temperature (°C) O[b]	Enthalpy (kJ/kg) I	Enthalpy (kJ/kg) O	Flow rate (kg/s)	Temperature (°C) I	Temperature (°C) O	Enthalpy (kJ/kg) I	Enthalpy (kJ/kg) O
Mt Amiata[c]	G,A	547	90	50	377	209	291	90	55	377	230
Larderello	D	1·3	160	95	2780	398	0·45	160	95	2780	398
Lago	G	0·7	125	100	2730	419	0·3	125	100	2730	419
Castelnuovo	G	0·4	115	100	2710	419	0·15	115	100	2710	419
Val di	G	8·3	95	75	398	314	2·5	95	80	398	335
Cecina[d]	D	53	95	70	398	293	14	95	75	398	314
Bulera[d,e]	G	1	120	40	1664	130	0·4	120	55	1664	190
Radicondoli[d]	G	70	95	65	398	272	47	95	70	398	293
Euganei Hills											
Albano Terme–Monteortone[f]	B,O	580	78	37	327	155	420	75	37	314	155
Montegrotto–Terme[f]	B,O	470	75	37	314	155	270	73	37	306	155
Battaglia–Terme[f]	B,O	110	64	37	268	155	18	62	37	260	155
Galzignano[f]	G	30	58	35	243	147	21	55	40	230	155
Vicenza[g] S. Donato	D	33·3	65	30	272	126	18	65	40	272	167
Milanese[g]	D	13·9	62	25	260	105	11	62	35	260	147
Ferrara[g]	D	111	95	35	398	147	35	95	45	398	188
Cesano[g]	O	30	130	70	540	293	5	130	75	540	314

[a] Types of use: A, air conditioning; B, balneotherapy; D, district heating; G, greenhouses; O, other.
[b] I, inlet; O, outlet.
[c] Agricultural drying plants here had not been completed in 1986.
[d] Projects due for completion in 1985.
[e] Stated data include electric generation (50 kWe) using a binary-cycle turbine.
[f] Space heating and sanitary hot water production for hotels.
[g] Projects due for completion between 1985 and 1987.

TABLE 7.18

Wells drilled to facilitate the direct heat use of geothermal resources in Italy up to 1 January 1985 (excluding those drilled for other purposes and subsequently used for the extraction of heat)[35]

Site	Date of drilling	Number of wells	Type of well[c]		Total metrage	Maximum temperature (°C)	Flowing enthalpy (kJ/kg)	Aggregate flow rate (kg/s)
			1	2				
Euganei Hills								
Albano Terme–Monteorteone[a]	1950–1984	150	P	P	60 000	84	332	580
Montegrotto–Terme[a]	1955–1984	77	P	P	27 000	80	335	470
Battaglia–Terme[a]	1960–1984	23	P	A[d]	5 800	74	310	130[e]
Galzignano[a]	1960–1984	8	P	A[d]	2 000	62	260	80[e]
Vicenza[b]	1983	1	P	P	2 150	68	285	35
	1979	1	I	—	2 500	64	268	55
San Donato								
Milanese[b]	1981	1	P	P	2 700	62	260	66
Ferrara[b]	1981	1	P	P	1 960	100	418	11
Cesano	1983	1	P	—	3 002	234[f]	d[f]	d[f]

[a] Data refer to wells producing currently.
[b] Projects due to be completed between 1985 and 1987.
[c] Types of wells: (1) P, production; I, injection; (2) P, pumped; A, artesian.
[d] Pumps are used in some to increase flow rate.
[e] Including pumped wells.
[f] d, dry well. Temperature refers to the bottom hole.

also to try partially to recharge the geothermal field as well as decelerate pressure drawdown in the more intensely exploited area. The relevant physics of flow and the work of R. C. Schroeder and his associates in 1982[50] have been discussed earlier in Chapter 5 (Section 5.3), where the one-dimensional vertical system approximation applicable at Larderello was described, in addition to the stable-isotope study of re-injection processes there made by F. D'Amore and others in 1987.[51] This latter was preceded by a study of the application of environmental isotopes as natural tracers in a re-injection experiment at Larderello made in 1981 by S. Nuti and his co-workers.[52] Re-injection testing initially involved the more marginal areas of the field, i.e. the less intensely exploited parts. Later similar work was done centrally, firstly in the 'superheated' zone characterized by fracture-governed permeability, and injection went on at a rate of 10–50 kg/s into the upper productive parts of the reservoir. The temperature is *ca.* 250°C, the pressure 5–7 bar and there is a high density of productive wells in the area with steam entries at depths between 400 and 600 m.

The re-injection began in January 1979 and went on, with a few short breaks, until April 1982. The surrounding wells were affected as regards fluid composition and production increases, but no breakthrough took place (in spite of the closeness of the wells—the nearest being only 150 m away); wellhead temperatures stayed practically constant. The inference is that a high vertical permeability exists, this permitting water to penetrate to depth and promoting an efficient fluid–rock heat exchange.[53]

The tracing techniques based upon variations in isotopic composition were applied also to the non-condensable gas contents and the quantity of injected water recovered as steam was calculated.[52] In fact, over 85% of re-injected water was recovered as superheated steam without any change in wellhead temperature of the productive wells. This verifies the hypothesis that the re-injected water goes down deep into the reservoir and contacts a large rock mass.

There is also good evidence, obtained by deep drilling in the central part of the Larderello field, of diffused fracturing down to at least 3 km depth. This also showed the existence of a vapour-dominated system with relatively low fluid pressure and high temperature (55 bar and over 300°C at 3 km depth). In fact, the deep structure of the Larderello field was the subject of a deep exploration programme commenced in 1974 by ENEL. A report on this in 1983 by F. Batini and his colleagues mentioned that seismic surveys have confirmed the existence of a deep 'K' horizon which is continuously present on all those seismic lines within the depth range 3–6 km.[54] The relevant wells demonstrated that the metamorphic basement comprises phyllites, micaschists and gneiss below Mio-Pliocene sediments, Cenozoic to Mesozoic allochthonous units (Tuscan and Ligurian Nappes), Verrucano sediments and tectonic slices composed of the former units and the underlying basement.

From the above discussion it is clear that, all in all, re-injection has proved to be the best way of extracting heat from the rocks in the parts of the reservoir which have been most depleted by past exploitation.

As for designing and making new power plants, it is clear that the shorter the time elapsing between identification of geothermal fluids and the beginning of their commercial utilization the better, as G. Allegrini and others indicated in 1985.[55] In Italy, investment for exploration and drilling comprises a considerable part of the total funding for a geothermal project, ranging from 43% for shallow reservoirs to 70% for deep ones.[31] An excellent review of the present status of geothermal activity in Italy was given by R. Carella and others in 1984.[56]

Clearly, construction time can be shortened by pre-ordering the relevant power plant before drilling is finished, but this must take into account the appropriate geothermal fluid characteristics. ENEL and the Italian electromechanical industry designed a small-capacity (15–20 MW) condensation-type geothermal power plant having such wide adaptability that it is designated 'universal'. It can be rapidly assembled and has almost constant efficiency. B. Billi and others in 1985 reported that then there were 13 such plants scheduled for installation in Italy, with five more under construction.[35]

7.3.5.1 Planned Progress in Italy from 1985 to 1995

The activities planned for the decade 1985–95 in Italy are intended to further the application of the country's considerable reserves of geothermal energy both to the generation of electricity and for direct uses. Potentially interesting regions will be explored and there will also be exploratory drilling work accompanied by the characterization of geothermal reservoirs together with the testing of geothermal fluids. In addition, exploitation drilling, field development and power plant or heat-supply plant testing will be undertaken in newly developed regions. Finally, there will probably be modernization of old power plants at Lardello. Some of the planned work is listed in Table 7.19.

At Larderello, the work will entail increasing production by drilling peripherally in the field and in accordance with permeable horizons which are thought to exist below the geothermal reservoir now being exploited. Centrally there will be renewal of old wells and drilling of new ones,

TABLE 7.19
The drilling and power plants planned for the decade 1985–1995 in Italy[35]

Region	Drilling		Power plants	
	Number of wells	Total metrage	Number of units	Total capacity (MW)
Larderello	45	100 000	8	360
Travala–Radicondoli	7	20 000	4	80
Mt. Amiata	75	250 000	7	140
Latera	10	22 000	3	45
Others[a]	43	108 000	2	30
Total	180	500 000	24	655

[a] Together with Latium and Campania, areas fringing Larderello and Travale–Radicondoli are included.

continuing re-injection of condensate water in an attempt partly to recharge the main reservoir, and modernization of old plants as well as the construction of new generating units and modernization and/or adaptation of the pipeline network. Five new units of 50–60 MW capacity each are scheduled to operate within a pressure range of between 5 and 15 bar. There will be investigations of the technical feasibility and economics of operating the power plants under conditions of load modulation so as to derive peak production of electricity. In addition, it is planned to experiment with a binary-cycle electricity generating unit having a 50-kW capacity and to install greenhouses in the Bulera area (80 000 km^2) as well as to emplace a district heating system in Castelnuovo (total volume 200 000 m^3).

At Travale–Radicondoli, deep exploration may reveal whether or not permeable horizons bearing high-temperature and high-pressure geothermal fluids occur under the geothermal reservoir now under exploitation. Associated drilling will be directed towards the northern and eastern boundaries of the field in order to gain insight into the down-faulted steps of the local geological structure. The general view is that the geothermal fluids potentially available here could contribute approximately 80 MW (cf. Table 7.16 above). A pipeline 40 km long is planned for the transportation of condensate water from the Travale power plants to Larderello for re-injection.

At Monte Amiata, drilling will be effected in several areas such as Bagnore, Piancastagnaio and Poggio Nibbio so as to increase geothermal fluid output from the reservoir corresponding to fractured carbonate formations. There will also be a deep drilling programme at Piancastagnaio aimed at exploitation of geothermal fluid occurring within permeable layers of the metamorphic basement some 1·5–2 km beneath the exploited reservoir. Altogether about 75 wells, each from 3 to 3·5 km long, will be drilled up to 1995, by which time geothermal fluids are expected to contribute approximately 140 MW (cf. Table 7.16). The greenhouse surface area is anticipated to rise from the existing 350 000 m² or so, and also to offer geothermal heat for driers for farm produce. Overall energy savings of approximately 80 000 OET/year are envisaged.

Finally, at Latera, a 4·5-MW exhausting-to-atmosphere plant was installed and put into service in November 1984. Further activities will involve determination of the characteristics of the power plant as regards the type of geothermal fluid available as well as surveys aimed at assessing production and re-injection work. Drilling will continue in order to recover fluid required to supply three condensing units of 15 MW capacity each and it is believed that, by the time the programme has been completed, 1300–1500 m³/h of residual water will become available. The possible use of its energy in agriculture and for other purposes will be studied.

In Italy, therefore, it can be seen from the foregoing remarks that there is a very full geothermal development programme planned up to 1995 and its implementation will require research both into resources assessment and the technology of deep exploration and use of geothermal fluids. This will be done by such bodies as ENEL, AGIP, IIRG (International Institute for Geothermal Research) of the CNR (National Research Council), some universities and other organizations. Related work will be completing surface prospection in new areas and speeding up drilling in promising zones, i.e. installing the 180 wells for a total of 500 000 m as indicated in Table 7.19. Deep drilling will be undertaken to depths as great as 5 km in some regions. Modernization of old power plants and the installation of new ones of standard size (15–20 and 50–60 MW) will take place, these making for improved efficiency and flexibility. The standard size will mean standardization of components of equipment as well as simplification of assembly and maintenance operations. Some artificial recharge of vapour-dominated hydrothermal fields will be attempted. Every effort is to be made to reduce the time lapse between discovering geothermal fluids and putting them into use. In sum, 900 MW should be installed by 1995, 655 MW of which will be newly constructed or modernized, 5×10^9 kW h/year becoming available for electrical utilization, with an energy savings equivalent to 350 000 OET annually for direct use (Table 7.20).

A number of other problems remain which could be tackled during the

TABLE 7.20
Existing and planned electricity production (MW) in Italy[35]

	Geothermal		Fossil fuels	
	Capacity (MW)	Use (GWh/year)	Capacity (MW)	Use (GWh/year)
January 1985				
In operation	4 592	2 840	36 592	127 785[a]
Under construction	60	—	8 908	—
Funded, not commenced	260	—	2 560	—
1995				
Total projected use	900	5 000	42 953	171 800
	Hydroelectric		Nuclear	
	Capacity (MW)	Use (GWh/year)	Capacity (MW)	Use (GWh/year)
January 1985				
In operation	17 465	45 182[a]	1 273	6 888[a]
Under construction	1 809	—	2 000	—
Funded, not commenced	800	—	6 000	—
1995				
Total projected use	29 441	78 260	12 873	58 572

[a] Provisional data.

decade under discussion. An important one relates to analysing reservoir pressure and decline curves in the Serrazzano geothermal field which is located a few kilometres west of Larderello. This has an interesting geological structure, Cretaceous flysch facies being underlain by Triassic to Palaeozoic carbonates with intercalated anhydrites or terrigenous beds. Carbonates and terrigenous beds constitute an anticlinal structure and, as the flysch facies are allochthonous, they can overlie directly either of these. At the top of the anticlinal structure, tension cracking and shear jointing occur and, together with fractures caused by overthrusting, constitute a favourable environment in the form of a permeable structural high. Indeed, natural surface steam manifestations once occurred along the high in question, but disappeared as extraction of geothermal fluids proceeded. Around 1930 drilling commenced and penetrated below the covering complex terrains; they ended in 1941, at which time there were seven productive wells yielding a total of 220 t/h of superheated steam to supply an electric power plant. In 1954 drilling recommenced and terminated in 1970, covering about 20 km^2. Now there are 19 wells producing some 300 t/h of superheated steam at a temperature ranging between 160 and 260°C at a pressure of ca. 5 atm. The content of non-condensable gases is approximately 3·5%. The pressure in the Serrazzano reservoir is monitored by means of 23 non-commercial wells which are usually shut in.

For years material balance principles have been applied to naturally found subterranean hydrocarbon gas reservoirs in order to estimate gas reserves. In the recent past the technique has been extended to vapour-dominated hydrothermal systems such as Serrazzano. Both steam and water co-exist in such systems and the resultant behaviour depends upon the fluid–rock heat exchange. Consequently, there must be an energy balance accompanying the material balance. Another approach used in the petroleum industry is that of decline curve analysis. The conventional linear p/z versus cumulative production material balance relationship is correct for closed single-phase gas reservoirs. Although this has not been proved for steam-producing systems with boiling water, P. Atkinson and others in 1978 noted that a line connecting available data points extrapolated back to zero production indicates an initial reservoir pressure approximating at least 40 atm.[57] Extrapolating the same data to zero reservoir pressure showed the total initial steam in situ to be ca. 170 million tons. An empirical curve-type matching technique was applied to production decline curves of wells in this reservoir and that for each well extrapolated to infinite production time in order to derive an estimate of total past and future production.

Summing these values for all of the producing wells in the reservoir, an estimated total production for past and future of 200 million tons was obtained. The agreement between estimated total production applying material balance principles and decline curve analysis is excellent, but this can be confirmed only by further field and theoretical work. In recent times, log-type curve analyses were performed on declining rate data for oil reservoirs producing under differing natural oil recovery methods, e.g. the solution–gas drive and natural water drive. The same approach is feasible with natural gas reservoirs and steam reservoirs. M. J. Fetkovich in 1973 introduced it in reservoir engineering literature and indicated that it is totally analogous to the log-type curve-matching procedure applied to constant-rate pressure-transient data.[58] Decline curves for Serrazzano producing wells were analysed using the Arps equation.[57,59]

The desirability of evaluating the effects of liquid vaporization on the p/z

behaviour of steam reservoirs was recognized ever since the technique was initially applied to a suitable geothermal steam system by H. J. Ramey Jr in 1970.[60] However, it was not until 1977 that a related study by W. E. Brigham and W. B. Morrow appeared in the literature.[61] They considered a closed porous reservoir comprising a steam cap overlying a liquid water table, steam production causing liquid vaporization. This phenomenon was taken as causing either the water level to fall so that there is no vapour present in the liquid region or the boiling to take place uniformly throughout the liquid region which permits the original steam–water interface to remain fixed. However, two significant aspects were neglected. One is liquid water influx (recharge) and the other is vapour pressure lowering as caused by liquid–solid adsorption or curved steam–water interfaces.

7.3.5.2 Strategy for Geothermal Development

As regards developing any geothermal field, an appropriate strategy must be devised, one starting with a background study providing an adequate historical perspective of the region concerned. Exploration is fundamental and appropriate techniques have been described in Chapter 3. An exciting new approach was announced by Sandia National Laboratories in Albuquerque, New Mexico, USA, in 1988.[62] This involves an almost complete down-hole seismic source which has been under development since the early 1980s. It generates seismic wave radiation patterns valuable in detecting geothermal energy sources as well as such subterranean fluids as oil and gas. The tool was aimed at basic geophysical research in continental scientific drilling programmes which are mapping the crust of the Earth; laboratory and field tests of the prototype have proved successful up to 5 km from the source. A framework must be erected, optimally under several aspects, for instance: decision-making which can be political and involve choosing between various options; thermo-dynamic, relating the resource to considerations of the availability of matter and energy; and organizational, entailing the working together of governmental agencies, geothermal coordinators and private companies.

As regards the organizational aspect, a recent illustration is the supply of both samples and data by Republic Geothermal Inc. to J. G. Zukin and his colleagues to examine uranium–thorium-series radionuclides in 300°C brines and reservoir rocks from two deep geothermal boreholes in the Salton Sea geothermal field in southeast California, USA.[63] The investigation in question comprised part of a study of the behaviour of these radionuclides as a natural analogue of possible radionuclide behaviour near a nuclear waste repository constructed in salt beds. The rock/brine concentration ratios varied from near-unity for isotopes of radium, lead and radon to approximately 50 000 for ^{232}Th. The high sorptivity of this latter is closely followed by that of ^{238}U and ^{234}U, implying that uranium is retained in the $+4$ oxidation state by the reducing conditions in the brines. A rather high solubility of ^{210}Pb and ^{212}Pb may be connected with the formation of chloride complexes. High radium solubility is certainly attributable to chloride complexing, together with lack of appropriate adsorption sites due to high salinity of the brines, and temperatures as well as reducing conditions which preclude formation of MnO_2 and $RaSO_4$. The ^{228}Ra/^{226}Ra ratios are about equal to those of their parents (^{232}Th/^{230}Th) in associated rocks and this indicates that radium equilibration in the brine–rock system is attained within the mean life of ^{228}Ra (8·3 years). The ^{224}Ra/^{228}Ra ratios in the brines are *ca.* 0·7, from which it may be inferred that either the brine composition is not homogeneous or radium equilibration in the brine–host rock system is incomplete within the mean life of ^{224}Ra (5·2 days) because the desorption of ^{224}Ra from the solid phase is impeded. The ^{228}Ac/^{228}Ra activity ratio in the brines is $\leq 0\cdot1$ and, from this ratio, a residence time of ^{228}Ac in the brine prior to sorption on to solid surfaces is estimated to be ≤ 70 min, which certainly shows the possibility of rapid removal of reactive isotopes from brines! The brine is much enriched in ^{226}Ra, at 2–3 dpm/g some 10^4–10^5 times the concentration of the parent ^{230}Th, but a 10% deficiency of this daughter radionuclide relative to its parent ^{230}Th occurs in the reservoir rocks. Material balance calculations for ^{226}Ra and ^{18}O imply that brines reside in the reservoir for 10^2–10^3 years, the Salton Sea geothermal field having formed 10 000–40 000 years BP, the porosity probably not exceeding 20%.[63]

Coordination is essential for geothermal work which includes so many multi-disciplinary activities. It will necessitate the collating and presenting of a vast amount of data, some encapsulated in standardized numerical mode and some incorporated into an appropriate geothermal-field and field-plant modelling system for reservoirs. Close cooperation between personnel at all levels is mandatory, although often not achieved. Implementing this is far from simple because of lack of a general consensus and also since compatibility with pre-existing and frequently highly structured governmental bodies has to be attained.

A critical problem in management is to determine who is in control of what, a matter which is simpler in cases where the project can be split up into a sequence of jobs capable of separate, or at least semi-independent, handling. In other types of power work, e.g. in hydroelectric schemes, the investigation of a site and parallel hydrological assessments can be finalized before any construction commences and the final power plant constrains neither the user nor the resource itself. But in geothermal development this is not possible, since both the investigation and the exploitation of the resource overlap, at least until such time as the resource parameters are measured, and afterwards during the manipulating and managing of the geothermal resource. Consequently, it is unsatisfactory to rely on an agency only involved in production to effect the drilling of investigation wells. Rather such drilling should be under the control and supervision of a suitable geothermal coordinating unit.

Utilizing the geothermal resource involves two phases, namely mining existing reservoirs and establishing permanent systems in equilibrium with natural input. The first need not necessarily be implemented, though it can be valuable in determining whether substantial reservoir modification will be required. Mining is an activity taking place when a particular hydrothermal system is undergoing artificial extraction of matter and energy at rates much higher than in the natural undisturbed system. If the geothermal reservoir is considered as a fixed container supplied almost at a constant rate from depth and having a constant resistance, the effect is a net withdrawal of geothermal fluid from it. Laboratory work demonstrates that the reservoir volume actually contracts to a smaller equilibrium shape, an effect produced in nature by a predominantly lateral inward net invasion of the reservoir by cold groundwater. However, this must be a long process, because it has not been observed, for instance at Wairakei after decades of operation.

As regards long-term exploitation, it may be assumed that there is a practically continuous hot-rock region at depth. But, if the natural hydrothermal systems resemble laboratory-scale models, the total energy of the system is smaller than in a corresponding non-volcanic zone (the recharge is mainly cold water) and also most of the energy is in the higher-level reservoirs. Although the matter is one for speculation, if the argument is valid then mining geothermal resources means removal of excess matter and energy over the natural replacement from high-level reservoirs. Ignoring the longer-term effects, i.e. those extending over thousands of years, geothermal reservoirs act as virtually fixed containers with recharge equal to that of the natural systems over the short term. In paving the way for their optimum exploitation, a clear definition of the roles and responsibilities of the participating governmental agencies is essential at the very start.

An innovative approach in interesting private investors would stimulate matters and also remove part of the burden from public utilities by replacing, at least to some extent, that equity for debt which a public-sector agency would incur otherwise.

A broader application of geothermal energy could be facilitated using smaller-scale power plants which might make the generation of geothermal power attractive in remote areas with low demand such as small island states.

As regards non-electrical uses of geothermal energy, additional applications should be sought, not least because they could accelerate the future development of the industry. All these aspects represent a challenge to power utilities to expand into new markets and simultaneously better the lot of mankind, especially the majority of the global population living in the developing countries. It is to be hoped that the necessary vision and hard work will be brought to bear successfully so that people with access to the geothermal resources of the Earth can benefit from their exploitation well into the 21st century.

REFERENCES

1. BARBIER, E., 1985. Geothermal energy in the context of energy in general and electrical power supply, national and international aspects. Section 1, UN seminar on the utilization of geothermal energy for electric power production and space heating, Florence, 14–17 May 1984. *Geothermics*, **14**(2/3), 131–41.
2. ANON., 1983. *Energy in Profile*, No. 5. Shell Briefing Service, London.
3. COLOMBO, U., 1983. Energy: industrial competition and technological development in New Zealand. *Proc. 2nd UN Symp. on Development and Use of Geothermal Resources, San Francisco, California, USA*, pp. 2359–63.
4. BARBIER, E., 1987. Geothermal energy in the world energy scenario. *Geothermics*, **15**(5/6), 807–19.
5. ANON., 1981. *Energy and Hydrocarbons*. ENI (Italian State Holding), Rome.
6. ANON., 1980. *Energy in Profile*, Shell Briefing Service, London.
7. ANON., 1982. *The Energy Spectrum*, No. 3. Shell Briefing Service, London.
8. ANON., 1984. *Energy in Profile*, No. 5. Shell Briefing Service, London.
9. ANON., 1985. *Energy in Profile*, No. 4. Shell Briefing Service, London.
10. ANON., 1983. *UN Statistical Yearbook*. United Nations, New York.
11. ANON., 1984. Nuclear energy, the international situation. *ENEA News Bulletin*, pp. 8–9. ENEA, Rome.
12. KANEKO, M., 1983. *National Outlook in Japan for Geothermal Energy Development*. Ministry of International Trade and Industry, Tokyo.
13. EZSCHEVICH, V. V., 1982. General report. In: *Seminar on the Medium-Term and Long-Term Prospects for the Electric Power Industry, London, 26–30 October 1981*. UN Economic Commission for Europe, Geneva, Switzerland.

14. Barbier, E. and Fanelli, M., 1983. Report to the International Institute for Geothermal Research, Pisa, Italy.
15. Di Pippo, R., 1985. Geothermal electric power, the state of the world—1985. *Int. Symp. on Geothermal Energy, Hawaii, August 1985*, International Volume, pp. 3–18. Geothermal Research Council.
16. Gudmundsson, J. S., 1985. Direct uses of geothermal energy in 1984. *Int. Symp. on Geothermal Energy, Hawaii, August 1985*, International Volume, pp. 3–18. Geothermal Research Council.
17. Donovan, P. R., 1984. *The Status of High Enthalpy Geothermal Exploration in the Developing Countries*. Report EP/SEM9/R84, United Nations Revolving Fund for Natural Resources Exploration (E).
18. Aidlin, J. W., 1975. United States law as it affects geothermal development. *Proc. 2nd UN Symp. on Development and Use of Geothermal Resources, San Francisco, California, USA*, pp. 2353–7.
19. Jessop, A. M., 1975. In: *Present Status and Future Prospects for Non-Electrical Uses of Geothermal Resources*, ed. J. H. Howard, pp. 94–5. Lawrence Livermore Laboratory, UCRL-51926.
20. Torfason, H., 1975. The law of Iceland as it affects geothermal development. *Proc. 2nd UN Symp. on Development and Use of Geothermal Resources, San Francisco, California, USA*, pp. 2435–7.
21. Alcaraz, A., 1975. In: *Present Status and Future Prospects for Non-Electrical Uses of Geothermal Resources*, ed. J. H. Howard, pp. 97–8. Lawrence Livermore Laboratory, UCLR-51926.
22. Dench, N. D., 1975. The law and geothermal development in New Zealand. *Proc. 2nd UN Symp. on Development and Use of Geothermal Resources, San Francisco, California, USA*, pp. 2359–62.
23. Surinkum, A., Saelim, B. and Thienprasert, A., 1986. Geothermal energy in Phrao Basin, Chiang Mai, Thailand. *Geothermics*, Sp. issue **15**(5/6), 589–95.
24. Lindal, B., 1977. Geothermal energy for space and process heating. In: *Geothermal Energy Technology*, ed. D. M. Considine, pp. 43–58. McGraw-Hill Book Co., New York.
25. Bodvarsson, G. and Zoëga, J., 1964. Production and distribution of natural heat for domestic and industrial heating in Iceland. *Proc. UN Conf. on New Sources of Energy*, Part 3, p. 452.
26. Garnish, J. D., Vaux, R. and Fuller, R. W. E., 1986. Applications and economics. In: *Geothermal Aquifers. Department of Energy R&D Programme 1976–1986*, ed. R. Vaux, pp. 72–100. Energy Technology Support Unit (Department of Energy), AERE, Harwell, Oxfordshire OX11 0RA, UK, 109 pp and appendices.
27. Jiga, 1985. *Géothermie Actualités*. Bordeaux, France.
28. Fanelli, M., 1986. Opening Speech, 1st Afro-Asian Geothermal Seminar, Chiang Mai, Thailand, 4–9 November 1985. *Geothermics*, Sp. issue **15**(5/6), 551–2.
29. Guha, S. K., 1986. Status of exploration for geothermal resources in India. *Geothermics*, Sp. issue **15**(5/6), 665–75.
30. Krishnaswamy, V. S. and Shankar, R., 1982. Geothermal resource potential of India. *Rec. GSI*, **III**(2), 17–40.

31. PRASERDVIGAI, S., 1986. Geothermal development in Thailand. *Geothermics*, Sp. issue **15**(5/6), 565–82.
32. BODVARSSON, G. S., PRUESS, K., STEFÁNSSON, V. and BJÖRNSSON, S., 1987. East Olkaria Geothermal Field, Kenya. 1: History match with production and pressure decline data. *J. Geophys. Res.*, **92**(B1), 521–39.
33. BODVARSSON, G. S., PRUESS, K., STEFÁNSSON, V. and BJÖRNSSON, S., 1987. East Olkaria Geothermal Field, Kenya. 2: Predictions of well performance and reservoir depletion. *J. Geophys. Res.*, **92**(B1), 541–54.
34. BARBIER, E., 1984. Eighty years of electricity from geothermal steam. *Geothermics*, **14**(4), 389–401.
35. BILLI, B., CAPPETTI, G. and LUCCIOLI, F., 1986. ENEL activity in the research, exploration and exploitation of geothermal energy in Italy. *Geothermics*, Sp. issue **15**(5/6), 765–79.
36. D'AMORE, F. and TRUESDELL, A., 1984. Helium in the Larderello geothermal steam. *Geothermics*, **13**(3), 227–39.
37. NUTI, S., 1984. Elementary and isotopic compositions of noble gases in geothermal fluids of Tuscany, Italy. *Geothermics*, **13**(3), 215–26.
38. BROWNE, P. R. L., 1979. Minimum age of the Kawerau geothermal field, North Island, New Zealand. *J. Volcanol. Geotherm. Res.*, **6**, 213–15.
39. DEL MORO, A., PUXEDDU, M., RADICATI DI BROZOLO, F. and VILLA, I. M., 1982. Rb–Sr and K–Ar ages on minerals at temperatures of 300–400°C from deep wells in the Larderello geothermal field (Italy). *Contrib. Min. Petrol.*, **81**, 340–9.
40. CAVARRETTA, G., GIANELLI, G. and PUXEDDU, M., 1982. Formation of authigenic minerals and their use as indicators of the physicochemical parameters of the fluid in the Larderello–Travale geothermal field. *Econ. Geol.*, **77**(5), 1071–84.
41. ATKINSON, P., BARELLI, A., BRIGHAM, W., CELATI, R., MANETTI, G., MILLER, F. G., NERI, G. and RAMEY, H. J., 1978. Well-testing in Travale–Radicondoli field. *Geothermics*, **7**, 145–84.
42. PATELLA, D., ROSSI, A. and TRAMACERE, A., 1979. First results of the application of the dipole electrical sounding method in the geothermal area of Travale–Radicondoli (Tuscany). *Geothermics*, **8**, 111–34.
43. CASTELLUCCI, P., MINISSALE, A. and PUXEDDU, M., 1983. Nature and tectonic setting of the Travale–Radicondoli basement in the Larderello geothermal field (Italy). *Mem. Soc. Geol. Ital.*, **25**, 237–45.
44. GERI, G., MARSON, I., ROSSI, A. and TORO, B., 1985. Crustal deformation and gravity changes during the first ten years of exploitation of the new Travale–Radicondoli geothermal field, Italy. *Geothermics*, **14**(2/3), 273–85.
45. ATKINSON, P., CELATI, R., CORSI, R., KUCUK, F. and RAMEY, H. J. JR, 1978. Thermodynamic behavior of the Bagnore geothermal field. *Geothermics*, **7**, 185–208.
46. ATKINSON, P. G., CELATI, R., CORSI, R. and KUCUK, F., 1980. Behavior of the Bagnore steam/CO_2 geothermal reservoir, Italy. *Soc. Petrol. Engrs J.*, **20**, 228–38.
47. CAVARRETTI, G., GIANELLI, G., SCANDIFFIO, G. and TECCE, F., 1985. Evolution of the Latera geothermal system. II: Metamorphic, hydrothermal mineral assemblages and fluid chemistry. *J. Volcanol. Geotherm. Res.*, **26**, 337–64.

48. BURNHAM, C. W., 1979. Magmas and hydrothermal fluids. In: *Geochemistry of Hydrothermal Ore Deposits*, 2nd edn, ed. H. L. Barnes, pp. 71–136. John Wiley & Sons, New York.

49. CAPPETTI, G., CELATI, R., CIGNI, U., SQUARCI, P., STEFANI, G. C. and TAFFI, L., 1985. Development of deep exploration in the geothermal areas of Tuscany, Italy. *1985 Int. Symp. on Geothermal Energy, Kailua-Kona, Hawaii*, pp. 303–9.

50. SCHROEDER, R. C., O'SULLIVAN, M. J., PRUESS, K. and CELATI, R., 1982. Reinjection studies of vapor-dominated systems. *Geothermics*, **11**(2), 93–119.

51. D'AMORE, F., FANCELLI, R. and PANICHI, C., 1987. Stable isotope study of reinjection processes in the Larderello geothermal field. *Geochim. Cosmochim. Acta*, **51**, 857–67.

52. NUTI, S., CALORE, C. and NOTO, P., 1981. Use of environmental isotopes as natural tracers in a reinjection experiment at Larderello. *Proc. 7th Workshop of Geothermal Reservoir Engineering, Stanford, California, 15–17 December*, pp. 85–9.

53. BERTRAMI, R., CALORE, C., CAPPETTI, G., CELATI, R. and D'AMORE, F., 1985. A three-year recharge test by reinjection in the central area of the Larderello field: analysis of production data. *1985 Int. Symp. on Geothermal Energy, Kailua-Kona, Hawaii*, pp. 293–8.

54. BATINI, F., BERTINI, G., GIANELLI, G., PANDELI, E. and PUXEDDU, M., 1983. Deep structure of the Larderello field: contribution from recent geophysical and geological data. *Mem. Soc. Geol. Ital.*, **25**, 219–35.

55. ALLEGRINI, G., GIORDANO, G., MOSCATELLI, G., PALAMÀ, A., POLLASTRI, G. G. and TOSI, G. P., 1985. New trends in designing and constructing geothermal power plants in Italy. *1985 Int. Symp. on Geothermal Energy, Kailua-Kona, Hawaii*, pp. 279–88.

56. CARELLA, R., VERDIANI, G., PALMERINI, C. G. and STEFANI, G. C., 1984. Geothermal activity in Italy: present status and prospects. In: UN Seminar on Utilization of Geothermal Energy for Electric Power Production and Space Heating, Florence, 14–17 May 1984. *Geothermics*, **15**(2/3), 247–54.

57. ATKINSON, P., MILLER, F. G., MARCONCINI, R., NERI, G. and CELATI, R., 1978. Analysis of reservoir pressure and decline curves in Serrazzano zone, Larderello geothermal field. *Geothermics*, **7**, 133–44.

58. FETKOVICH, M. J., 1973. Decline curve analysis using type curves. SPE Paper 4629 presented at the 48th Annual Fall Meeting of the Society of Petroleum Engineers of AIME, Las Vegas, Nevada, USA.

59. ARPS, J. J., 1945. Analysis of decline curves. *Trans. AIME*, **160**, 228.

60. RAMEY, H. J. JR, 1970. A reservoir engineering study of The Geysers geothermal field. Evidence Reich and Reich, Petitioners vs Commissioner of Internal Revenue. 1969 Tex Court of the United States, 52, T.C. No. 74.

61. BRIGHAM, W. E. and MORROW, W. B., 1977. p/z behavior for geothermal steam reservoirs. *Soc. Petr. Eng. J.*, **17**(6), 407–12.

62. ANON., 1988. 'In Brief' item on Sandia National Laboratories. *Geotimes*, **33**(3), 27.

63. ZUKIN, J. G., HAMMOND, D. E., KU, T.-L. and ELDERS, W. A., 1987. Uranium–thorium series radionuclides in brines and reservoir rocks from two deep geothermal boreholes in the Salton Sea Geothermal Field, southeastern California. *Geochim. Cosmochim. Acta*, **51**(10), 2719–31.

Glossary

'The joy of discovery is certainly the liveliest that the mind of man can ever feel.'

CLAUDE BERNARD

Apparent resistivity: In interpreting electric surveys, this is a quantity calculated from the values of current strength and potential difference with the aid of a geometric factor which depends upon electrode array and distances. If the relevant substance were isotropic and uniform, then its apparent resistivity would be equal to the specific resistivity of this infinitely thick layer. Practically, this means that to produce the same measured signal under the same conditions and at the same instant in time as the recorded signal, the requisite resistivity of a homogeneous Earth comprises the apparent resistivity. Problems may arise with the concept when the analogy between an apparent resistivity computed from geophysical observations and the true resistivity structure of the sub-surface is drawn too tightly.

In electromagnetic applications, there are several definitions of the parameter, but those in common use do not always show the best behaviour. Many of the features of the apparent resistivity curve which have been interpreted as physically significant with one definition disappear when another definition is substituted. It is misleading to compare the detection or resolution capabilities of different field systems or configurations only on the basis of the apparent resistivity curve.

Benioff zone: At destructive plate boundaries, crust is being consumed by being forced down under island arc–ocean trench systems or mountain belts with cold lithosphere descending into the asthenosphere as deep as 700 km. These usually curvilinear features comprise subduction or Benioff zones (after the seismologist Hugo Benioff). They are recognized by a zone of earthquake hypocentres from shallow (below deep trench and outer arc) to deep in location.

Bouguer anomaly: An anomaly arising after the application of the Bouguer reduction, although the term is applied also to the anomaly remaining after a total topographic reduction for Hayford's zones A–O or after application of both the Bouguer reduction and a terrain correction factor. The former is sometimes termed a modified Bouguer anomaly.

Bouguer reduction: A correction allowing for the effect of the mass between the surface of the Earth and the datum plane, generally the geoid, or, in water-covered areas, for the effect of the deficiency of mass between sea-level and the sea-floor. The mass or deficiency is assumed to be identical with an infinite horizontal plate.

British thermal unit (Btu): That amount of heat required in order to raise the temperature of 1 lb of water through $1°F$ ($= 251·997$ cal or $1055·06$ J).

Calorie: Unit of quantity of heat, namely that required to raise the temperature of 1 g of water through $1°C$.

Circulation system: A sub-class of the convective geothermal system occurring in low-porosity/fracture permeability environments in areas of high-to-normal regional heat flow. Circulation systems can develop also in regions devoid of young igneous intrusions and may derive from the deep circulation of meteoric water in the thermal regime of conductive regional heat flow. The temperature actually reached by the water will depend upon the magnitude of this heat flow and also the depth to which circulation takes place. Usually, higher temperatures will be found at shallower depths in regions of elevated conductive heat flow.

Conductive geothermal system: This category of geothermal system is characterized by a thermal regime solely due to conduction, frequently in the steady state. A working fluid either exists already or must be introduced. (See also Geothermal systems.)

Convective geothermal system: This category of geothermal system is characterized by natural circulation of a working fluid so that most of the heat transfer is effected by circulating fluids rather than by conduction. Convection tends to increase the temperature in the upper part of such a system while temperatures in the lower part decline. There are two sub-classes, namely hydrothermal and circulation. (See also Geothermal systems.)

Core of the Earth: The part of the planetary interior below the Gutenberg discontinuity, i.e. extending from 2900 km depth to the centre of the Earth. An outer liquid core has been distinguished from the inner solid one which starts at a depth of 5000 km. The latter has a density perhaps as high as $14.5 \, g/cm^3$ and a mainly nickel–iron (NiFe) composition. Its temperature is estimated as exceeding 2700°C and its pressure probably exceeds 3.5×10^6 bar.

Cristobalite: A crystalline polymorph of silica formed by the inversion of tridymite at 1470°C.

Crust of the Earth: That portion of the planet overlying the Mohorovičić discontinuity and divisible into sial and sima. Three layers are distinguishable within it, namely the topmost one made up of sedimentary rock and about 1 km thick, an intermediate one about 2.7 km thick and composed either of consolidated sedimentary materials or of modified bottom layer, and the base some 5 km thick and comprising basic igneous matter.

Darcy (D): A standard unit of permeability named after Henri Darcy, a 19th-century French water engineer. 1 darcy is equivalent to the passage of $1 \, cm^3$ of fluid of 1 centipoise viscosity flowing in 1 second under a pressure differential of 1 atmosphere through a porous medium having an area of cross-section of $1 \, cm^2$ and a length of 1 cm. A millidarcy (mD) is 0.1% of a darcy.

Darcy's Law: A derived equation for the flow of fluids involving the assumption that such flow is laminar and that inertia can be neglected. The numerical formulation of the law is used generally in studies of gas, oil and water production from subterranean formations. For instance, in gas flow, the flow velocity is proportional to the product of the pressure gradient and the ratio between permeability and density divided by the viscosity of the gas.

Delta (δ): The isotopic composition of a water sample is expressed in terms of the per-thousand deviation of the isotopic ratio from that of a standard, usually SMOW (standard mean ocean water), and the data are given as delta units defined by

$$\delta \, (\permil) = \frac{(R_{sa} - R_{st})}{R_{st}} \times 1000$$

in which $R_{sa} = {}^{18}O/{}^{16}O$, ${}^{13}C/{}^{12}C$, D/H, etc., in the unknown (sample) and R_{st} = the corresponding ratio in the standard.

Dike: A discordant sheet of igneous rock, i.e. one which cuts across the bedding or structural planes of the host rock. Dikes may be composite or multiple, sometimes occurring in swarms either radial or parallel in pattern. They may be found also in ring form. Occasionally, they extend for hundreds of kilometres.

Direct current: An electric current of constant direction of flow.

Dyne: The cgs system unit of force. That force which, acting upon a mass of 1 g, will impart to it an acceleration of $1 \, cm/s^2$. 1 dyne $= 10^{-5}$ newton (N).

Doublet: A geothermal system employing two wells, hot water being extracted from one and cooled water recharging to the aquifer through the other. The extraction point in the aquifer must be separated from the re-injection point to prevent rapid 'short-circuiting' by the cooled water. In the absence of this phenomenon and in terms of an infinite homogeneous and confined aquifer, the cold front would eventually reach the extraction point anyway. In practice, re-injected water warms up a little in passing through the rock and as many as three or more passes may take place before the outlet temperature starts to fall appreciably. A typical separation distance in an aquifer is approximately 1 km. There are two feasible approaches, namely to drill wells 1 km apart or to drill them near each other and deviate them so as to attain the requisite separation within the aquifer. Occasionally, one vertical and one deviated well may be drilled. Well spacing is usually selected so as to attain a useful resource lifespan of 20–30 years. Since aquifers are heterogeneous, a hydraulic boundary may be set up within one and actually impede the flow of water through it. Compared with a single well, doublets have a restricted area of influence, therefore closer packing, and a bigger exploitation of the resource is possible. The recharging work obviates problems stemming from a continuously decreasing reservoir pressure, and also the re-injection well offers an environmentally acceptable means for the disposal of the water, which may be a brine.[1]

Down-hole pumps: Once a well has been drilled, the water pressure in the aquifer usually causes the water or brine to rise and overflow may result. Where this happens, a surface pump suffices to induce circulation in a doublet system and assist re-injection. Sometimes a thermo-siphon effect alone is enough to produce adequate flow in summer. In the UK, no geothermal borehole has been drilled in which water reaches the surface. Such cases necessitate lowering a pump down-hole, i.e. into the well. As this pumps it lowers the water level, producing a drawdown, and hence lowers the pressure at the well bore adjacent to the aquifer. The pump must always be maintained below the water level. It has been found that electrical submersible down-hole pumps vary in reliability. US on-land experience over 17 years in a field with more than 600 electrical submersible pumps operational recorded an average life of just over two years for such pumps.[1]

Eigenvalues: These pertain to free oscillations which are of interest in earthquake engineering. Free vibrations of a system for which $\mathbf{P}$, the column vector of external forces, equals $\mathbf{M\ddot{x}_0} = 0$ and $\mathbf{y} = \mathbf{x}$, and $\mathbf{M\ddot{x}} + \mathbf{Kx} = 0$ lead to a situation in which a structure is said to vibrate in one of its natural modes when its time-dependent free displacements can be written

$$\mathbf{x}(t) = z_n \Theta_n(t)$$

where n denotes the order of the mode, z_n is the shape of the mode which does not vary with time, and Θ_n is a scalar function. That is to say, the system oscillates in one of its natural modes if the base remains motionless, all of its masses describe a synchronous motion and the shape of the configuration is independent of time (although its magnitude varies with t). Substitution of the penultimate into the last equation produces, on separation, the following expression:

$$(\mathbf{K} - \omega_n^2 \mathbf{M})z_n = 0$$

in which ω_n is a constant for the system. Characteristic roots ω_n^2 are known as critical values or eigenvalues.

Electric conductance: This term refers to the ease with which an electric current is able to flow through a medium and it is the reciprocal of resistance. Its symbol is G and its unit is the siemens: $1\,S = 1\,\Omega^{-1}$.

Electrode array: In the resistivity method, various systems of two current
  electrodes (C1 and C2) are used for d.c. supply to the ground and two or
  more measuring electrodes (P1, P2, etc.) are employed for noting the
  consequent potential differences. There are several possibilities. One is
  the Wenner array which has four equidistant electrodes in line: C1–P1–
  P2–C2 with distances P1–P2 = $\frac{1}{3}$(C1–C2). A second is the Schlumberger
  array which has a similar linear configuration, but distances P1–P2 $\leq$
  $\frac{1}{3}$(C1–C2). The dipole–dipole array has the measuring dipole P1–P2 in
  any position with respect to the current dipole C1–C2. Finally, the Lee
  array may be mentioned. This is a five-electrode configuration, C1–P1–
  P0–P2–C2, with distances C1–P1 = P1–P2 = P2–C2.

Enthalpy: Heat content, H. A thermodynamic property of a substance given
  by $H = U + pV$, where U is the internal energy, p is the pressure and V is
  the volume. Thus it is the expression of the energy contained in a system
  and may be used to denote the quantity of heat within steam or water
  produced from a geothermal source. The unit is the joule (1 J = 0·239 cal).
  A high-enthalpy (high-caloric or high-temperature) system is one
  producing steam or water of which the temperature is high enough to
  permit application to generation of electricity, i.e. in excess of 150°C. A
  low-enthalpy (low-caloric or low-temperature) system is one in which
  the temperature is below 150°C.

Erg: A cgs system unit of work or energy. The work effected by a force of
  1 dyne acting through a distance of 1 cm. 1 erg = 10^{-7} J.

Four-arm dipmeter: A logging tool designed to measure the strike and dip of
  bedding planes. It has four pad-type electrodes arranged in a co-planar
  orthogonal pattern. The pads are pressed against the well bore by a
  controllable force and make measurements which depend upon the
  electric conductivity of the rock. The reference pad, 1, is magnetically
  orientated and two independent calipers assess the diameter of the
  borehole between pads 1 and 3, and 2 and 4. Because of the design, the
  four-arm dipmeter becomes aligned with the long axis of a non-circular
  well.

Fumarole: A steam vent found where there is intense heat, but little water.
  As well as steam, gases such as carbon dioxide and hydrogen sulphide
  are emitted at high temperatures. Fumaroles occur in volcanic regions,

often in lava and, in addition, in geyser areas where they normally lie on higher ground than flowing springs, for instance in the Yellowstone National Park, Wyoming, USA, which has only 10 to 15 at high temperatures up to 138°C. The hottest and most spectacular from 1925 to 1927 was Black Growler, which emitted steam at this temperature from a vent located at the bottom of a 13 m-diameter, 1 m-deep crater. By 1931 it became inactive, but it has sometimes rejuvenated subsequently, activity apparently relating to alterations in the quantities of near-surface subterranean water. The composition of gases discharged from this fumarole was 99·6% steam, as might be expected, the remainder comprising mostly carbon dioxide (0·386%) together with minute quantities of hydrogen sulphide (0·01%), hydrogen (0·002%), nitrogen–argon (0·002%) and methane.

Geoid: Equipotential surface of the Earth at mean sea-level.

Geoisothermic plane: A plane of equal temperature within the Earth.

Geopressurized reservoirs: A category of conductive geothermal system entailing low-enthalpy confined aquifers in high-porosity/permeability sedimentary sequences in regions of normal to slightly higher heat flow. Such geopressurized reservoirs represent a particular case of sedimentary aquifers in which the pore fluids sustain pressures exceeding that of the water column, i.e. greater than the hydrostatic pressure. Actually, these fluids carry much of the total overburden, i.e. of the lithostatic pressure, values of the order of 100 MPa being typical. Such reservoirs are confined by impervious and low-conductivity shales both above and below them. Contrary to normal sedimentary reservoirs in which both porewater and heat are expelled during diagenetic compaction, geopressurized zones represent thermal traps. An optimum instance of a geopressurized zone is given by the northern Gulf of Mexico basin in the USA, both on- and off-shore. Geopressurized reservoirs are found to occur in the depth range of 3–7 km and also contain large amounts of natural gas.

Geotherm: A function which relates temperature to depth within the Earth.

Geothermal: Of the surface and sub-surface manifestations of the heat of the Earth at which the temperature exceeds that ambient in the adjacent surroundings, e.g. volcanoes, thermal fluids, hot springs and so on.

Geothermal anomaly: An area with heat flow characteristics which differ from those of the surrounding region. Both geological and hydrological factors can act to produce such a positive anomaly which denotes a concentration of thermal energy into a geothermal system. The most significant geothermal anomalies relate to ascending magma.

Geothermal gradient: Temperature gradient, the rate of increase of temperature with depth in the Earth with the following units: $1 \text{ K/m} (= 1°\text{C/m}, °\text{C/100 m or }°\text{C/km})$.

Geothermal province: A heat flow province, i.e. a geographical area possessing similar heat flow characteristics.

Geothermal step: Increase of depth related to a $1°\text{C}$ increase in temperature with the following unit: $\text{m/K} = \text{m/}°\text{C}$.

Geothermal systems: Locations at which concentrations of the internal thermal energy within the Earth occur in sufficient strength to constitute resources of geothermal energy. They fall into two major categories, namely convective and conductive. The former are characterized by natural circulation of a working fluid which transfers heat and may be steam or water, or a mixture of both. The latter operate by means of thermal regimes entailing conduction only, often in the steady state. A working fluid is either present, e.g. from deep-seated aquifers in sedimentary basins, or it must be supplied. Convective geothermal systems may be sub-divided into the hydrothermal type situated in high-porosity/permeability environments, and the circulation type situated in low-porosity/fracture permeability ones having high-to-normal regional heat flow. Conductive geothermal systems may be sub-divided into low-temperature/low-enthalpy aquifers located within high-porosity/permeability sedimentary sequences, including geopressurized zones, in regions of normal or slightly elevated heat flow and hot dry rock in a high-temperature/low-permeability environment.

Geothermal wells: In drilling these wells, exactly the same procedure is followed as with oil and gas on-land, the relevant rig embodying two principal parts, namely the main hoist and the rotating table. A square or hexagonal sliding tube, the kelly, is lowered into the table and its rotation transferred to the drill string, a series of hollow tubes (drill pipes) carrying the drilling bit at its lower end. As the depth increases, additional lengths of drill pipe are inserted into the top of the string below the kelly so that drilling proceeds in stages according to the length of the drill pipe, this depending upon the operating length of the kelly, usually just below 10 m. Drilling mud from mud pumps is fed into the hollow kelly using a flexible pipe. This mud is a specialized fluid and its density can be adjusted. It descends through the drill pipe and passes across the drilling bit, thereafter ascending to the surface again through an annular space between the wall of the well and the drill pipe. Once up it is screened and filtered and then may be re-used. Its purposes are to support the wall of the open hole and control any gas or fluid pressures in formations encountered by alteration of its density, to cool the drilling bit and to lift drilling cuttings to the surface. The necessary weight is applied to the drilling bit by the drill string, which may include sections of heavy-walled tube termed collars. During deepening of the drill hole, the weight on the drill bit must be adjusted and the hoist is retained attached to the kelly to achieve this end. The hoist brake must be handled carefully so as to follow the kelly downwards and not remove the weight from the bit. When the drill bit becomes worn, it must be replaced. This is accomplished by withdrawing the drill string in a reversal of the procedure mentioned above, a routine operation called 'tripping'. The work is done by taking out sections comprising three lengths of drill pipe, these then being stacked vertically at a corner of the drilling platform.

At suitable stages during drilling, a long assembly of steel casing is lowered into the borehole and the annular space between it and the hole filled with cement. This prevents the hole collapsing. Once such a length of casing has been installed, the drilling goes on further with a reduced diameter of drill bit (since this latter must pass through the casing). Particular care is taken when drilling through formations of especial geothermal significance. Then the drilling mud is often changed to one which will do less damage to the aquifer.

Of course, as drilling proceeds, observations are made. Cuttings are analysed after they are separated from the drilling mud, and thus changes in the formation are identified. Depths and thicknesses of formations are inferred from data regarding the drill string length and the delay time for the cuttings to get to the surface. Occasionally, drilling may be arrested in order to carry out specific tests.

Geophysical logging necessitates lowering relevant instruments one by one into the borehole: thus a suite of logs is accumulated. The wire line supporting them enables depths to be measured as well as providing a means of transmitting their signals. Measurements are usually made at selected intervals in the borehole. They record characteristic features such as porosity, density and thickness of various strata. The extent to which such logging is effected relates to the nature of the borehole itself, i.e. whether it is an exploratory one or intended for production. In exploratory boreholes, logging must be carried out immediately prior to each casing stage.

A drill stem test (DST) can be effected when an interesting formation is reached and this both samples and measures the pressure of water in it, also offering preliminary evidence of formation permeability. Normally, a test of the potential water yield is made by means of a 'gas-lift' to pump out the water. The gas, often nitrogen, is either pumped down the drill pipe to return up the annular gap or vice versa. Rising gas bubbles reduce the density of the water in the well as well as the hydrostatic pressure in the aquifer adjacent to it, and thus induce flow. Rates up to 30 litre/s can be sustained and even higher rates imposed over short time intervals. These surge the well and remove any drilling mud which has succeeded in penetrating the formation concerned. Several hours of steady pumping, together with measurement of pressure recovery between spells of pumping, facilitate estimation of formation transmissivity (permeability × thickness) near the well.

Hiring a rig is very expensive; for example, in 1985 the operating expenses were approximately £7000 daily. Therefore work is done round the clock although, where geothermal reservoirs are exploited to supply space heating to nearby buildings, overnight noise could be a problem. This results from the activity of the diesels powering the rig, banging together of pipes as the drill string is lengthened or taken out of the borehole and possible squealing of the hoist brake as it permits the block to follow the kelly. However, much effort has been put into minimizing this noise; this has been quite successful in UK.[1]

Geothermics: That branch of geophysics dealing with the internal heat and temperature of the Earth.

Geothermometer: Initially, a word used for a mercury thermometer based upon the overflow principle and designed for use in boreholes. Now an abbreviation for geochemical thermometer, a number of isotopic indicators of sub-surface temperatures in relation to the origins of fluids in geothermal systems.

Geothermometry: The measurement of temperatures within the Earth.

Geyser: Essentially a hot spring, but characterized by periodic thermo-dynamic and hydrodynamic instability. For a geyser to arise several conditions must be fulfilled. Firstly, it must have a source of heat and also it must have a place in which to store water while this heats sufficiently to erupt. Then it must possess an opening of optimum size out of which hot water can be ejected. Finally, there must be subterranean channels connected with it so that fresh water can re-supply it after eruption has occurred. The power of eruption reflects the fact that transformation of 1 g of water into steam effects as much work as detonating 1 g of an explosive. This spectacular phenomenon actually takes place when part of the stored water becomes unstable, i.e. its heat content attains some critical level of distribution. There may be abrupt and vigorous steam generating within the geyser relatively close to its surface opening. Where there is intense heat, but little water, a fumarole results. In the case where hot water fills with mud, a mud pot is found.

Geysers are uncommon, occurring in a few widely separated and highly localized regions such as Yellowstone National Park, Wyoming, USA, where there are some 200 active ones. However, all are found in volcanic regions and contain large quantities of rhyolite from which spouting proceeds.

Gravity anomaly: The difference between an observed value of the acceleration due to gravity and a theoretical (or normal) one. Differences in acceleration due to gravity are measured using a gravimeter and gravity stations incorporating such instruments often comprise a grid, each of the stations constituting a base station. A latitude correction is introduced to allow for the effects of latitude on the second-order derivative of the potential gravity of a normalized Earth and a topographic correction allows for differences in elevation above or below mean sea-level of a station. In addition, there is a free-air reduction correction taking into account its elevation and assuming that there is no mass between it and the surface of reference, i.e. as if in 'free air'.

In marine gravimetry, the same correction is combined with twice the attraction of an infinite slab of water of an elevation equal to the distance between the instrument and mean sea-level. The Bouguer correction allows approximately for the effect of the mass between the surface of the Earth and the datum plane, usually the geoid, or, in water-covered regions, for the effect of the mass deficiency between mean sea-level and the sea-floor. This mass deficiency is assumed to be identical with an infinite horizontal plate. The attraction of an infinite (Bouguer) plate is independent of the distance to the gravity station. The effect of topography after application of Bouguer reduction (see above) is termed terrain correction. The anomaly after application of the Bouguer reduction is termed the Bouguer anomaly (see above). A modified Bouguer anomaly is that remaining after application of total topographic reduction for Hayford's zones A–O and topographic and isostatic reduction for Hayford's zones 1–18 (see below). An isostatic anomaly results from the application of an isostatic reduction (see Isostasy). A regional gravity map shows only the gradual changes of gravity and an isanomalic or contour map shows lines connecting points of equal anomaly. Isogal and isogam are alternative words for isanomalic contour.

Hayford zone: A sub-division, one of many made of the globe for facilitating calculation of the topographic and isostatic reductions around a gravity station.

Heat: Energy possessed by a substance in the form of kinetic energy of atomic or molecular translation, rotation or vibration. The heat contained by a body is the product of its mass, temperature and a specific heat capacity. It may be expressed as joules (SI units), calories or British thermal units.

Heat capacity: That quantity of heat required to raise the temperature of a body by 1 K with the following unit: joules/Kelvin (J/K) $(1 \, \mathrm{J/K} = 239 \times 10^{-3} \, \mathrm{cal/°C})$. (See also Specific heat capacity.)

Heat conduction: The transfer of heat as the result of molecular interaction. This is the major form of heat transfer in rocks.

Heat convection: The transfer of heat through the motion of deformable media.

Heat-current density: Heat flow, i.e. the quantity of heat which flows through a unit area in a unit time. Where heat transfer by conduction is concerned, it may be expressed as the product of temperature gradient and thermal conductivity. The relevant unit is watt/metre^2 (1 W/m$^2 =$ 0·0239 mcal/cm^2 s^{-1}).

Heat flow provinces: That linear correlations between heat flow, q_0, and surface heat production, A_0, in plutonic and metamorphic terrains from around the world exist is well established and the following expression is applicable:

$$q_0 = q^* + DA_0$$

in which D is a function of the depth extent of the upper crustal zone of radionuclide enrichment and q^* represents the heat flow contribution from beneath this zone. Those regions in which the relationship is satisfied are known as heat flow provinces.

In the case of basement rocks (granite plutons and Early Palaeozoic metasediments) in southern Britain, heat flow and heat production data seem to be related in a linear manner and data were interpreted in terms of a classic q_0–A_0 heat flow province with $q^* = 27$ mW/m^2 and $D = 16·6$ km.[2,3] Subsequently, P. C. Webb and others in 1987 claimed that additional q_0–A_0 determinations made during an assessment of the hot dry rock geothermal potential of the UK made the current data set irreconcilable with the concept of a single UK heat flow province.[4] Their interpretation of the available information from two voluminous granite batholiths accords with geochemical evidence that depth distributions of heat production differ according to magma type.

Heat flow rate: The quantity of heat which passes across a certain surface in unit time. 1 W = 239 mcal/s.

Heat flow unit (HFU): A practical unit of heat current density in the lithosphere. 1 HFU = 10^{-3} W/m^2.

Heat radiation: Transfer of heat by the emission of electromagnetic waves.

Hot dry rock (HDR): This constitutes a heat resource accessible using modern drilling technology and devoid of a working fluid because of the low permeability of the deep-seated rocks. Extraction may be possible by introduction of an appropriate artificial fluid to create a circulation system. Obviously, boreholes are vital and, since drilling them entails costs which increase almost exponentially with depth, the temperature–depth curve must be assessed prior to commencing such an exercise.

Hot spring: A spring created when the quantity of subterranean water is too great to be converted into steam by a magma, the heated water flowing upwards until emergence.

Hot-water system: This term may be applied to a geothermal resource in which water temperatures are usually below the boiling point at atmospheric pressure, ranging from 50°C upwards. Water from such systems is similar to the adjacent groundwater and also to the surface waters of nearby regions. In fact, it often appears as hot springs.

Hydrothermal processes: These are associated with igneous activity and involve heated or superheated water. The latter is very active and can break down silicates as well as dissolving substances normally considered as insoluble. There are two major possible consequences, namely alteration processes, e.g. kaolinization and serpentinization, and deposition processes responsible for ores, for instance of copper, lead and zinc.

Hydrothermal system: This is a sub-class of the convective geothermal system occurring in high-porosity/permeability environments related to shallow, young silicic intrusions. Almost all geothermal systems so far developed for the commercial production of electric power fall into this sub-class. It comprises two sub-divisions depending upon the heat: discharge balance found. These are vapour-dominated systems and liquid-dominated, the latter being the more common.

Hyperthermal field: A wet geothermal resource producing pressurized water at a temperature exceeding the boiling point. Boiling water may emerge as a fluid at the surface, but a fraction flashes into steam. Alternatively, a dry geothermal resource producing saturated or slightly superheated steam at pressures above atmospheric. Both types fall into the category of hydrothermal, convective, liquid-dominated systems.

Hyperthermal region: One in which the geothermal gradient exceeds 80°C/km.

Induced polarization (IP) method: The same electrode array and a similar set-up to those employed in resistivity surveys are utilized and the IP effect measured. Monitoring of either the slow decay of voltage in the ground after a short current pulse (time domain method) or of low frequency variations of earth impedance (frequency domain method) is carried out.

Isenthalpic: That situation existing when the enthalpy of a flowing steam and water mixture entering a well below ground includes liquid which was originally so at a distance, but boiled on its way. The consequence may be the creation of a two-phase zone locally surrounding the feed point of the well.

Isostasy: The phenomenon by which the masses between the Earth's surface and the mean sea-level in land regions and the mass deficiencies between sea-level and the ocean floor in water-covered regions are compensated, in general, by masses deeper within the planet with reversed density signs. Thus is achieved a theoretical balance of all large parts of the terrestrial crust as though these were floating on a denser underlying layer. Consequently, areas of less dense crustal material rise topographically above areas of denser material.

Isothermic plane: A plane of equal temperature within the Earth.

Liquid-dominated system: A sub-division of the hydrothermal system sub-class of the convective geothermal system in which both steam and water usually exist in varying ratios, the chemical composition of the geothermal fluids characterizing the system. Examples include Imperial Valley, California, USA, and Wairakei and Broadlands in New Zealand.

Lithosphere of the Earth: The rigid outermost shell of the Earth which translates coherently during plate movements. Seismologically, it constitutes material lying above the upper mantle low-velocity zone and, from a rheological standpoint, it is the material above the thermally controlled depth where ductile behaviour becomes predominant (as assessed, for example, in terms of effective viscosity or creep strength). However, none of these definitions is unambiguous. The low-velocity zone is absent in a number of cratonic areas and the thickness and properties of plates vary with age and tectonic province. Also decouplings of a 'flake' of a plate over another one along sub-horizontal planes are not rare. Finally, the rheological properties of the lithosphere (however defined) are a function of the process used to determine them, which indicates their dependence upon magnitude and rate of stress and strain. In fact, a rheological stratification of the lithosphere may reflect the stresses originating from the deformation process or be a function of the properties of the actual materials involved. G. Ranalli and D. C. Murphy discussed the phenomenon in 1987.[5]

Magnetotelluric method: This entails measurement of components of electric and magnetic fields due to fluctuating telluric currents and facilitates the acquisition of deep structural information.

MAGSAT: This is a NASA satellite which was launched on 30 October 1979 and descended on 11 June 1980. It had an initial orbit with apogee of 561 km and perigee of 352 km which decayed to a more circular shape (apogee ≤ 400 km) as the mission approached completion. Magnetic field measurements were effected using a triaxial set of fluxgate magnetometers for directional (vector) field components and a caesium-vapour magnetometer for the absolute (scalar) field. The latter operated sporadically and was used mainly to calibrate the vector instruments. MAGSAT was the first satellite to provide a global survey of tri-vector magnetic fields and subsequent maps. Since it was designed in part for solid-Earth magnetic field studies, its lower-altitude orbit gave the most accurate and highest-resolution satellite surveying of the geomagnetic field so far available. The measured field includes contributions from the terrestrial core field (of magnitude up to *ca.* 50 000 nT), from geomagnetic transient fields from external ionosphere/magnetosphere current systems (up to hundreds of nanotesla (nT)) and from magnetic-anomaly fields produced from magnetization in the Earth's crust (up to a few tens of nanotesla). Extraction of the crustal anomaly fields by data pre-processing by NASA at the Goddard Space Flight Center resulted in observed anomalies with magnitudes up to ± 20 nT and spatial resolution of 200–250 km to map crustal magnetic sources. This resolution is twice as good as the first satellite data available for preliminary mapping of crustal anomalies, from the absolute-field measuring and high-altitude POGO (Polar Orbiting Geophysical Observatory) missions of 1965 to 1971. It may be added that NASA is proposing to launch an even higher-resolution Geopotential Research Mission, GRM (GRAVSAT/MAGSAT, satellite from the Space Shuttle in about 1990. This would comprise a satellite pair in circular orbit around the Earth at an altitude of only 160 km, measuring both gravity field to an accuracy of 1 mGal and vector and scalar magnetic fields to an accuracy of 2 nT.

Mantle of the Earth: The part of the planetary interior between the Mohorovičić and Gutenberg discontinuities, i.e. from about 35 km down to 2900 km. Densities range from 3·3 g/cm³ at the Moho to 5·7 g/cm³ at the Gutenberg discontinuity. The upper mantle is believed to be composed mainly of garnet, spinel or plagioclase lherzolite depending upon the depth. This is supported by the nature of ultramafic xenoliths carried to the surface by alkali basalts and kimberlites, and of ultramafic orogenic masifs such as those of western Europe thought to have been directly emplaced from the underlying upper mantle.

Mass transfer of heat: That phenomenon resulting from the movement of material possessing a higher temperature than its surroundings, e.g. plutonism, volcanism and thermal springs.

Micro-earthquake: An earthquake of magnitude 2 or less on the Richter scale; the limit of 2 is arbitrary.

Mise-à-la-masse method: This is a resistivity technique whereby one current electrode is inserted into the accessible portion of a conducting body such as a mineral deposit in a borehole, and the second electrode is placed at a great distance. The potential electrodes are shifted around with the object of mapping the potential differences present around the conducting body. From such data the shape and dimensions of the conducting body can be assessed.

Mofette: A type of fumarole.

Moho, Mohorovičić discontinuity: A seismically defined outer layer within the Earth which separates the crust from the mantle and has an average depth of 35 km.

Mud pots: Almost waterless hot springs, what water is present being mixed with fine, insoluble mineral matter, mostly clays, to form a viscous, variously coloured wet mud of differing consistency. The colours are due to a sulphur shortage because an abundance of this element would transform the iron oxides into pyrite, a grey mineral. Where the mud is unusually viscous, the feature becomes a mud volcano. Usually, mud pots occur in areas where the considerable hydrogen sulphide content of acidic spring waters has decomposed the surface rocks. They may be single or grouped and are found in most geothermal regions such as the

Yellowstone National Park, Wyoming, USA, where some are as much as a hectare in area. A famous one called Ngapuna Tokatoru near the Puarenga Geyser in the Whakarewarewa area in New Zealand has surface patterns which alter seasonally.

Mud volcano: If a mud pot emits highly viscous mud, blobs of it may be thrown out and coalesce to form a cone or mound. This is a mud volcano and hard-baked, almost extinct ones have often been found in geothermal regions.

Nano-earthquake: This word is sometimes used for a small-scale micro-earthquake.

Noise: There is no standard terminology for the various types of noise and hence the expression 'ambient noise' can be used to refer to everything attributable to the environment of a recording experiment. Ambient noise includes: (a) low-amplitude 'background noise' regarded as ergodic and attributed to wind, moving water, micro-seismic activity and far-off, incoherently scattered, cultural noise; (b) time-stationary noise from 'static ambient noise sources' arising from fixed machinery or localized off-line activity; (c) 'noise bursts' from moving ambient noise sources that vary on a time-scale of a few seconds, often vehicular in origin; (d) 'noise spikes' from electrical sources such as atmospheric discharge of piezo-electric impulses.

Orogeny: A period of mountain building leading to the formation of those intensely deformed belts constituting mountain ranges. Belts of such deformed rocks may be termed orogens.

Potassium–argon dating: A method of age determination based upon the accumulation of stable argon-40 produced *in situ* by the radioactive decay of potassium-40 by electron capture and positron emission with a decay energy of 1.51 MeV: 11% of potassium-40 atoms decay by electron capture to an excited state of argon-40 which de-excites by emission of a γ-ray with an energy of 1.46 MeV. Also 0.16% of the decays are by electron capture directly to the ground state of argon-40. Positron decay takes place only 0.001% of the time and is succeeded by two annihilation γ-rays with a combined energy of 1.02 MeV. Decay to stable calcium-40 is favoured by 88.8% of the potassium-40 atoms and occurs with negatron emission directly to the ground state with a decay energy of

1·32 MeV. Consequently, only about 11·2% of the potassium-40 atoms decay to argon-40.

Recommended decay constants are:

(1) For potassium-40 to argon-40: $0·581 \times 10^{-10} \, a^{-1}$;
(2) For potassium-40 to calcium-40: $4·962 \times 10^{-10} \, a^{-1}$;

and therefore the total decay constant for potassium is $5·543 \times 10^{-10} \, a^{-1}$, corresponding to a half-life of $1·250 \times 10^9 \, a$.

Radioactive (radiogenic) heat: That heat arising from radioactive decay comprising the principal heat source in the lithosphere.

Resistivity: That resistance offered by a cube of unit dimensions of a material to the flow of electric current. Its unit is the ohm metre, $\Omega\,m$ (i.e. $\Omega\,m^3/m^2$).

Rubidium–strontium dating: A method of age determination based upon the accumulation of strontium-87 produced *in situ* by the radioactive decay by β-emission of rubidium-87. The relevant half-life has been measured often, but with contradictory results. The most recent value is $48·8 \times 10^9 \, a$, corresponding to a decay constant for rubidium-87 of $1·42 \times 10^{-11} \, a^{-1}$.

Samarium–neodymium dating: These are rare earth elements linked in a parent–daughter relationship by the α-decay of samarium-147 to stable neodymium-143 with a half-life of $106 \times 10^9 \, a$, the decay constant being $6·54 \times 10^{-12} \, a^{-1}$. They occur in many rock-forming silicate, phosphate and carbonate minerals, and the decay scheme mentioned is useful in dating terrestrial rocks, stony meteorites and lunar rocks. Also the growth of radiogenic neodymium-143 and strontium-87 together give a new insight into the geochemical evolution of planetary objects and the genesis of igneous rocks.

Self-potential (SP) or spontaneous polarization method: This involves the measurement of natural electrical fields due to electrochemical effects around sulphide ore bodies or to the physical effects of flowing fluids such as groundwater.

Semi-thermal region: One in which the geothermal gradient is between 40 and 80°C/km.

Sial: A rock layer underlying all continents which ranges from granitic at the top to gabbroic at the base. Its thickness varies from 30 to 35 km and its name derives from the main constituents, namely silica and alumina.

Sill: A sheet of igneous rock conforming to bedding or structural planes in a host rock. Sills may be composite or multiple and sometimes are quite extensive.

Sima: The outer shell of the Earth. Under continents it underlies the sial, but under oceans it directly underlies oceanic water. Once considered basaltic in nature, specific gravity ≈ 3, it has more recently been regarded as peridotitic in composition, specific gravity $\approx 3\cdot3$.

SLAR, side-looking airborne radar: Radar itself is a radio detection and ranging method for obtaining images of the surface of the Earth and/or locating and identifying distant objects. Transmitted radio waves, normally in the 3 cm band, are scattered by material objects and the returning waves are recorded. By emitting pulses of high-frequency waves, it is possible to measure accurately the distances of such objects. SLAR is a variant of ordinary radar, comprising an all-weather, day and night, airborne radar especially capable of imaging large areas of terrain and yielding photograph-like pictures. It can be used to produce planimetric, geomorphic, geological and other maps in areas obscured as regards visible or infra-red methods by clouds or atmospheric haze, but it cannot give such detailed data as ordinary aerial photography. Physiographical features such as mountains and plains are easily recognized by both approaches and objects as small as a few metres across can be identified. Variations in tone found in radar imagery stem from the reflectivity characteristics of a region rather than from the factors of brightness and colour significant in photography. As a variety of wavelengths is available to SLAR operators, that selected is critical in that the reflectance of a particular surface varies with the wavelength employed. Also the penetration of a surface increases with wavelength.

Specific heat capacity: That quantity of heat necessary to elevate the temperature of a unit mass of a material by 1°. It has the following unit: $J/kg\,K$ ($1\,J/kg\,K = 239\,cal/g\,°C$).

Spheroid: This word is used in geodesy to describe a surface of revolution approaching an ellipsoid of revolution with slight flattening. In fact, the equilibrium surface of a fluid Earth is a spheroid.

Steam: Water above its boiling point and in the gaseous state. The white clouds emitted from geysers comprise droplets of liquid water condensed from steam, a phenomenon also seen when kettles boil.

Subduction zone: A region on Earth in which convergent movement of plates occurs, the subducting (oceanic) one being thrust under the over-riding (obducting) one to produce orogeny on the continental side and island arcs and deep sea trenches on the oceanic side. Subduction zones constitute some of the most peculiar localities on the planet. Their first manifestation was the deep-focus earthquake plane discovered by Wadati in 1935 and investigated further by Benioff in 1954, this being a zone in which the lithospheric material descends into the deep mantle. Structures and processes in such subduction zones were the subjects of an interdisciplinary symposium (No. 1) at the 18th General Assembly of the International Union of Geodesy and Geophysics at Hamburg in the FRG in August 1983.[6]

Televiewer: The borehole televiewer is an ultrasonic logging tool providing high-resolution data regarding borehole elongation and natural fracture distributions.

Telluric currents: Natural electrical currents in the Earth caused by variations in its magnetic field and by its rotation through external (ionospheric) electrical fields. They may show rapid variation in intensity and also in direction during magnetic storms. Apart from during these latter, rather rare, events, the relevant electric field intensities are of the order of 1–10 mV/km. They may disturb self-potential (SP) measurements.

Terrain correction: A term used for the effect of topography after application of the Bouguer reduction.

Terrestrial heat flow: The transfer of heat in the Earth which is usually directed towards the surface and tends to equalize temperature differences.

Thermal conductivity: The quantity of heat transferred across a unit area in a unit time as the result of a unit temperature gradient with the following unit: W/m K (1 W/m K $= 2\cdot39$ mcal/cm s °C).

Thermal diffusivity: A measure of the rate of penetration of heat through conduction into a medium. It is expressed as the thermal conductivity divided by the product of specific heat capacity and density of the material penetrated.

Thermistor: Highly accurate thermometer appropriate for employment in boreholes and based upon the relationship between electrical resistance and temperature.

Thermocouple: Highly accurate thermometer appropriate for employment in boreholes and based upon the thermo-electric effect in a closed circuit comprising two or more electrical conductors.

Time domain: The induced polarization method is applied in mineral prospection and its effects measured by a similar outfit and the same electrode array that is used for resistivity surveys. Observations are made on either the slow decay of voltage in the ground following a short current pulse or low-frequency variation of earth impedance. The first constitutes the time domain method and the second the frequency domain one.

Tomography: A recent seismological development dependent upon high-speed digital computation is tomography, a technique long used in radio astronomy and referring to the numerical reconstruction of an image of internal structure from data expressed as a functional (line integrals or Radon transform) of the function (parameter) to be imaged.[7] In simplest form, it was used to infer the core of the Earth and has been generalized for computer algorithms to the block perturbation method of K. Aki and others.[8] Tomographic algebraic reconstruction illustrates both the frustrations and dilemmas of seismologic exploration of the interior of the Earth and a classical geophysical tomography problem is to reconstruct a velocity profile for a part of the Earth such that the travel-times agree with measurements. This can be done by solving a large, sparse, least-squares problem. Traditionally, such problems in geophysical tomography have been approached by row-action methods, e.g. the algebraic reconstruction technique of P. Gilbert in 1972.[9] Also the conjugate gradient algorithm of M. Hestenes and E. Stiefel in 1952 has been applied.[10] In 1987 J. A. Scales discussed tomographic inversion versus this latter, applying the conjugate algorithm method to the least-squares solution of large sparse systems of travel-time equations.[11] It proved to be fast, accurate and suitable for the sparsity of the matrix. A quite comprehensive discussion of tomography was given by K. E. Bullen and B. A. Bolt in 1985.[12]

Transmissivity: The product of permeability or hydraulic conductivity and thickness of the reservoir rock.

Travertine: A type of calcareous tufa deposited by some hot springs in volcanic areas.

Triple junction: The place of meeting of two plate boundaries or three plates.

Tufa: Calcareous tufa comprises masses of calcium carbonate deposited from a solution of calcium bicarbonate.

Uranium and thorium series: The decay series arising from uranium-238 and -235 as well as from thorium include many isotopes with different chemical properties and mean lives which can cause significant radioactive disequilibria in groundwaters or geothermal waters as well as possibly in solid phases closely associated with such waters. The decay of uranium-238 initiates the uranium series ending in stable lead-206 by emission of eight α-particles and six β-particles. The decay of uranium-235 produces the actinium series ending with stable lead-207 after emission of seven α- and four β-particles. The decay of thorium-232 results in the emission of six α- and four β-particles with the formation of stable lead-208 as its end product. Forty-three isotopes of 12 elements are created as intermediate daughters in these decay series, but none is a member of more than one series. That is to say, each decay chain always leads to the formation of a specific isotope of lead. The half-lives of all three parents, uranium-238, uranium-235 and thorium-232, are all much longer than those of their respective daughters. Consequently, these decay series satisfy the prerequisites for the establishment of secular equilibrium.

Vapour-dominated system: Steam predominates in this sub-division of the hydrothermal sub-class of the convective geothermal system. Examples include The Geysers in California, USA, and Larderello in Italy.

Watt: The energy expended per second by an unvarying electric current of 1 ampere flowing through a conductor of which the ends are maintained at a potential difference of 1 volt (1 W = 1 J/s).

Working fluid: A naturally circulating fluid, water or steam, or a mixture of both, which transfers heat in a convective geothermal system.

REFERENCES

1. GARNISH, J. D., VAUX, R. and FULLER, R. W. E., 1986. *Geothermal Aquifers, Department of Energy R&D Programme 1976–1986,* ed. R. Vaux, pp. 19–21. Energy Technology Support Unit for the Department of Energy, AERE, Harwell, Oxfordshire OX11 0RA, UK, 109 pp. and appendices.
2. RICHARDSON, S. W. and OXBURGH, E. R., 1978. Heat flow, radiogenic heat production and crustal temperatures in England and Wales. *J. Geol. Soc. London,* **135,** 323–37.
3. RICHARDSON, S. W. and OXBURGH, E. R., 1979. The heat flow field in mainland UK. *Nature (London),* **282,** 565–7.
4. WEBB, P. C., LEE, M. K. and BROWN, G. C., 1987. Heat flow–heat production relationships in the UK and the vertical distribution of heat production in granite batholiths. *Geophys. Res. Lett.,* **14**(3), 279–82.
5. RANALLI, G. and MURPHY, D. C., 1987. Rheological stratification of the lithosphere. *Tectonophysics,* **132,** 281–95.
6. KOBAYASHI, K. and SACKS, I. S. (eds), 1985. Structures and processes in subduction zones. Selected papers from the Interdisciplinary Symposium No. 1, 18th General Assembly of the International Union of Geodesy and Geophysics, Hamburg, 22–24 August 1983. *Tectonophysics,* Sp. issue **112**(1–4), 1–561.
7. DEANS, S. R., 1983. *The Radon Transform and Some of its Applications.* John Wiley & Sons, New York.
8. AKI, K., 1967. Scaling law of seismic spectra. *J. Geophys. Res.,* **72,** 1217–31.
9. GILBERT, P., 1972. Iterative methods for the three-dimensional reconstruction of an object from projections. *J. Theor. Biol.,* **36,** 105–17.
10. HESTENES, M. and STIEFEL, E., 1952. Methods of conjugate gradients for solving linear systems. *Nat. Bur. Standards J. Res.,* **49,** 409–36.
11. SCALES, J. A., 1987. Tomographic inversion via the conjugate gradient method. *Geophysics,* **52**(2), 179–85.
12. BULLEN, K. E. and BOLT, B. A., 1985. *An Introduction to the Theory of Seismology.* Cambridge University Press, London, 499 pp.

Geothermal Miscellanea

'Quod si cui mortalium cordi et curae sit, non tantum inventis haerere, atque iis uti, sed ad ulteriora penetrare; atque non disputando adversarium, sed opere naturam vincere; denique non belle et probabiliter opinari, sed certo et ostensive scire; tales tanquam veri scientiarum filii, nobis se adjungant.'

SIR FRANCIS BACON (*Novum Organum*, Prefatio, 1620)

There are 72 publications in the following list, mostly very recent. The majority deal with matters considered peripheral to the book, discussing speculative questions such as the origins and histories of geothermal fields and the sources of their fluids, upper crustal structure and crustal energetics as well as forecasting low-thermal water resource with isostatic methods and describing researches which are not yet economically significant. Specialized aspects of stable and radioactive (radon) isotope geochemistry and crystallography (Ca-pyriboles) as well as engineering operations in geothermal well-drilling are covered by some of them. Others relate to planning, e.g. future tasks entailed in producing geothermal energy in Hungary, possible exploitation of such energy in the Bohemian Massif, hot dry rock developments in Japan and a feasibility study for a 55-MW generator in the Los Humeros area of Mexico. A few refer to abstruse matters such as a statistical study of the geothermal field and the so-called technological types of geothermal systems. A number of interesting contributions to the first EEC/US workshop on geothermal hot dry rock technology and to the 5th international symposium on water–rock interactions, both held in 1986, have been included. In this manner, very localized matters not covered in the text are made accessible to the interested reader, for instance the origin of hot springs at Polichnitos in Greece, the geochemistry of thermal springs in Monghyr District, Bihar,

India, the classification of karstic and karstic–thermal springs at West Bükk in Hungary, and geothermal resource investigations at Twentynine Palms in California. Other topics are the extent of geothermal fields such as Larderello and Los Humeros, their structure (for instance that of Las Derrumbadas) and the history of their geothermal fluids, e.g. those of Krafla and Reykjanes in Iceland. A survey of the geothermal industry in Iceland is listed. In addition, descriptions of researches proceeding on apparent resistivity measurement and silica geothermometry are given. There are also a few treating continuing work in areas already adequately discussed earlier—for instance, two on the further development of the San Kamphaeng geothermal region in Thailand, contributions being cited from the 5th regional congress on the geology, mineral and energy resources of southeast Asia which took place in 1986. Another instance is a paper on the geothermal resources of the United Kingdom, namely low-enthalpy resources in deep sedimentary basins and hot dry rock in radiothermal granites. A fourth relates to the Qualibou Calders work on St Lucia in the West Indies. Although some of the themes discussed in the references below are well outside the scope of this book (which is earth science-orientated rather than concerned with, say, engineering or statistics), they are sufficiently relevant to be mentioned and this is done by grouping the appropriate papers under a general heading or under the names of the 23 countries involved (listed alphabetically).

General

1. ALTHAUS, E. and EDMUNDS, W. M., 1986. Geochemical research in relation to hot dry rock geothermal systems. In: Proc. 1st EEC/US workshop on geothermal hot dry rock technology, May 28–30 1986, Brussels, Belgium, ed. J. D. Garnish. *Geothermics*, **16**(4), 451–8.
2. FOURNIER, R. O., 1987. Comments based on hydrologic and geochemical characteristics of active hydrothermal systems. In: *Coupled Processes Associated with Nuclear Waste Repositories, September 18–20 1985, Berkeley, CA, USA*, ed. C.-F. Tsang, pp. 733–7. Harcourt Brace Jovanovich, Orlando, FL, USA.
3. GARNISH, J. D., 1987. Introduction: background to the workshop. In: Proc. 1st EEC/US workshop on geothermal hot dry rock technology, May 28–30 1986, Brussels, Belgium, ed. J. D. Garnish. *Geothermics*, **16**(4), 323–9.
4. HAENEL, R., PEARSON, R. A. and PINE, R. J., 1987. Borehole measurements in the field of hot dry rock experiments. In: Proc. 1st EEC/US workshop on geothermal hot dry rock technology, May 28–30 1986, Brussels, Belgium, ed. J. D. Garnish. *Geothermics*, **16**(4), 433–9.
5. O'SULLIVAN, M. J., 1985. Aspects of geothermal well test analysis in fractured reservoirs. In: Single and multi-phase fluid flow through heterogeneous

permeable materials, November 18–25 1985, Hamilton, New Zealand. Pt 3: ed. I. G. Donaldson, *Transport in Porous Media*, **2**(5), 497–517.

Canada

1. JONES, A. G. and GARLAND, D. G., 1984. Preliminary interpretation of the upper crustal structure beneath Prince Edward Island. In: 7th Workshop on electromagnetic induction in the Earth and the Moon, August 15–22 1984, Ile Ife, Nigeria, ed. M. Menvielle. *Annales Geophysicae, Ser. B, Terrestrial and Planetary Physics*, **4**(2), 157–63.

China

1. XUQUAN, W., 1987. Forecasting of low-thermal water resource with isostatic method. *Hydrogeology and Engineering Geology*, **1987**(5), 41–4; 54 (Chinese summary).
2. YAO ZUJIN, 1986. The comparison on distribution between current geothermal system and Mesozoic ore deposits in South China. In: *5th Int. Symp. on Water–Rock Interactions, August 8–17 1986, Reykjavik, Iceland*, chairperson B. Hitchon, pp. 644–7.

Czechoslovakia

1. HAZDROVA, M., 1986. Possibilities and ways of exploitation of geothermal energy in the Bohemian Massif. In: *5th Int. Symp. on Water–Rock Interactions, August 8–17 1986, Reykjavik, Iceland*, chairperson B. Hitchon, pp. 248–50.
2. JURANEK, J. and HARNOVA, J., 1986. Correction of the chemistry of calcium in hydrogeothermal pressure systems; MINEQUA computer program. *Vodni Hospodárstvi, Rada B*, **36**(1), 21–4 (Russian, English summary).
3. JURANEK, J. and PULKRÁBKOVÁ, Z., 1987. 'MINEQUA'; Part II, Carbonate equilibria in the thermal water of the Podhájska geothermal system. *Geothermics*, **16**(5/6), 555–65.

France

1. GERARD, A. and KAPPELMEYER, O., 1987. The Soultz-sous-Fôrets Project. In: Proc. 1st EEC/US workshop on geothermal hot dry rock technology, May 28–30 1986, Brussels, Belgium, ed. J. D. Garnish. *Geothermics*, **16**(4), 393–9.
2. MOSNIER, J., 1987. The work of the Orleans Laboratory of Applied Geophysics. In: Proc. 1st EEC/US workshop on geothermal hot dry rock technology, May 28–30 1986, Brussels, Belgium, ed. J. D. Garnish. *Geothermics*, **16**(4), 405–7.

Germany (Federal Republic)

1. KAPPELMEYER, O. and JUNG, R., 1987. HDR experiments at Falkenberg/ Bavaria. In: Proc. 1st EEC/US workshop on geothermal hot dry rock

technology, May 28–30 1986, Brussels, Belgium, ed. J. D. Garnish. *Geothermics*, **16**(4), 375–92.

2. SAUER, K., 1985. The use of geophysical methods to heighten the security in the discovery and protection of thermal and mineral springs. In: *Geothermics, Thermal–Mineral Waters and Hydrogeology*, ed. E. Romijn, pp. 167–80. Theophrastus Publ., Athens, Greece.

3. WOHLENBERG, J. and KEPPLER, H., 1987. Monitoring and interpretation of seismic observations in hot dry rock geothermal energy systems. In: Proc. 1st EEC/US workshop on geothermal hot dry rock technology, May 28–30 1986, Brussels, Belgium, ed. J. D. Garnish. *Geothermics*, **16**(4), 441–5.

Greece

1. FYTIKAS, M. and KAVOURIDIS, T., 1985. Geothermal areas of Sousaki–Loutraki. In: *Geothermics, Thermal–Mineral Waters and Hydrogeology*, ed. E. Romijn, pp. 19–34. Theophrastus Publ., Athens, Greece.

2. PAPASTAMATAKI, A. and LEONIS, C., 1985. The origin of Polichnitos hot springs. In: *Geothermics, Thermal–Mineral Waters and Hydrogeology*, ed. E. Romijn, pp. 141–9. Theophrastus Publ., Athens, Greece.

3. SCHROEDER, B., 1985. Structural evolution and mineral, thermal springs of the Neogene basin at the Isthmus of Corinth, Greece. In: *Geothermics, Thermal–Mineral Waters and Hydrogeology*, ed. E. Romijn, pp. 111–19. Theophrastus Publ., Athens, Greece.

Hungary

1. SCHEUER, G., 1985. Classification of the karstic and karstic–thermal springs of West Bükk. *Mernok-Geologiai Szemle*, **34**, 193–209.

2. SZILAS, A. P., 1985. Geothermal energy production in Hungary; present situation and tasks for the future. *Foldtani Kutatas*, **28**(3), 71–4.

Iceland

1. GUDMUNDSSON, J. S. and PÁLMASON, G., 1987. The geothermal industry in Iceland. *Geothermics*, **16**(5/6), 567–73.

2. SVEINBJORNSDOTTIR, A. E., COLEMAN, M. L. and YARDLEY, B. W. D., 1986. Origin and history of hydrothermal fluids of the Reykjanes and Krafla geothermal fields, Iceland. *Cont. Min. Petrology*, **94**(1), 99–109.

India

1. MUKHERJEE, A. L., 1985. Geochemistry and hydrology of thermal springs in Monghyr District, Bihar, India. In: *Geothermics, Thermal–Mineral Waters and Hydrogeology*, ed. E. Romijn, pp. 151–63. Theophrastus Publ., Athens, Greece.

Israel

1. BEIN, A. and ARAD, A., 1986. Fresh and saline thermal water–rock interaction in analogous carbonate aquifers. In: *5th Int. Symp. on Water–Rock Interactions, August 8–17 1986, Reykjavik, Iceland,* chairperson B. Hitchon, pp. 43–4.

Italy

1. CECCARELLI, A., 1987. The southern boundary of Larderello geothermal field. *Geothermics,* **16**(5/6), 501–13.
2. LAPENNA, V. A., SATRIANO, C. and PATELLA, D., 1987. On the methods of evaluation of apparent resistivity under conditions of low message-to-noise ratio. *Geothermics,* **16**(5/6), 487–504.

Japan

1. CHIBA, H., 1986. Chemistry and stable isotope study of fluids from the Hacchobaru geothermal system, Japan. In: *5th Int. Symp. on Water–Rock Interactions, August 8–17 1986, Reyjkavik, Iceland,* chairperson B. Hitchon, pp. 127–9.
2. KATAGIRI, K. and ANDO, S., 1987. Exploration for geothermal resources and drilling of geothermal wells. *Tsuchi-To-Kiso JSSMFE* (Soil Mechanics and Foundation Engineering), **35**(6 Series (353)), 25–31.
3. KURIYAGAWA, M., 1987. Hot dry rock geothermal energy in Japan; developments by MITI and NEDO. In: Proc. 1st EEC/US workshop on geothermal hot dry rock technology, May 28–30 1986, Brussels, Belgium, ed. J. D. Garnish. *Geothermics,* **16**(4), 401–3.

Jordan

1. RIMAWI, O., 1988. Hydrochemistry and groundwater system of the Zerka Ma'in-Zara thermal field, Jordan. *J. Hydrol.,* **98**(1–2), 147–63.

Mexico

1. ARRENDONDO FRAGOSOS, J., 1987. Gravimetric study to delineate the extent of the geothermal field of La Caldera de Los Humeros, Puebla. *Geotermia,* **3**(1), 53–63 (English summary).
2. BALLINA LÓPEZ, H. R., 1987. A geoelectric resistivity study on the eastern area of Los Azufres, Michoacan. *Geotermia,* **3**(2), 145–54 (English summary).
3. BIGURRA PIMENTEL, E., 1987. Natural potential of geothermal energy. *Geotermia,* **3**(2), 171–80 (English summary).
4. CAMPOS-ENRIQUEZ, J. O. and GARDUNO-MONROY, V. H., 1987. The shallow structure of Los Humeros and Las Derrumbadas geothermal fields, Mexico. *Geothermics,* **16**(5/6), 539–54.

5. CENDEJAS, F. A., 1987. Feasibility studies of a 55-MW generator in the Los Humeros geothermal field, Puebla. *Geotermia*, **3**(1), 77–92.
6. CHAVEZ, R. E., 1987. An integrated geophysical study of the geothermal field of Tule Chek, BC, Mexico. *Geothermics*, **16**(5/6), 529–38.
7. DÍAZ MOLINARI, M., 1987. Evaluation of the dynamic elastic properties in the area of the machine housing and cooling tower for the Los Azufres geothermal center, Michoacan. *Geotermia*, **3**(1), 65–76 (English summary).
8. HERRERA FRANCO, J. J. and CASTILLO HERNANDEZ, D., 1987. Geothermal resources in Jalisco. *Geotermia*, **3**(1), 3–18 (English summary).
9. MILÁN VALDEZ, M., 1987. Important geologic aspects of a geothermal exploration exploration operation in Pathe Field, Hildago and Queretaro. *Geotermia*, **3**(1), 31–9.
10. PALMA GUZMÁN, H. and BIGURRA PIMENTEL, E., 1987. Geoelectric analysis around the proposed well AZ-51 in Los Azufres, Michoacan. *Geotermia*, **3**(1), 41–52 (English summary).
11. ROSAS ELGUERA, J., 1987. Subsurface structures as indicators in geothermal exploration. *Geotermia*, **3**(1), 19–29 (English summary).

New Zealand

1. PROL-LEDESMA, R. M., 1986. Hydrothermal alteration patterns and fluid inclusions in Rotokawa geothermal field (drill hole RK-6), New Zealand. In: *5th Int. Symp. on Water–Rock Interactions, August 8–17 1986, Reyjkavik, Iceland*, chairperson B. Hitchon, pp. 448–51.

St Lucia

1. WOHLETZ, K., KEIKEN, G., ANDER, M., GOFF, F., VUATAZ, F. and WADGE, G., 1986. The Qualibou Caldera, Saint Lucia, West Indies. *J. Volcanol. Geotherm. Res.*, **27**(1/2), 77–115.

Switzerland

1. MAZOR, E., DUBOIS, J. D., FLUCK, J. and JAFFÉ, F. C., 1988. Noble gases as tracers identifying geothermal components in regions devoid of surface geothermal manifestations; a case study in the Baden springs area, Switzerland. *Chem. Geol., Isotope Geoscience Section*, **72**(1), 47–61.

Thailand

1. RAMINGWONG, T. and PRASERDVIGAI, S., 1986. Development of San Kamphaeng geothermal energy project, Thailand. In: 5th Regional Congress on geology, mineral and energy resources of southeast Asia, April 9–13 1986, ed. G. H. Teh. *Bull. Geol. Soc. Malaysia*, **19**, 627–41.
2. SERTSRIVANIT, S., TANTISUKRIT, C. and PRASERDVIGAI, S., 1986. Magnetic

spectrum of the San Kamphaeng geothermal area, northern Thailand. In: 5th Regional Congress on geology, mineral and energy resources of southeast Asia, April 9–13 1986, ed. G. H. Teh. *Bull. Geol. Soc. Malaysia,* **19**, 587–95.

United Kingdom

1. BESWICK, A. J., BARON, G. and GARNISH, J. D., 1987. Drilling for dry hot rock reservoirs. In: Proc. 1st EEC/US workshop on geothermal hot dry rock technology, May 28–30 1986, Brussels, Belgium, ed. J. D. Garnish. *Geothermics,* **16**(4), 429–32.
2. DOWNING, R. A. and GRAY, D. A., 1986. Geothermal resources of the United Kingdom. *J. Geol. Soc., London,* **143**, 499–507.
3. ROBERTS, P. J., LEWIS, R. W., CARRADORI, G. and PEANO, A., 1987. An extension of the thermodynamic domain of a geothermal reservoir simulator. *Transport in Porous Media,* **2**(4), 397–420.

United States of America (General)

1. MORRIS, C. W., CAMPBELL, D. A. and PETTY, S., 1987. Analysis of geothermal wells in naturally fractured formations with rate-sensitive flow. *SPE Formation Evaluation,* **2**(4), 567–72.
2. WHEATALL, M. W. and MARSDEN, S. S., 1986. The streaming potentials generated by simulated wet steam. In: *5th Int. Symp. on Water–Rock Interactions, August 8–17 1986, Reykjavik, Iceland,* chairperson B. Hitchon, pp. 623–5.

Alaska

1. SASADA, M., 1987. Research on magma-hydrothermal systems in Continental Scientific Drilling Program. *Tsuchi-To-Kiso JSSMFE* (Soil Mechanics and Foundation Engineering), **35** (6 Series (353)), 7–12 (Japanese, English summary).

California

1. ELDERS, W. A., 1987. Research drilling in an active geothermal system; Salton Sea Scientific Drilling Project (SSSDP). In: AAPG Annual Convention with division SEPM/EMD/DPA. *AAPG Bulletin,* **71**(5), 552–3.
2. ELDERS, W. A., 1986. The Salton Sea Scientific Drilling Project; an investigation of an active hydrothermal system in the Colorado River delta of California. In: *5th Int. Symp. on Water–Rock Interaction, August 8–17 1986, Reykjavik, Iceland,* chairperson B. Hitchon, pp. 193–6.
3. GOLDSTEIN, N. E., 1987. Pre-drilling data review and synthesis for the Long Valley Caldera, California. *EOS, Trans. Am. Geophysical Union,* **69**(3), 43–5.
4. MARTIN, J. J. and ELDERS, W. A., 1986. Geothermal resource investigations at Twentynine Palms, California. In: *Geology Around the Margins of the Eastern*

San Bernardino Mountains, ed. M. A. Kooser, Vol. 1, pp. 85–94. Publications of the Inland Geological Society, Inland Geological Society, Redlands, CA, USA.
5. WILLIAMS, A. E. and OAKES, C. S., 1986. Isotopic and chemical variations in hydrothermal brines from the Salton Sea geothermal field, California. In: *5th Int. Symp. on Water–Rock Interactions, August 8–17 1986, Reykjavik, Iceland*, chairperson B. Hitchon, p. 636 (abstract).
6. YAU, Y. C., PEACOR, D. R. and ESSENE, E. J., 1986. Occurrence of wide-chain Ca-pyriboles as primary crystals in the Salton Sea geothermal field, California. *Cont. Min. Petrology*, **94**(1), 127–34.

Hawaii

1. THOMAS, D. M., 1986. The hydrothermal system associated with the Kilauea East Rift Zone, Hawaii. In: *5th Int. Symp. on Water–Rock Interactions, August 8–17 1986, Reykjavik, Iceland*, chairperson B. Hitchon, pp. 569–72 (abstract).

Michigan

1. VUGRINOVICH, R., 1987. Regional heat flow variations in the northern Michigan and Lake Superior region determined using the silica heat flow estimator. *J. Volcanol. Geotherm. Res.*, **34**(1–2), 88815–24.

Nevada

1. LYLES, B. F., 1985. *Time-Variant Hydrogeologic and Geochemical Study of Selected Thermal Springs in Western Nevada*. Master's Thesis, University of Nevada, Reno, NV, USA, 203 pp.
2. WENDELL, D. E., 1985. *Geology, Alteration and Geochemistry of the McGinness Hills Area, Lander County, Nevada*. Master's Thesis, University of Nevada, Reno, NV, USA, 123 pp. (Thermal springs.)

Washington

1. BOGART, L. E. and READDY, L. A., 1987. Importance of fault mapping to mineral/geothermal exploration; relationship to fluid migration and ore formation, northwest Washington. In: *Proc. 5th Thematic Conference on Remote Sensing for Exploration Geology; Technology for a Competitive World, September 29–October 2 1986, Reno, Nevada, USA*, Vol. 1, chairperson J. J. Cook, pp. 395–404. Environ. Res. Inst. Mich., Ann Arbor, MI, USA.

USSR

1. BELODED, V. D., 1986. Characteristics of geothermal energy. In *Geotekhnologiya Toplivno-energeticheskikh Resursov Sbornik Nauchnykh Trudov*, ed. E. B. Chekalyuk. Izd. Naukova Dumka, Kiev, Ukrainian SSR, pp. 74–82.

2. DYAD'KIN, Y. D. and GENDLER, S. G., 1986. Technological types of geothermal systems. In *Geotekhnologiya Toplivno-energeticheskikh Resursov Sbornik Nauchnykh Trudov*, ed. E. B. Chekalyuk. Izd. Naukova Dumka, Kiev, Ukrainian SSR, pp. 38–53.

3. IL'IN, V. A., 1986. Active geothermal processes and crustal energetics. *Int. Geol. Rev.*, **28**(11), 1262–8 (trans.).

4. KABO, V. A., MUSIN, Y. A. and IDRISOVA, S., 1987. Peculiarities in changes of radon concentration in thermal waters of northern Kirghizia in connection with seismicity. *Iz. Akad. Nauk Kirghizskoy SSR* (Kyrgyz SSR Ilminder Akademiyasynin), **1987**(1), 84–9.

5. SARDOROV, S. S. and KAZIYEV, K. S., 1986. Statistical study of the geothermal field. *Dokl. Akad. Nauk SSR*, **289**(5), 1083–8.

6. VASIL'YEV, V. A., ZAYCHIK, L. I., PEREDERIY, A. L. and RABOVSKIY, V. B., 1986. Estimation of underground circulation systems extracting heat from dry rocks with man-made permeability, construction methods of hydrofractures. In *Geotekhnologiya Toplivno-energeticheskikh Resursov Sbornik Nauchnykh Trudov*, ed. E. B. Chekalyuk. Izd. Naukova Dumka, Kiev, Ukrainian SSR, pp. 62–70.

Vanuatu

1. BATH, A. H., 1986. Mixed sources and reactions of hydrothermal springs in Efate, Vanuatu island arc. In: *5th Int. Symp. on Water–Rock Interaction, August 8–17 1986, Reykjavik, Iceland*, chairperson B. Hitchon, pp. 39–42.

Yugoslavia

1. SAJKOVIC, I., 1984. Contribution of geoelectrical sounding to the study of geological features of the Klokot Banja Tertiary Basin. *Vesnik, Zavod za Geoloska i Geofizicika Istrazivanja, Serija B: Inzenjerska Geologija i Hidrogeologija*, **20**, 137–46 (English summary).

Companies of Geothermal Interest

Alfa-Laval (Swedish pump manufacturers)
American Oil Shale Corporation
Ansaldo Componenti SpA (Genoa manufacturers)
APV (UK pump manufacturers)
Atkins, W. S. & Partners

Batelle-Northwest (US company)
Ben Holt Company (CA, USA)
Bozzetto (scale inhibitor manufacturer)

Carrier (US pump manufacturers)
Centrilift (US pump manufacturers)
Coury & Associates (US company)

Eurosyn (scale inhibitor manufacturer)

Floway (US pump manufacturers)

GEA (West German pump manufacturers)
Geothermal Resources International
Gibb, Sir Alexander & Partners (UK company)

Hoare Lee and Partners

Jackson, Bryan (US pump manufacturers)

KSB (West German pump manufacturers)

Man & Neunert (West German pump manufacturers)
Merz & McLellan & Virkir Ltd
Mitsubishi (scale inhibitor manufacturer)
Monsanto (scale inhibitor manufacturer)
Montedison–Arca (scale inhibitor manufacturer)

Nadar (scale inhibitor manufacturer)

Peirless (US pump manufacturers)
Pfizer (scale inhibitor manufacturer)
Pompes Guinard (French pump manufacturers)
Procedo (scale inhibitor manufacturer)

Reda (US pump manufacturers)
Republic Geothermal Inc.

Sandia National Laboratories, NM, USA
Sproule Associates Ltd
SRI International, California, USA
Sulzer (Swiss pump manufacturers)
Systems, Science and Software, La Jolla, CA, USA

Vicarb (French pump manufacturers)

Westinghouse (US pump manufacturers)

York (US pump manufacturers)

Zanussi (Italian pump manufacturers)

APPENDIX 3

Organizations of Geothermal Interest

GEOTHERMAL RESOURCES COUNCIL, PO Box 1350, Davis, California 95617, USA.

Executive Director: David N. Anderson. Founded 1972. Members, 1350; Staff, 8. Budget, $600 000. Publications: *Bulletin*, 11/year, *Transactions of Annual Meetings*, *Membership Roster* (annual) and (irregularly) Special Reports. Regional groups, 8. Library, 1750 volumes. Telex: 822410. Convention/Meeting: Annual.

GEOTHERMAL WORLD INFO CENTER, 17304 Village, No. 17, Camarillo, California 93010, USA.

Contact Alan Tratner. Founded 1972. Maintains Speakers' Bureau. Publications: *Geothermal Energy Magazine* (monthly), *Geothermal World Directory* (annually), occasional research reports. Also known as *Geothermal World*. Convention/Meeting: Annual.

AMERICAN GEOLOGICAL INSTITUTE, 4220 King Street, Alexandria, Virginia 22302, USA.

Publishes monthly a *Bibliography and Index of Geology* containing references to geothermal topics, the Chief Editor being Sharon Tahirkheli as of 1988. The work is photocomposed from citations in GeoRef, a database produced at the American Geological Institute. As well as constituting the source of the AGI Bibliography, GeoRef is searchable on-line and is used to produce special bibliographies and year-end journal indices. Toll-free number: (800) 336-4764.

INTERNATIONAL INSTITUTE FOR GEOTHERMAL RESEARCH and INTERNATIONAL SCHOOL OF GEOTHERMICS, Piazza Solferini 2, 56100 Pisa, Italy. Director: Mario Fanelli. Both bodies collaborate with the CENTRE FOR ENERGY STUDIES in Genoa, Italy.

SOCIETÀ GEOLOGICO ITALIANA, l'Istituto di Geologia e Paleontologia dell'Università, 00185 Rome, Italy.

THE GEOLOGICAL SOCIETY, LONDON, Burlington House, Piccadilly, London W1V 0JU, UK.

Founded 1807, received Royal Charter 1825 for 'investigating the mineral structure of the earth'. Promotes all aspects of geological science and has THE GEOLOGISTS' ASSOCIATION in its building. President: B. E. Leake. Receives papers from time to time from the British Geological Survey, a component body of the National Environment Research Council which transmits to it government funds to assemble, manage, interpret and communicate a National Geosciences Database for the benefit of the nation. This Survey was founded in 1835.

DEUTSCHE GEOLOGISCHE GESELLSCHAFT, 3000 Hannover 51, Alfred-Bentz-Haus, Federal Republic of Germany.

BUREAU DES RECHERCHES GÉOLOGIQUES ET MINIÈRES (BRGM), Paris, Tour Mirabeau, 39–43 quai André-Citroën 75739, France.

THE ROYAL GEOLOGICAL AND MINING SOCIETY OF THE NETHERLANDS (KNGMG), PO Box 157, 3300 AH Dordrecht, The Netherlands.

APPENDIX 4

World Geothermal Localities

(*YNP* = *Yellowstone National Park, Wyoming, USA*)

Abaya, Ethiopia
Acholi, Uganda
Aeolian volcanic islands, Italy
Ahuachapán, El Salvador
Akananda, India
Alaska, USA
Albano Terme, Italy
Aluto, Ethiopia
Amatitlan, Guatemala
Amphoe Phrao, Thailand
Arad-Timisoara, Roumania
Atiamuri, New Zealand
Auburn, New York State, USA

Bagnore, Italy
Bakreswar, India
Batong-Buhay, Philippines
Battaglia Terme, Italy
Bay Islands, India
Bead Geyser, YNP
Beehive Geyser, YNP
Beowawe Geyser, Iceland
Beppy, Japan
Berlin, El Salvador
Bimbandh, Bihar, India

Black Growler fumarole, YNP
Bogoria Lake, Kenya
Bohemian Massif, Czechoslovakia
Brawley, California, USA
Broadlands, New Zealand
Bulera, Italy
Buranga Hot Springs, Sempaya Valley, Uganda

Cambay, Gujarat, India
Camborne HDR, Cornwall, UK
Campi Flegri Caldera, Italy
Carnmenellis granite, Cornwall, UK
Casamicciola Hot Springs, Ischia
Castelnuovo, Italy
Castle Geyser, YNP
Cerro Prieto, Mexico
Cesano, Italy
Chilas, Pakistan
Chinamara, El Salvador
Corbetti, Ethiopia
Coso Hot Springs, California, USA
Cretail (Paris)

Daisy Geyser, YNP
Damavand, Mt, Iran
Dixie Valley, Nevada, USA

East Mesa, Imperial Valley, California, USA
East Olkaria, Kenya
Eburru, Kenya
El Tatio, Chile
Euganei Hills, Italy

Falkenberg, Bavaria, F.R. Germany
Fang geothermal area, Thailand
Fenton Hill HDR, New Mexico, USA
Ferrara, Italy

Gabbro, Italy
Geyser Valley, California, USA

Geysir geyser area, Iceland
Giant Geyser, YNP
Great Geyser, Iceland

Hachimantai volcanic region, Japan
Hannah, Mt, California, USA
Hatchobaru, Japan
Heber, California, USA
Hilo, Hawaii
Hunza, Pakistan

Ischia volcanic island
Isthmus of Corinth, Greece
Ixtlan de los Hervores, Mexico

Jalisco, Mexico
Jewel Geyser, YNP

Kawo-Kamojang, Indonesia
Kawerau, New Zealand
Ketetahi, New Zealand
Kilauea, Hawaii, USA
Kinnaur District, India
Kirghizia, USSR
Kizildere, Turkey
Kokot Banja Tertiary Basin, Yugoslavia
Korako, New Zealand
Krafla, Iceland
Krisuvik, Iceland
Kurikome, Japan
Kyle Hot Springs, Nevada, USA

La Fortuna, Costa Rica
Lago, Italy
Lagoni Rossi, Italy
Lahendong, Indonesia
La Leccia, Italy
La Primavera, Jalisco, Mexico
La Soledad, Jalisco, Mexico

La Soufrière, St Vincent
La Union, Costa Rica
Larderello, Italy
Las Derrumbadas, Mexico
Las Hornillas, Costa Rica
Lassen, Mt, USA
Latera, Italy
Limagne, France
Linow Lake volcanic area, Indonesia
Lipari volcanic island
Lone Star Geyser, YNP
Los Alamas HDR, USA
Los Azufres, Michoacan, Mexico
Los Humeros, Pueblo, Mexico
Los Negritos, Michoacan, Mexico
Lund, Sweden

Maku-Khoy, Azerbaijan, Iran
Mammoth Lakes, California, USA
Marchwood borehole, UK
Matsukawa, Japan
McCall, Glady well, Louisiana, USA
Meaux, Paris
Melun, Paris
Merksplas, Holland
Metanopoli, Italy
Milos, Greece
Minahasa volcanic region, Indonesia
Mofete, Italy
Mokai, New Zealand
Molinetto, Italy
Momotombo, Nicaragua
Monghyr District, Bihar, India
Mon Pelée, Martinique
Monte Amiata, Italy
Montecerboli, Italy
Montegrotto Terme, Italy
Monterotondo, Italy
Morbo bath, Italy
Moyutag, Guatemala

Námafjall, Iceland
Ngatamariki, New Zealand
Ngawha Springs, New Zealand
Nisyros, Greece

Ohaki, New Zealand
Old Faithful Geyser, YNP
Onikobe, Japan
Oradea-Satu Mare, Roumania
Orakei-Korako, New Zealand
Otake, Japan

Palabukan-Ratu, Java, Indonesia
Pannonian Basin, eastern Europe
Paratunka, USSR
Paris Basin, France
Pathe, Mexico
Pauzhetka, USSR
Petropavlovsk-Kamchatskiy, USSR
Phrao Basin, Thailand
Phlegraean Field, Italy
Piancastagnaio, Italy
Plume Geyser, YNP
Po Valley, Italy
Podhájska, Czechoslovakia
Poggio Nibbio, Italy
Pohutu Geyser, New Zealand
Pontian geothermal aquifer, Roumania
Prabatki-salak, Java, Indonesia
Puarenga Geyser, New Zealand
Puchuldiza, Chile
Puga Valley, Himachal Pradesh, India
Puhima thermal area, Hawaii

Radicondoli, Tuscany, Italy
Reporoa, New Zealand
Reyhotshver Geyser, New Zealand
Reykjanes, Iceland
Riverside Geyser, YNP
Roosevelt Hot Springs, Utah, USA
Rosemanowes HDR, Cornwall, UK

Rotokawa, New Zealand
Rotorua-Whaka, New Zealand

Sabalon volcano, Azerbaijan, Iran
Sahand volcano, Azerbaijan, Iran
Salak, Java, Indonesia
Salar de Empexa, Bolivia
Salbardi Hot Springs, India
Salton Sea, California, USA
San Donato Milanese, Italy
San Jacinto, Nicaragua
San Kampaeng, Thailand
San Marcosgarai, Jacinto, Mexico
San Martino, Italy
San Vincent, El Salvador
San Vito, Italy
San Miguel, Azores
Santa Maria, Guatemala
Sasso Pisano, Italy
Seismic Geyser, YNP
Serranzano, Italy
Simla, India
Sousaki-Loutraki, Greece
Soultz-sous-Forets, Alsace, France
South Geyser, Iceland
Southern Negros Island, Philippines
St Lucia, Caribbean
Steamboat Springs, Nevada, USA
Strokkur Geyser, Iceland
Surajkind, Bihar, India
Suriri, Chile
Svartsengi, Iceland

Takinoue, Japan
Tattapani Hot Springs, India
Tattapani-Jhor, India
Tauhara, Wairakei, New Zealand
Te Kopia, New Zealand
Tengehong, Yunnan, China
Thana District, India
The Geysers, California, USA

Thisted, Denmark
Tianjin, China
Tikitere-Taheke, New Zealand
Tokaanu-Waihi, New Zealand
Tongonang, Philippines
Travale, Tuscany, Italy
Travale-Radicondoli, Tuscany, Italy
Tule Chek, Mexico

Umhavre-Khed, Maharashtra, India
Upper Geyser Basin, YNP
Upper Pannonian, Hungarian reservoir
Uxhaver Geyser, Iceland

Val di Cecina
Vallonsordo, Italy
Volcanoes National Park, Hawaii
Vicenza, Italy
Vulcanello, Vulcano volcanic island, Italy

Waikiti, New Zealand
Waimangu, New Zealand
Waiotapi-Reporoa, New Zealand
Wairakei, New Zealand
Waitangi, New Zealand
Washington State, USA
Wessex Basin, UK
West Bükk, Hungary
West Coast, India
West Pennines Basin, UK
West Lancashire Basin, UK
Whakarewarewa Geyser Basin, New Zealand
White Dome Geyser, YNP
White Island Volcano, New Zealand

Yangbajain, Tibet
Yellowstone National Park, Wyoming, USA

Zambian Hot Springs
Zerka Ma'in-Zara thermal field, Jordan
Zunil, Guatemala

Author Index

Numbers in italic type indicate those pages on which references are given in full

459

Subject Index